T0142488

Studies in Systems, Decision and Control

Volume 260

The series “Studies in Systems, Decision and Control” (SSDC) covers both new developments and advances, as well as the state of the art, in the various areas of broadly perceived systems, decision making and control–quickly, up to date and with a high quality. The intent is to cover the theory, applications, and perspectives on the state of the art and future developments relevant to systems, decision making, control, complex processes and related areas, as embedded in the fields of engineering, computer science, physics, economics, social and life sciences, as well as the paradigms and methodologies behind them. The series contains monographs, textbooks, lecture notes and edited volumes in systems, decision making and control spanning the areas of Cyber-Physical Systems, Autonomous Systems, Sensor Networks, Control Systems, Energy Systems, Automotive Systems, Biological Systems, Vehicular Networking and Connected Vehicles, Aerospace Systems, Automation, Manufacturing, Smart Grids, Nonlinear Systems, Power Systems, Robotics, Social Systems, Economic Systems and other. Of particular value to both the contributors and the readership are the short publication timeframe and the world-wide distribution and exposure which enable both a wide and rapid dissemination of research output.

** Indexing: The books of this series are submitted to ISI, SCOPUS, DBLP, Ulrichs, MathSciNet, Current Mathematical Publications, Mathematical Reviews, Zentralblatt Math: MetaPress and Springerlink.

More information about this series at http://www.springer.com/series/13304

Alla G. Kravets · Alexander A. Bolshakov · Maxim V. Shcherbakov
Editors

Cyber-Physical Systems: Industry 4.0 Challenges

Editors
Alla G. Kravets
Volgograd State Technical University
Volgograd, Russia

Maxim V. Shcherbakov
Volgograd State Technical University
Volgograd, Russia

Alexander A. Bolshakov
Peter the Great St. Petersburg Polytechnic University
St. Petersburg, Russia

ISSN 2198-4182 ISSN 2198-4190 (electronic)
Studies in Systems, Decision and Control
ISBN 978-3-030-32650-0 ISBN 978-3-030-32648-7 (eBook)
https://doi.org/10.1007/978-3-030-32648-7

This Springer imprint is published by the registered company Springer Nature Switzerland AG
The registered company address is: Gewerbestrasse 11, 6330 Cham, Switzerland

Preface

This book is devoted to the study of breakthrough technologies for the industrial cyber-physical systems' implementation as a basic of the Fourth Industrial Revolution (Industry 4.0). This book defines the activities in the industrial technologies field that leverage the exploitation of IoT and cyber-physical systems development capabilities in an embedded system's context. The authors study the activities which allow for the developing new and innovative concepts combining various sensing, actuating, communication, processing system's functions including smart systems' integration and system of systems convergence by addressing a large scale of applications.

The book also describes the intelligent control and implementation of these technologies in the different industrial sectors such as energy, buildings, mobility, health, urban development, in order to realize the novel hardware or software, and algorithms, functionalities, and features. The authors describe the development of industrial cyber-physical systems, considering the increasing complexity of the interaction and the importance of ensuring safety during their work.

The book analyzes the requirements for cybersecurity systems, including the protection of confidential information, the integrity of basic information resources, as well as blocking attacks and preventing possible threats. The authors highlight the problem of ensuring the cybersecurity of industrial cyber-physical systems as the most important and consider different approaches to its solution.

Also, this book dwells with the analysis of the direction in engineering education, which helps the person who received it in creating cyber-physical systems. The authors describe the impact of education in the universities on the performance of industrial complexes, the organization of education for the development of cyber-physical systems.

Edition of the book is dedicated to the 120th Anniversary of Peter the Great St. Petersburg Polytechnic University and technically supported by the Project Laboratory of Cyber-Physical Systems of Volgograd State Technical University.

Volgograd, Russia — Alla G. Kravets
St. Petersburg, Russia — Alexander A. Bolshakov
Volgograd, Russia — Maxim V. Shcherbakov
August 2019

Contents

Cyber-Physical Systems Intelligent Control

Industrial Cybersecurity

IoT for Industrial Cyber-Physical Systems

Cyber-Physical Principles of Information Processing in Ultra-Wideband Systems

Sergey Chernyshev

Abstract Information processing in ultra-wideband systems based on the cyber-physical principles of a neuron, in a distributed structure of which impulses propagate, is considered. The processing of ultra-wideband pulses carrying information is carried out on distributed structures of irregular lines, in which the wave parameters change. Such processing can be carried out both on the "pass" and on the "reflection", and it is possible that filters are connected in parallel, which are matched with various ultra-wideband signals.

Keywords Ultra-wideband system · Signal processing · Neuron · Distributed lines · Filters synthesis · Transient characteristic

1 Introduction

Ultra-wideband (UWB) systems, as follows from the definition of DARPA (USA), call the system of transmitting or extracting information that has a relative frequency bandwidth of more than 50%. Such systems have the maximum capacity.

In accordance with the basic Shannon formula, the capacity with a theoretical infinite increase in the frequency band reaches its limiting value, where and are the average signal power and the spectral power density of white noise. The transmission speed of ultra-wideband (UWB) signals can reach tens of billions of pulses per second. The transmission of UWB messages takes microseconds. Such maximum achievable characteristics of UWB systems make them attractive for many applications: in data transmission systems, medical systems, security systems, radar systems, etc. [1–11]. The IEEE 802.15.4a/b standard provides for ultra-wideband wireless communication with high speed. The recently published IEEE STD 802.15.6-2012 includes IR-UWB and FM-UWB technologies. In Russia and the United States pay attention to the development of UWB technology [12, 13]. The processing of information transmitted by UWB signals encounters difficulties associated with the

S. Chernyshev (✉)
Bauman Moscow State Technical University, 5, 2-nd Baumanskaya str., Moscow 105005, Russia
e-mail: chernshv@bmstu.ru

A. G. Kravets et al. (eds.), *Cyber-Physical Systems: Industry 4.0 Challenges*, Studies in Systems, Decision and Control 260,
https://doi.org/10.1007/978-3-030-32648-7_1

impossibility of applying analog-digital conversion due to the large width of their frequency spectrum, which can reach tens of gigahertz. This necessitates the use of special methods for processing information using cyber-physical principles.

2 Cyber-Physical Principles of Neurons

Cyber-physical systems computing is distributed throughout the system. The most vivid example is the human brain. The brain is an unsurpassed intellectual system in nature. Considerably inferior in speed to modern computers, the brain, however, demonstrates significantly higher intellectual capabilities. In this regard, it is an urgent task to use for processing information, as far as possible, new principles based on the properties of the neuron system of the brain.

A neuron [14], like a brain cell, is (Fig. 1) attached to the supply line 1 (axon), has a tree formation consisting of branches 2 (dendrites) connected to a common adder 3 (cell body). The permission or prohibition of the generation of pulses (excitation or inhibition) in a neuron is carried out using synapses 4, controlled by terminal fibers 5, coming from the axons of other neurons.

The impulses generated in dendrites are folded in the body of a neuron with certain time delays [15], forming an analog signal in the form of a complex sequence of pulses entering the axon for transmission to other neurons or to the executive nerve endings. The duration of such pulses is about 1.5 ms, although their geometric length is about 3 cm. Such a geometric length of a pulse in free space corresponds to a pulse duration of 0.1 ns. The speed of these pulses along the axon in the aquatic environment is about 20 m/s, and the speed of propagation of the pulse in free space is 300 million m/s. These data show that the speed of propagation of pulses along the axon is less than the speed of propagation of pulses in free space at about 15 million times. Such a propagation could take place in a guiding electromagnetic system located in a medium with a relative permittivity equal to 2.25×10^{14} (for

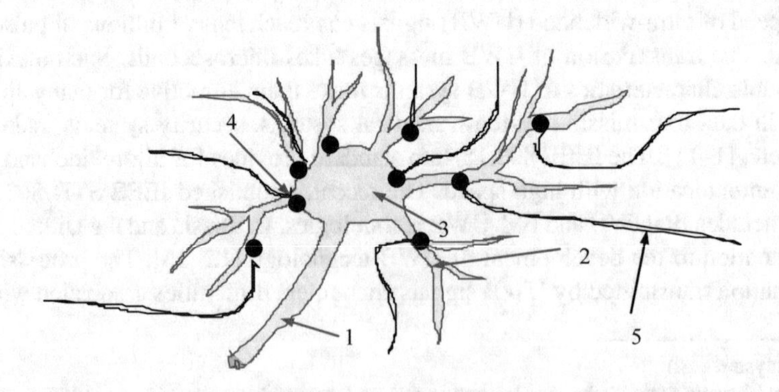

Fig. 1 Neuron

free space it is equal to 1), into which an equivalent impulse that has a duration of 1.5 ms in free space falls. The geometric length of such equivalent pulses in free space should be about 450 km. If the dendrite is represented in the form of a distributed line along which the impulse propagates, then the processes occurring in this line will be completely identical to the processes occurring in some equivalent air-filled distributed line. With a signal duration of 0.1 ns, the upper frequency of the pulse spectrum is 10 GHz, and such a signal should be attributed to the ultra-wideband type.

3 Cyber-Physical Principles of Information Processing on Distributed Lines

Such principles of processing ultra-wideband information can be applied by creating distributed structures on irregular transmission lines [16, 17], like dendrites, in which cyber-physical principles are implemented. In such lines, the wave parameters vary along the longitudinal coordinate. Information processing in them is distributed, as in cyber-physical systems.

Figure 2 shows the configuration of a distributed irregular transmission line, which is a cascade connection of segments with different wave resistances R_i (i = 1, 2, 3, 4, …, N + 1 having such lengths that on each of them an equal time delay of the wave occurs. In Fig. 2 is designated: a_I-the falling line, b_{II}-the passed wave. Such a distributed connection has filtering functions due to the fact that as a result of reflections of an electromagnetic wave between inhomogeneities in the form of resistance leaps, the shape of the transmitted wave differs from the incident wave and the reflected wave also appears. The electromagnetic T-wave incident on the input of such a connection in the form of an ultra-wideband pulse, the duration of which does not exceed the propagation time over a separate section. Due to such a short duration, the frequency band occupied by such a signal is ultra-wide, which determines the specificity of its propagation. Each i-th leap is characterized by a normalized wave scattering matrix of the form

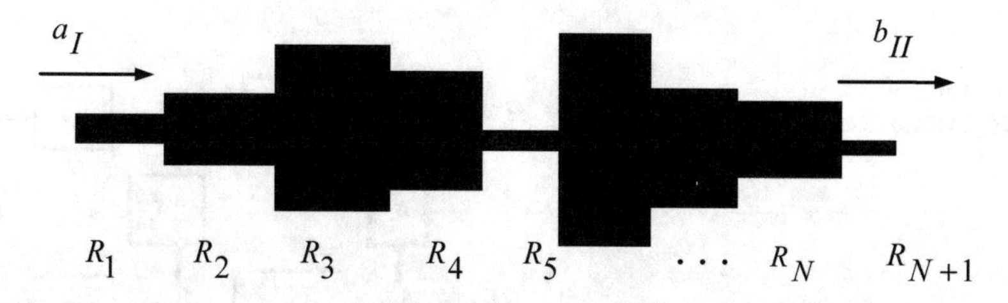

Fig. 2 Configuration irregular transmission line

$$\begin{bmatrix} S_i & \sqrt{1-S_i^2} \\ \sqrt{1-S_i^2} & -S_i \end{bmatrix}$$

where $S_i = (R_{i+1} - R_i)/(R_{i+1} + R_i)$ is the local wave reflection coefficient from the junction of the sections of lines with different wave resistances.

The electromagnetic wave passes sectors between the jumps with time T. Due to the equal delay of the wave on all sections, the z-transfer function of these sections can be written as z^{-1}, where z is the operator of Z-transformation [16].

The reflections of the pulses between the sections create feedbacks on each of them. In this regard, such an irregular distributed line can be considered as a discrete control system with numerous feedbacks. Figure 3 shows one link of such a discrete system.

Z-the transfer function "per pass" through two adjacent leaps connected by a regular segment can be written as

$$S_{21}^{i,i+1}(z) = \frac{\sqrt{1-S_i^2}\sqrt{1-S_{i+1}^2} \cdot z^{-1}}{1+S_i S_{i+1} z^{-2}}.$$

The denominator of this expression is determined by the feedback arising in the connection due to the oscillation of the wave between the two inhomogeneities, and the double reflection of this wave determines the second inverse power of the parameter z. The propagation of a wave through a junction gives a factor z^{-1} in the numerator. If we consider a cascade connection (Fig. 2), then the wave transfer coefficient through it takes the form:

$$S_{21}(z) = \frac{\prod_{i=1}^{N}\sqrt{1-S_i^2} \cdot z^{-N}}{[1+S_1 S_{2,N}(z)z^{-2}][1+S_2 S_{3,N}(z)z^{-2}]\cdots[1+S_{N-1}S_N z^{-2}]}, \quad (1)$$

where $S_{i,N}(z)$ is the wave coefficient of reflection from a cascade connection consisting of the i-th leap and the other leaps, up to the N-th. As can be seen from (1), repeated wave reflections between inhomogeneities lead to the appearance in the

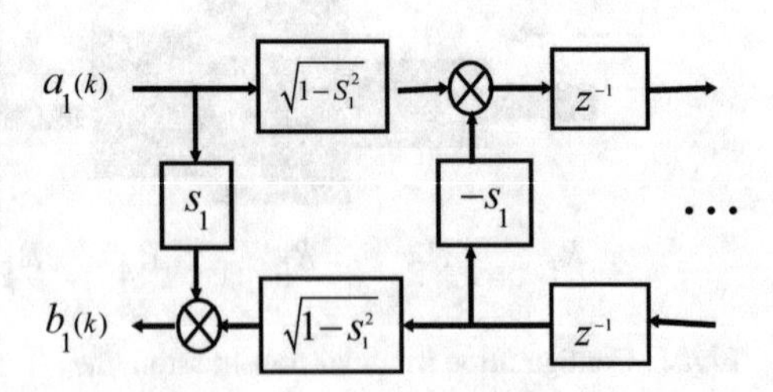

Fig. 3 Link of cyber-physical discrete system

denominator of numerous factors, including the second inverse power of the parameter z. Expanding the brackets of the factors of the denominator, we find that it is a polynomial of degree 2N:

$$S_{21}(z) = \frac{\prod_{i=1}^{N} \sqrt{1 - S_i^2} \cdot z^{-N}}{\left[1 + w_1 z^{-2} + w_2 z^{-4} + \cdots + w_N z^{-2N}\right]}, \tag{2}$$

where the coefficients w_n depend on the local wave reflection coefficients and are calculated as

$$w_n = \sum_{m=1}^{n} \sum_{p=1}^{n} \cdots \sum_{l=1}^{n} \sum_{i=1}^{n-l} S_i S_{i+l} \sum_{k=i+l+1}^{n-m} S_k S_{k+m} \sum_{j=k+m+1}^{n-p} S_j S_{j+p} \cdots, \tag{3}$$

where $l + m + p + \cdots = n$. The number of indices $l, m, p, \ldots$ is equal to the whole number obtained from division $N/2$, respectively, the same pairs of multiplications are included as much as possible in (3). This expression was obtained as a result of inductive analysis of the constituent parts of the coefficients w_n.

Z-image of the transmitted wave is associated with the Z-image of the incident wave:

$$b_{II}(z) = S_{21}(z) a_I(z).$$

The use of reverse Z-transformation allows you to find a temporary form of the transmitted wave:

$$b_{II}(k) = \sum_{i=1}^{\infty} G_{21}(i - k) a_I(i).$$

where $G_{21}(k)$ is the discrete impulse response of the line per pass.

To solve the problem of synthesizing a line for a given transmitted wave $b_{II}(k)$ with a known incident wave $a_I(k)$, obtaining expression (3) is useful because the number of these coefficients in the denominator (2) is limited. Therefore, if we find the dependence of these coefficients on the values $b_{II}(k)$ and $a_I(k)$, then, by solving a system of n equations of the form (3), we can find the desired wave reflection coefficients from leaps.

Z-transfer function "on reflection" has the following form:

$$S_{11}(z) = \frac{S_1 + S_{2,N}(z) z^{-2}}{\left[1 + S_1 S_{2,N}(z) z^{-2}\right]}. \tag{4}$$

Z-image of the reflected wave is associated with the Z-image of the incident wave:

$$b_I(z) = S_{11}(z)a_I(z).$$

Opening the polynomial $S_{2,N}(z)$ in (4), and dividing the numerator polynomial by the denominator polynomial, we obtain the following expression:

$$b_I(z) = (C_0 + C_1 z^{-2} + C_2 z^{-4} + C_3 z^{-6} + \cdots)a_I(z).$$

The coefficients of the series are determined by the formulas

$$C_0 = S_1,\quad C_1 = (1 - S_1^2)S_2,\quad C_2 = (1 - S_1^2)(1 - S_2^2)S_3 - (1 - S_1^2)S_1 S_2^2,$$
$$C_3 = (1 - S_1^2)(1 - S_2^2)\cdot(1 - S_3^2)S_4 - (1 - S_1^2)(1 - S_2^2)S_1 S_2 S_3 - (1 - S_1^2)(1 - S_2^2)S_2 S_3^2 + (1 - S_1^2)S_1^2 S_2^3, \ldots,\quad C_i = \cdots.$$

After applying the reverse Z-transformation allows you to find the relationship between the reflected and incident waves:

$$b_I(k) = C_0 a_I(k) + C_1 a_I(k-2) + C_2 a_I(k-4) + C_3 a_I(k-6) + \cdots.$$

The developed algorithm for the synthesis "on reflection" [16, 17] allows you to determine all the necessary reflection coefficients $S_i,\ i = 1, 2, 3, \ldots$ for given wave values $b_I(k)$ and $a_I(k)$.

Synthesized filters can be used both in series and in parallel.

Figure 4 shows an example of a parallel link. In such a combination, these filters are the original dendrites of the cyber-physical system. The chains of such links create an information processing system for various ultra-wideband signals carrying information.

As an example, we will give a filter intended for the correlation processing of the ultra-wideband signal shown in Fig. 5. Using the developed algorithm for the synthesis of the filter matched with the signal shown in Fig. 5, the configuration of the matched microstrip filter shown in Fig. 6 was synthesized. In this figure, W is the width of the microstrip, and x is the length. The distributed cyber-physical

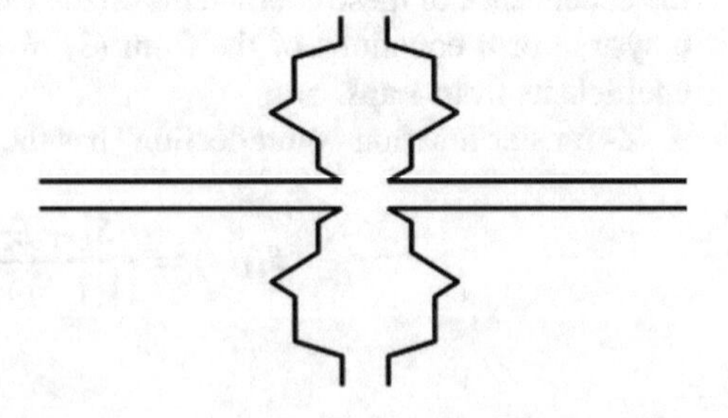

Fig. 4 Part of the parallel connections system

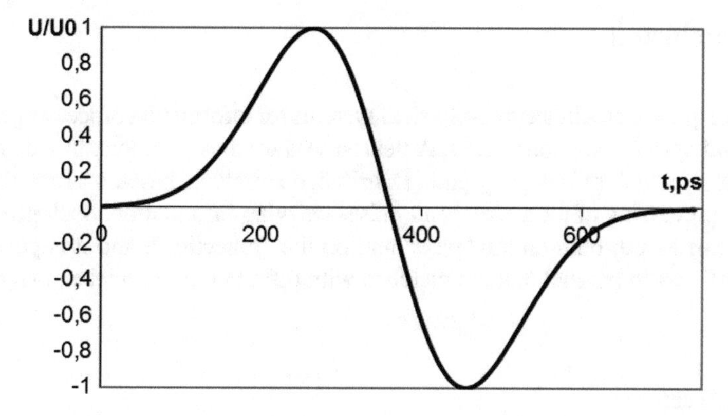

Fig. 5 Single ultra-wideband signal

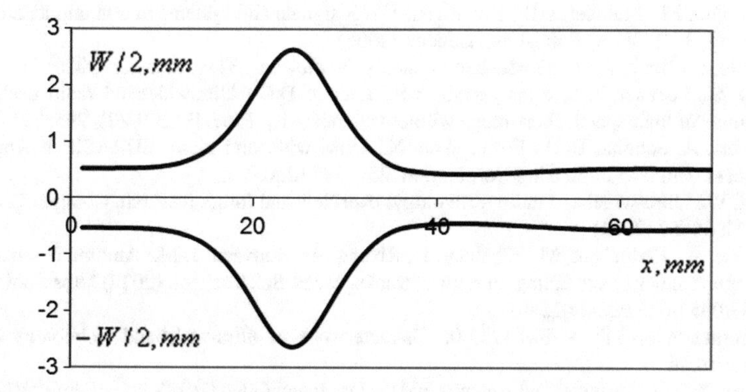

Fig. 6 Distributed filter configuration

system implemented in such a filter provides the correlation processing of the ultra-wideband signal. The wave reflected from such a filter is a correlation function of an ultra-wideband signal.

The smooth nature of changes in the distributed wave parameters in such a line is explained by the small sampling interval of the signal, which was equal to 10 ps. If the sampling interval is longer, a jump-like character of the change in wave parameters may appear, such as shown in Fig. 2. But in any case, the sampling interval must meet the condition of Kotelnikov.

4 Conclusion

The principles of creating cyber-physical systems for information processing in ultra-wideband systems are considered. A neuron was used as a sample, in a distributed structure of which pulses propagate. Distributed structures based on irregular lines provide processing of ultra-wideband pulses carrying information. Such processing can be carried out both on the "pass" and on the "reflection", and it is possible to connect filters in parallel that are matched with different ultra-wideband signals.

References

1. Taylor, J.D. (ed.): Introduction to Ultra-Wideband Radar Systems, 670p. CRC Press, Inc. (1995)
2. Ghavami, M., Michael, L.B., Kohno, R.: UWB signals and systems in communication engineering, 354p. Wiley Publishing, London (2006)
3. Aiello, R., Batra, A.: Ultra wideband systems. In: Newnes, 341p. Elsevier (2006)
4. Roy, S., Foerster, J.R., Somayazulu, V.S., Leeper, D.G.: Ultrawideband radio design: the promise of high-speed, short-range wireless connectivity. Proc. IEEE **92**(2), 295–311 (2004)
5. Robert, A. Scholtz, D.M., Pozar, Won, N.: Ultra wideband radio. EURASIP J. Appl. Sig. Process. Hindawi Publishing Corporation, 252–272 (2005)
6. Hirt, W.: Ultra-wideband radio technology: overview and future research. Comput. Commun. **26**(1), 46–52 (2003)
7. Mroue, A., Heddebaut, M., Elbahhar, F., Rivenq, A., Rouvaen, J.-M.: Automatic radar target recognition of objects falling on railway tracks. Meas. Sci. Technol. (2012) https://doi.org/10.1088/0957-0233/23/2/025401
8. Recomendation ITU-R SM.1755-0: Characteristics of ultra-wideband technology (2006). www.itu.int
9. Liang, X.L.: Ultra-wideband Antenna and design. IntechOpen (2012). https://doi.org/10.5772/47805
10. Isaev, V.M., Semenchuk, V.V., Meshchanov, V.P., Shikova, L.V.: UltraWuidebsnd fixed phase shifters based on coupled transmission lines with stubs. J. Commun. Technol. Electron. **60**(6), 566–571 (2015)
11. Meschanov, V.P., Metelnikova, I.V., Tupikin, V.D., Chumaevskaya, G.G.: A new structure of microwave Ultra-wideband differential shifter. IEEE Trans. Autom. Control **MTT-42**(5), 762 (1994)
12. Russian SCRF Decision № 09-05-02 dated 15 Dec 2009
13. US Federal Communications Commission (FCC) Decision № FCC 02-48 from 14 Feb 2002
14. Hubel, D.H.: Eye, brain, vision. 2nd ed., 256p. Henry Holt and Company (2012)
15. Chernyshev, S.L., Chernyshev, A.S.: Modeling of the neural network elements taking into account the properties of the neuron as a biological object. Nauka i obrazovanie, 11, https://doi.org/10.7463/1114.0743838 (2014)
16. Chernyshev, S.L.: Analysis and synthesis of smooth irregular transmission lines in the time domain. In: Instrument Making, pp. 52–54. Bulletin Bauman Moscow State Technical University (1994)
17. Chernyshev, S.L.: Analysis and synthesis of UWB filters and shapers in time domain. In: 2013 IEEE International Conference on Microwave Technology & Computational Electromagnetics, Proceedings, pp. 127–130. Qingdao, China (2013)

The Model of Reliability of Dublated Real-Time Computers for Cyber-Physical Systems

V. A. Bogatyrev, S. M. Aleksankov and A. N. Derkach

Abstract The article is devoted to the impact of recovery strategies and organizing migration of virtual resources on the reliability of fault-tolerant embedded two-machine computing systems. This computer is focused on using cyber-physical systems, which are critical to the continuity of the controlling computational process. Fault tolerance of a computer system is realized in the case of migration of a computational process from a failed computer to a working one. The computational process should not be interrupted after failures. The Markov models of reliability are proposed. Embedded two-machine onboard systems are critical to the continuity of the computational process. Systems include the failure criterion such as loss of continuity of the computational process without the implementation of recovery.

Keywords Virtualization · Virtual machines · Clusters · Reliability · Fault tolerance · Non-stationary availability factor

1 Introduction

The most important objective for embedded computing facilities of cyber-physical systems is to increase their reliability and fault tolerance to ensure high overall, informational and functional system security [1–3]. Cyber-physical systems are characterized by the operation of real-time systems with criticality for maintaining control continuity [4, 5]. The using of clustering and virtualization technologies contributes to the achievement of high and stable indicators of performance, reliability and fault tolerance [4–6] of computer systems, including control systems, which do not allow

V. A. Bogatyrev (✉) · A. N. Derkach
Faculty of Software Engineering and Computer Systems, Saint-Petersburg National Research University of Information Technologies, Mechanics and Optics, Saint-Petersburg 197101, Russian Federation
e-mail: vladimir.bogatyrev@gmail.com

S. M. Aleksankov
Research Institute Mashtab, 5A Kantemirovskaya Street, Saint-Petersburg 194100, Russian Federation

A. G. Kravets et al. (eds.), *Cyber-Physical Systems: Industry 4.0 Challenges*, Studies in Systems, Decision and Control 260,
https://doi.org/10.1007/978-3-030-32648-7_2

interruptions in the computational process. High fault tolerance of computational processes [7, 8] during virtualization is achieved by the migration of virtual machines between physical nodes of the cluster. Migration accelerates the process of reconfiguring the computational system and increases the reliability and security of critical real-time systems [9]. During virtualization, several virtual machines use one pool of physical resources [10], in which they migrate when failures occur. Migration of virtual resources and computational processes in real-time systems can be used to preserve the continuity of the computational process after failures of physical nodes.

2 Mathematical Model of the Duplicated Real-Time Computing System

Reliability of computational systems focused on the use of embedded in cyber-physical systems can be achieved on the basis of Virtual Machine (VM) migration technology. The fault tolerance technology with maintaining the continuity of a recovery computer system with cluster architecture after the failure of its physical resources is implemented during the migration.

The technology «Fault tolerance» supports two copies of the VM in the operational memory on different physical computers of the recovery system. The computational process continues on the backup computer after the failure of the physical computer that realizes control of the cyber-physical system [11]. VM virtual disk images are stored on dedicated or distributed data storage with synchronous data replication. The backup computer must maintain an up-to-date copy of the RAM [11, 12] of the active VM. The «Fault tolerance» technology is supported by the products: VMware Fault Tolerance, Kemari for Xen and KVM [11, 12].

Recovery time after failures depends on the structure and amount of data storage, which can be shared or localized for each computer. In case of failures, only RAM, virtual processor registers and VM virtual device states are transferred if the cluster has shared storage [13–15]. Information from hard disks is transferred additionally if there are localized data storages in the cluster.

The purpose of the study is to increase the reliability of the system, which does not allow interruptions in the computing process. The system must be able to perform the required functions, to support the operability of the required resources and ensure necessary conditions for their operation. The condition for the required functioning of the duplicated computer systems is to ensure the continuity of the computational process in the event of normal operation and during recovery after failures. Recovery is relevant in this cluster if the backup nodes continue to implement the functions of the system. The computer system enters a state of unrecoverable failure in case of interruption computations (control in a cyber-physical system). Such problem appears as a result of failures when it cannot control the system during a possible recovery of computers.

An object of the study is Fault tolerance duplicate system with a cluster architecture. The cluster contains two computers and local storage devices connected to it. The servers are connected through a switch. The data storage system is presented as local storage for each physical node (hard disks). «Fault tolerance» technology supports synchronous data replication between local storages [16–19].

The duplicate system contains two computers (main and backup), each of them is connected to a local storage device (hard disk) [20]. Each computer has access only to the attached hard disk.

To control cyber-physical systems, individual computers can be combined into clusters of duplicated systems. Machine-to-machine exchange and distribution of requests between computers can be organized through redundant communication tools, including through aggregated channels [21–23].

In case of failure of the primary physical computer fault tolerance duplicate system initiates the launch of the VM's copy on the backup server without interrupting the calculations.

The choice of service strategy of two-machine computer systems is largely determined by the characteristics of the purpose and operation of the cyber-physical system. The limitations of the possibilities for recovering from the failures and the organization of data and virtual resources migration are taken into account. Consider several computer system recovery strategies.

Strategy A is an operational recovery beginning immediately after the failure of the physical resources of computers at any stage of the computer system degradation. Interrupt calculations are allowed in the strategy. During constructing a computer system model, it is assumed that failures are detected instantly and control affects the decrease in reliability of the system [24].

Strategy B is a non-operational recovery performed after resource failures causing the computer system to become inoperable or the state of the specified limit level of the cluster's effectiveness decrease. Interrupts of the computational process are allowed. Strategy B is analyzed in [24].

The service strategies investigated in [24] are characterized by the ability to restore the system regardless of the idle time (interruption) of the computational process. Idle after failures is associated with the migration of virtual resources and actual data, as well as reconfiguration and recovery of physical nodes.

Strategy C—operational recovery of duplicated computer system with the inadmissibility of the computational process discontinuity while the computational process restores and migrates to a backup server. Such limitations are characteristic of cyber-physical real-time control systems, for which interruptions of the computational process may lead to loss of control over the object (for example, a technological process or vehicles) or loss of critical information.

The state and the transition diagram of a fault-tolerant cluster with the recovery strategy C is shown in Fig. 1. The diagram shows the failure and recovery rates, respectively, of the computer λ_0 and μ_0; disk λ_1, μ_1; switch λ_2, μ_2. The actual data replica loads with the intensity μ_3 on the restored disk (synchronization of the distributed storage system).

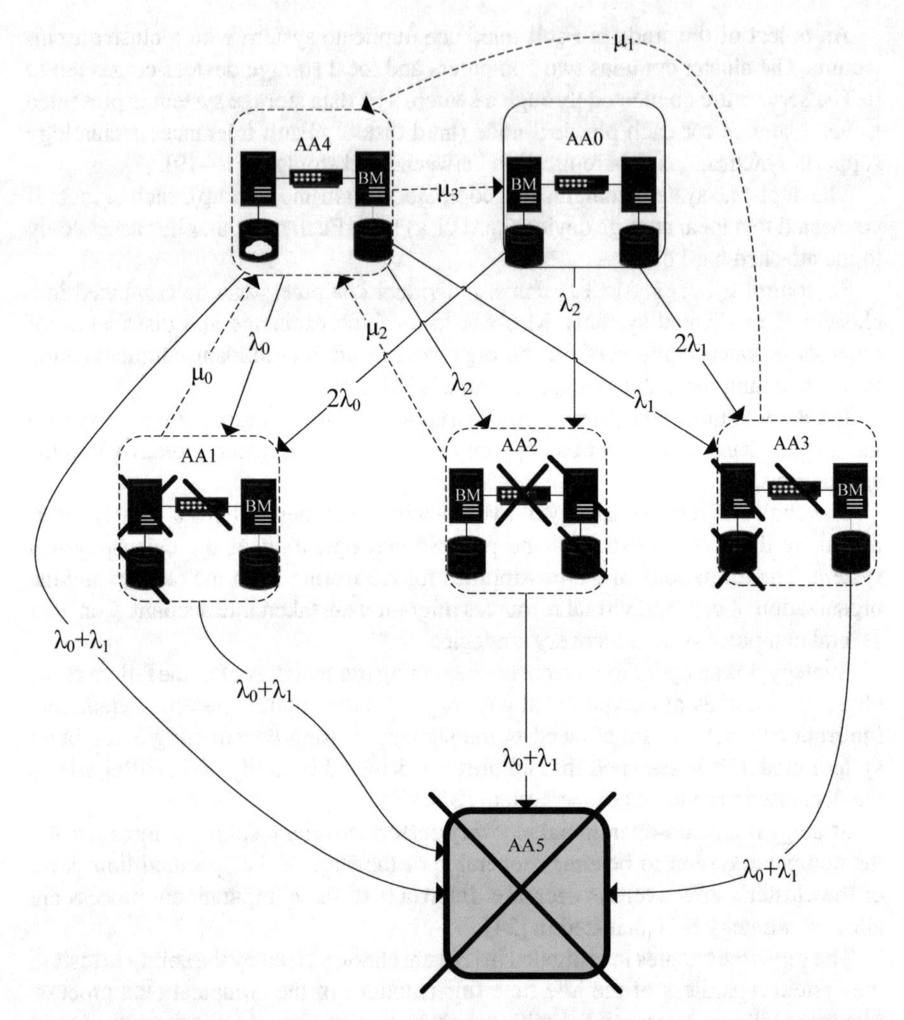

Fig. 1 State and transitions graph of the cluster with the requirement of ensuring the continuity of the computational process

The system of differential equations in accordance with the state and transitions diagram in Fig. 1 (strategy C) has the form:

$$
\begin{aligned}
P_0'(t) &= -(2\lambda_0 + \lambda_2 + 2\lambda_1)P_0(t) + \mu_3 P_4(t),\\
P_1'(t) &= -(\lambda_1 + \lambda_0 + \mu_0)P_1(t) + 2\lambda_0 P_0(t) + \lambda_0 P_4(t),\\
P_2'(t) &= -(\lambda_1 + \lambda_0 + \mu_2)P_2(t) + \lambda_2 P_0(t) + \lambda_2 P_4(t),\\
P_3'(t) &= -(\lambda_1 + \lambda_0 + \mu_1)P_3(t) + \lambda_1 P_4(t) + 2\lambda_1 P_0(t),\\
P_4'(t) &= -(2\lambda_0 + \lambda_2 + 2\lambda_1 + \mu_3)P_4(t) + \mu_1 P_3(t) + \mu_0 P_1(t) + \mu_2 P_2(t),\\
P_5'(t) &= -(\lambda_1 + \lambda_0)(P_1(t) + P_2(t) + P_3(t) + P_4(t)).
\end{aligned} \tag{1}
$$

The results of calculations and comparisons of non-stationary availability factors of a computer system with the considered service strategies are presented in the form of graphs.

3 Calculation Results

The results of reliability calculations based on the model equations are presented in Fig. 2, in which the curves 1 and 2 reflect the results of the non-stationary availability factor's evaluations for service strategies A and B, respectively. The curve 3 corresponds to the probability that the computer system will work until the computing process is discontinued during its migration to the backup server (strategy C).

Calculations are performed at failure rates $\lambda_0 = 1.115 \times 10^{-5}$, $\lambda_1 = 3.425 \times 10^{-6}$, $\lambda_2 = 2.3 \times 10^{-6}$ 1/h, and recovery rates $\mu_0 = 0.33$ 1/h, $\mu_1 = 0.17$ 1/h, $\mu_2 = 0.33$ 1/h, $\mu_3 = 1$ 1/h, $\mu_4 = 2$ 1/h, $\mu_5 = 0.5$ 1/h.

The impact significance on the requirements for ensuring the continuity of the computational process during its migration to the backup computer is confirmed by calculations. The greatest probability to find the system in working condition in the strategy A. The strategy C shows better results than the strategy B.

The presented calculations do not consider the influence of control. Models [25] that provide for the availability of operational and test control for duplicated computing systems, taking into account the optimal frequency of activation of control, are used to account for this effect.

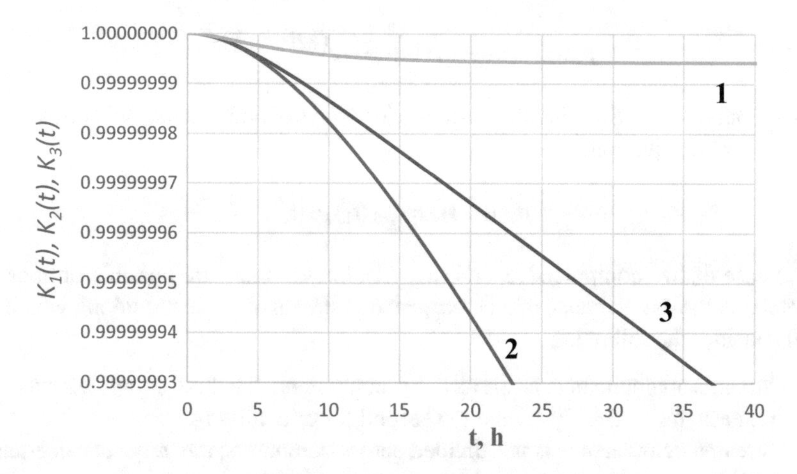

Fig. 2 Non-stationary coefficients applied to recovery strategies A, B, C

4 Analysis of the Possibility of Timely Requests' Execution

For real-time cyber-physical systems, the reliability (stability) of operation depends not only on the readiness of the system structure to perform the required functions while ensuring the continuity of the computational process, but also on the timeliness of fulfilling delay-critical requests.

Structural reliability is characterized by the availability factor defined above. Timeliness of calculations characterizes the functional reliability of the system. Timeliness of calculations is the probability not to exceed the delay in servicing the request (the established maximum allowable time t).

The flow of service requests to computer nodes of cyber-physical systems, as a rule, is heterogeneous and contains requests of different criticality to the execution time.

For the considered two-machine cluster, unworkable and workable states are possible, in which the input request stream can be executed on two or one of the computers.

Consider a case in which the input stream is heterogeneous. In this case, there are requests for two criticality gradations for the waiting time t_1 and t_2 ($t_1 < t_2$), arriving with intensities Λ, Λβ and performing for the same average time v.

In the case of readiness (operability) of one computer out of two and the lack of service priorities, the probability of requests' execution with waiting criticality (maximum allowable waiting time) t_1 and t_2 ($t_1 < t_2$) we calculate, respectively, as:

$$p_1(t_1) = 1 - \Lambda(1+\beta)ve^{\left(\Lambda(1+\beta)-\frac{1}{v}\right)t_1}, \quad (2)$$

$$p_2(t_2) = 1 - \Lambda(1+\beta)ve^{\left(\Lambda(1+\beta)-\frac{1}{v}\right)t_2}. \quad (3)$$

The probability that the criticality requests for delays of both t_1 and t_2 will be executed in time by the equation:

$$p_{12}(t_1, t_2) = p_1(t_1)p_2(t_2). \quad (4)$$

In case of two computers' operability (readiness), there are possible options for organizing the maintenance of a heterogeneous stream of requests, among which we will consider the following:

- B1 Streams maintenance is divided by computers: the first computer executes requests for criticality t_1, and the second for criticality t_2.
- B2 Streams maintenance is not divided across computers; any request can equally well be sent to the queue of the first or second computers.
- B3 Servicing of streams of criticality to t_2 delays is performed exclusively by the second computer, and criticality to delays of t_1 is divided between two computers with probability g.

For the service organization option B1, the probability of executing requests with a waiting criticality (maximum allowable waiting time) t_1 and t_2 ($t_1 < t_2$) is calculated, respectively, as:

$$p_1(t_1) = 1 - \Lambda v e^{\left(\Lambda - \frac{1}{v}\right)t_1}, \tag{5}$$

$$p_2(t_2) = 1 - \Lambda\beta v e^{\left(\Lambda\beta - \frac{1}{v}\right)t_2}. \tag{6}$$

The probability of timeliness of servicing requests of both threads p_{12} (t_1, t_2) is calculated by (4).

Probability for service option B2:

$$p_1(t_1) = 1 - \frac{1}{2}\Lambda(1+\beta) v e^{\left(\frac{1}{2}\Lambda(1+\beta) - \frac{1}{v}\right)t_1}, \tag{7}$$

$$p_2(t_2) = 1 - \frac{1}{2}\Lambda(1+\beta) v e^{\left(\frac{1}{2}\Lambda(1+\beta) - \frac{1}{v}\right)t_2}. \tag{8}$$

The probability of timeliness is calculated by (4).

Probability for service option B3:

$$p_1(t_1) = g\left[1 - g\Lambda v e^{\left(\Lambda g - \frac{1}{v}\right)t_1}\right] + (1-g)\left[1 - \Lambda(\beta + 1 - g) v e^{\left(\Lambda(\beta+1-g) - \frac{1}{v}\right)t_1}\right], \tag{9}$$

$$p_2(t_2) = 1 - \Lambda(\beta + 1 - g) v e^{\left(\Lambda(\beta+1-g) - \frac{1}{v}\right)t_2}. \tag{10}$$

The probability of timeliness of service requests of two streams of different criticality to the expectation is calculated by (4).

4.1 The Example of Calculating the Probability of Timely Execution of Requests

Let us determine the probability of the timeliness of the redundant service in case of heterogeneity of the input stream, which includes requests for two grades of criticality for the waiting time t_1 and t_2 ($t_1 < t_2$).

In the calculations, we assume that $v = 0.1$ s, $t_1 = 0.2$ s, $t_2 = 0.4$ s.

The dependence of the probability of requests' timely execution, which have different criticalities p_{12} (t_1, t_2) on the intensity of the requests' stream Λ, is presented in Fig. 3. In Fig. 3 at $\beta = 0.5, 0.8, 1.5$, the curves 1–3 correspond to the variant of organizing the service of streams with divided service B1, and the curves 4–6 with combined service B2. Calculations show the expediency of combined service

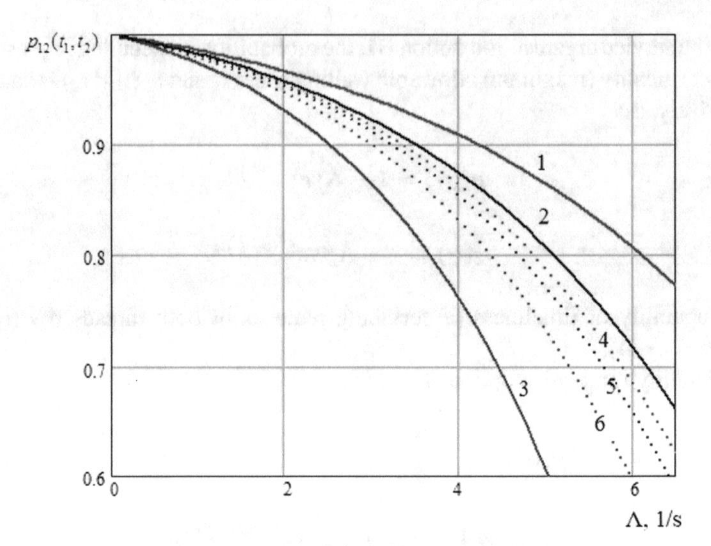

Fig. 3 The dependence of the probability of timely execution of requests from the intensity of the input stream

streams, which have different criticality, at low intensity of the second flow (less critical to the expectation) relative to the first.

For the service strategy B3, the influence on the probability of timely servicing the requests of two streams p_{12} (t_1, t_2) of the requests' intensity Λ and the fraction g of the first stream's requests (criticality t_1) sent to the first computer queue is shown in Fig. 4, in which, at $\beta = 0.5$, the curves 1–5 correspond to g = 1, 0.8, 0.5, 0.2. The presented dependences show the importance of stream control (change g) and the impact on the timeliness of servicing requests for the total stream of requests that have different criticality to delays.

The dependence of the timely servicing probability the requests of the two streams p_{12} (t_1, t_2) from the fraction g sent to the first computer is shown in Fig. 5, in which $\beta = 0.5$ the curves 1–4 correspond to the intensity of the input stream $\Lambda = 1, 2, 3, 4$ 1/s.

The presented dependences allow us to establish the significance of the distribution effect of streams that have different criticality to delays on the probability of timely execution of the requests' total stream with different criticalities. The analysis performed allows us to conclude that there is an optimal share of requests distributed to computers that have retained their functionality, depending on their criticality to delays. The share maximizes the probability of timely execution of total stream requests.

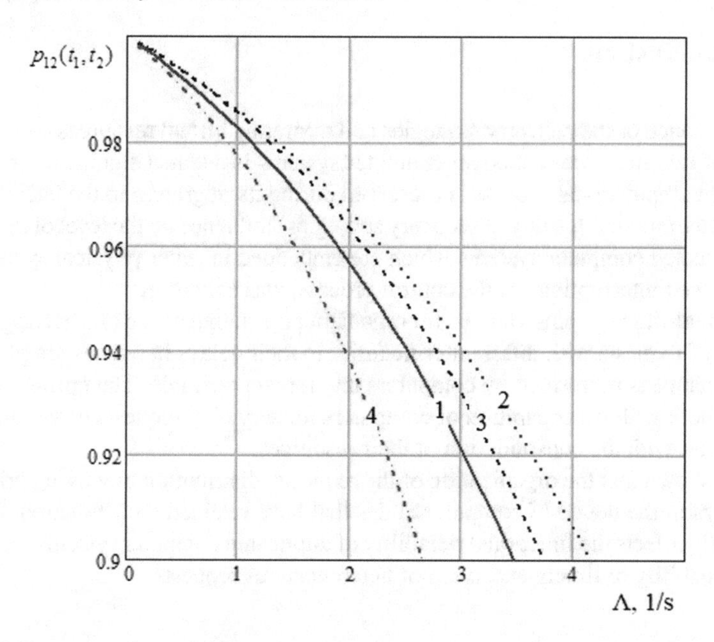

Fig. 4 Influences on the probability of requests timely servicing of two streams with requests' intensity Λ, depending on the fraction g of requests' first stream sent to the queue of the first computer

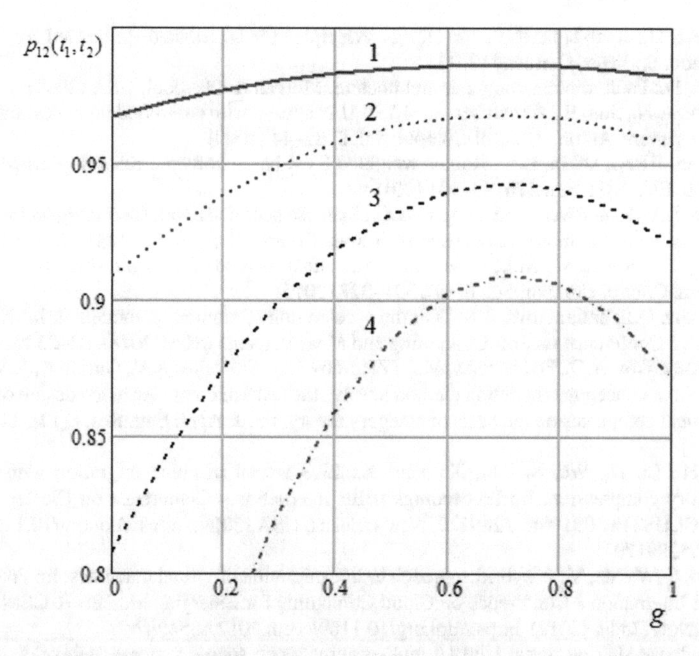

Fig. 5 Dependence of the probability of timely servicing of requests for two streams p_{12} (t_1, t_2) on the fraction g, requests sent to the first computer, with different intensity of requests Λ

5 Conclusions

The influence of the recovery strategies and migrating virtual resources on the reliability of two-machine redundant computer systems is deemed significant, provided that the computational process is preserved during its migration to the backup computer after failures. A study of recovery strategies' influence on the level of reliability of duplicated computer systems which are embedded in cyber-physical systems, do not allowed interruptions of the control process, was conducted.

For real-time systems, options for organizing the maintenance of a heterogeneous stream of requests with different criticalities to their delays in queues are proposed. Maintenance is performed by computers that remain operable. The options are presented both with the separation of computers for servicing requests of varying criticality, and with the consolidation of their resources.

It is shown that the organization of the requests' distribution of varying criticality to delays in the queue of computer nodes that have retained their functionality significantly affects the functional reliability of duplicated computer systems, including the probability of timely execution of heterogeneous requests.

References

1. Kopetz, H.: Real-Time Systems: Design Principles for Distributed Embedded Applications, 2nd edn. Springer, Germany (2011)
2. Sorin, D.: Fault tolerant computer architecture. Morgan & Claypool, USA (2009)
3. Dudin, A.N., Sun, B.: A multiserver MAP/PH/N system with controlled broadcasting by unreliable servers. Autom. Control Comput. Sci. **5**, 32–44 (2009)
4. Coolen, F.P.A., Utkin, L.V.: Robust weighted SVR-based software reliability growth model. Reliab. Eng. Syst. Saf. **176**, 93–101 (2018)
5. Utkin, L.V., Zaborovsky, V.S., Popov, S.G.: Siamese neural network for intelligent information security control in multi-robot systems. Autom. Control Comput. Sci. **8**(51), 881–887 (2017)
6. Aliev, T.I., Rebezova, M.I., Russ, A.A.: Statistical methods for monitoring travel agencies. Autom. Control Comput. Sci. **6**(49), 321–327 (2015)
7. Kutuzov, O.I., Tatarnikova, T.M.: On the acceleration of simulation modeling. In: XXI International Conference on Soft Computing and Measurements (SCM'2018), 23–25 May 2018
8. Korobeynikov, A.G., Fedosovsky, M.E., Zharinov, I.O., Shukalov, A.V., Gurjanov, A.V.: Development of conceptual modeling method to solve the tasks of computer-aided design of difficult technical complexes on the basis of category theory. Int. J. Appl. Eng. Res. **6**(12), 1114–1122 (2017)
9. Jin, H., Li, D., Wu, S., Shi, X., Pan, X.: Live virtual machine migration with adaptive memory compression. In: Proceedings IEEE International Conference on Cluster Computing (CLUSTER'09). Art. 5289170, New Orleans, USA (2009). https://doi.org/10.1109/clustr.2009.5289170
10. Sahni, S., Varma, V.: A hybrid approach to live migration of virtual machines. In: Proceedings IEEE International Conference on Cloud Computing for Emerging Markets (CCEM), 12–16, Bangalore, India (2012). https://doi.org/10.1109/ccem.2012.6354587
11. Knowledge sharing portal UNIX/Linux-systems, open source systems, networks, and other related things. http://xgu.ru/wiki/Kemari. Last accessed 25 Mar 2019
12. Dittner, R., Rule, D.: The Best Damn Server Virtualization Book Period, 2nd edn. Syngress, USA (2011)

13. Zhu, Jun, Jiang, Zhefu, Xiao, Zhen: Optimizing the performance of virtual machine synchronization for fault tolerance. IEEE Trans. Comput. **12**(60), 1718–1729 (2011)
14. Agrawal, S.: Hardware virtualization towards a proficient computing environment. Int. J. Innov. Appl. Stud. **2**(3), 528–534 (2013)
15. Khaled, Z.I., Hofmeyr, S., Iancu, C., Roman, E.: Optimized pre-copy live migration for memory intensive applications. In: International Conference for High Performance Computing, Networking, Storage and Analysis, Article 40 (2011)
16. Chandak, A., Jaju, K., Kanfade, A.: Dynamic load balancing of virtual machines using QEMU-KVM. Int. J. Comput. Appl. **6**(46), 10–14 (2012). (0975-8887)
17. Adamova, K.: Anomaly detection with virtual service migration in cloud infrastructures. Master thesis. 263-0800-00L (2012)
18. Liang, Hu, Zhao, Jia, Gaochao, Xu, Ding, Yan: HMDC: live virtual machine migration based on hybrid memory copy and delta compression. Appl. Math. Inf. Sci. **7**(2L), 639–646 (2013)
19. Soni, G., Kalra, M.: Comparative study of live virtual machine migration techniques in cloud. Int. J. Comput. Appl. **14**(84), 19–25 (2013). (0975-8887)
20. Ageev, A.M.: Configuring of excessive onboard equipment sets. J. Comput. Syst. Sci. Int. **4**(57), 640–654 (2018)
21. Bogatyrev, A.V., Bogatyrev, S.V., Bogatyrev, V.A.: Analysis of the timeliness of redundant service in the system of the parallel-series connection of nodes with unlimited queues. In: 2018 Wave Electronics and its Application in Information and Telecommunication Systems (WECONF) (2018)
22. Bogatyrev, V.A: On interconnection control in redundancy of local network buses with limited availability. Eng. Simul. **16**(4), 463–469 (1999)
23. Bogatyrev, V.A.: Increasing the fault tolerance of a multi-trunk channel by means of inter-trunk packet forwarding. Autom. Control Comput. Sci. **33**(2), 70–76 (1999)
24. Bogatyrev, V.A., Aleksankov, S.M., Derkach, A.N.: Model of cluster reliability with migration of virtual machines and restoration on certain level of system degradation. In: 2018 Wave Electronics and its Application in Information and Telecommunication Systems (WECONF) (2018)
25. Bogatyrev, V., Vinokurova, M.: Control and safety of operation of duplicated computer systems. Commun. Comput. Inform. Sci. **700**, 331–342 (2017)

Data Storage Model for Organizing the Monitoring of POS-Networks Processes and Events

Dmitriy Kozlov, Natalia Sadovnikova and Danila Parygin

Abstract The chapter describes the main problems of monitoring the POS-networks operation. The task of developing a new method of storing and processing data of the POS-devices network for effective monitoring of events and processes occurring in it is considered. The main problems associated with storing data about the topology of the POS-network, its structural elements, events and processes occurring in it are formulated. Storage methods and data processing algorithms that can solve actual problems when solving the problem of monitoring a large number of POS devices are described.

Keywords Monitoring · Information system · POS networks · Graph models · Event flow · Data model · Database

1 Introduction

Today the system of electronic payments is a complex structure. It contains many heterogeneous subsystems and elements. The information protection in the performing payment transaction process requires stricter control and threat detection in the operational mode. The system can prevent the leakage of critical information if the speed of detection and response to a cyber-threat are fast enough. The problem is aggravated by the constantly increasing topological complexity of networks, which form software and hardware POS subsystems.

There are a lot of methods to combat cyber threats in POS networks [1]. The introduction of a unified information system for the POS-networks administration and its integration with the proactive monitoring service can be one of the solutions that

D. Kozlov (✉) · N. Sadovnikova · D. Parygin
Volgograd State Technical University, Lenin av., 28, Volgograd 400005, Russian Federation
e-mail: mrdiko4@gmail.com

N. Sadovnikova
e-mail: npsn1@ya.ru

D. Parygin
e-mail: dparygin@gmail.com

A. G. Kravets et al. (eds.), *Cyber-Physical Systems: Industry 4.0 Challenges*, Studies in Systems, Decision and Control 260,
https://doi.org/10.1007/978-3-030-32648-7_3

significantly reduce the risks. This service should allow the POS-network operator to collect and provide information about existing threats and network status in real-time. The monitoring service allows us to quickly detect security threats [2], effectively eliminate them, and in some cases even prevent their occurrence [3]. It is possible to detect cyber threats and timely informing decision-makers (operators of the POS network administration system) about them [4] with the help of modern technologies, such as real-time databases and real-time web [5].

The operator in real-time will have access to all events and processes occurring in the POS network when implementing the monitoring service. However, this will increase the effectiveness of detection of threats, only if the operator will be able to correlate each process and their events, with a particular device. The complexity of the events classification task is associated with several factors:

- A large number of events. It is simply impossible for a person to analyze such a volume of information in real-time.
- A large number of different types of processes occurring in the system. Classify an event with a particular process of a certain type often requires a detailed analysis of the event parameters and presents a challenge for the operator.
- The complex topology of modern POS-networks. Often, an event belongs to a process that occurs not in one device but in several connected devices, or even in a group of devices containing both connected and unconnected devices. In this case, the correlation of a particular event that occurred on a particular device with the process that encompasses a group of devices taking into account the topology of the network formed by a given group of devices is a nontrivial task.

Therefore, the organization of effective monitoring is possible only if there is a tool to filter the events flow, associate events with a specific process and with all the associated POS-network structure elements. A cognitive representation method of the POS-network topology and events associated with its elements was proposed to implement it [6]. This approach allows to solve the tasks and significantly reduce the time to identify threats, their analysis, and prevention.

However, a large number of POS-network elements and connections between them, and an order of magnitude more events generated by the POS-network in order to build an effective monitoring system require nontrivial approaches to organizing data storage and processing. In addition, it was demonstrated in [6] that the cognitive mapping of the POS-network in the form of a special type graph with the display of real-time events on it is the most effective from the point of view of implementation monitoring the POS-networks operation. This method of presentation imposes additional requirements on the filtering data speed on existing topological connections between the POS-network elements.

Thus, it is necessary to develop a model for storing and processing POS-network data to build an effective monitoring system. The data model is the set of data structures and their processing operations [7]. Consequently, it is necessary to determine the structure of POS-network elements, the relations between them and the main possible operations on these structures.

2 POS-Network Administration and Monitoring System Architecture

At present, each customer who purchases and implements the POS administering system sets the requirements to the simplicity of access to actual information, preferring the cloud systems. At the same time, the systems themselves should subject to the requirements of high performance, security, information sharing between different levels of user responsibility, as well as reliability and clarity.

Recently, technologies are being actively developed that can help in building cloud solutions that meet the requirements presented above. These technologies include so-called Real-time web technologies [8], Event Sourcing [9], and real-time databases [10].

As shown in [4], the most effective approach for organizing cloud-based real-time data flow monitoring is the integration of real-time databases into the system operation process. In this case, it is not necessary to store the entire domain data layer in this database, but only the events that occur in the system that have the necessary information for its immediate presentation to a user. The described architecture can be illustrated by the UML component diagram (see Fig. 1).

The figure shows that the use of a real-time database that stores events that occur in the system does not negate the presence of a database that stores in it the entire domain data layer. This database can also be relational for organizing high-speed queries and a more rigorous description of the domain data layer.

In turn, the NoSQL database RethinkDB [11] was chosen to implement the real-time database for event storage. Data are stored in JSON format documents. This allows us to store and process events that have a flexible format, which can be changed during the system lifecycle. In addition, RethinkDB allows notifying clients about changes in data using the WebSocket protocol. This architecture allows to open a messaging channel directly from the browser, and both the server and the client can be the initiator of sending a message. Today, JSON format is the de facto standard for exchanging data on the web. Using the JSON format to store events also allows us to generate, notify, process events with virtually no additional transformations.

3 The Relational Data Model of the POS-Network Administration System

POS-network is a set of software and hardware solutions that perform payment transactions or involved in their provision. Hardware solutions include payment terminals, PIN pads, readers and other various model devices. In addition, there are models of devices, which capable of performing several of the listed functions or all at the same time. There are also software solutions running on personal computers and taking on some of the terminal's functions. In the future, each of the hardware solutions will be referred to as a physical device (PD).

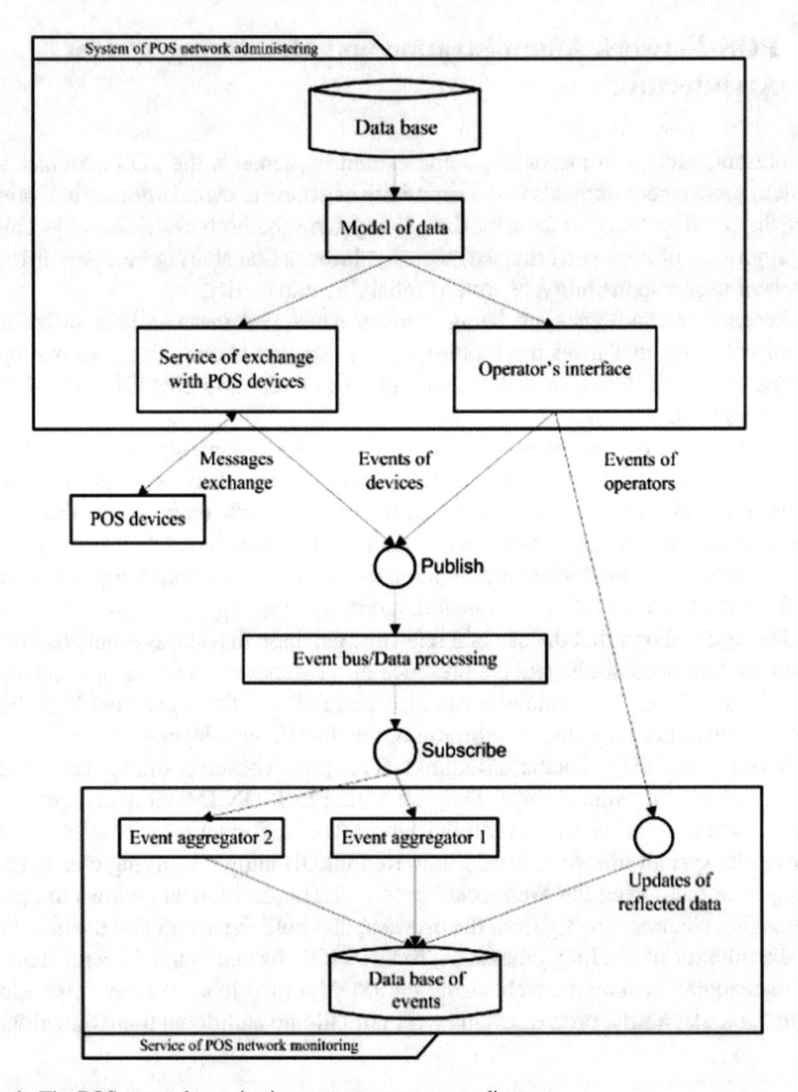

Fig. 1 The POS-network monitoring system components diagram

Often it is possible to physically connect several physical devices. For example, connecting a terminal and a PIN pad or PIN pad and reader. One physical link connects two devices: the control device and the controllable device. Several connected devices form a device complex. The structure of devices complex is almost always hierarchical and tree-like. The device's complex has one main device—the root of the complex. Each controllable device has only one control device.

The physical device has a hardware implementation and is used to conduct financial transactions at trade points. Each trade point has its own geographical address.

Each of the physical devices has its own software. For example, the terminal correct operation needs the software that provides to perform payment operations in its serving acquirer bank. Such software will be called a logical device (LD). Moreover, it is important to remember that there are devices capable of working with several acquiring banks (multi-banking) or with several currencies. In this case, work with each of the banks and with each of the currencies requires a special instance of software running on the device. It follows that one or more logical devices correspond to one physical device.

Thus, each of the logical devices is connected with the acquirer bank. This connection is carried out using a bank account that is available to the owner of the physical device. The bank account holder will be referred to as the Customer (the owner of the account/accounts in the acquiring bank).

Merchant is a legal entity that carries out its activities in the sphere of selling goods and providing services. Merchant can accept payment cards using POS-terminals. Each logical device is connected to the corresponding Merchant. One Merchant can operate at one or several trade points. It is also possible that several Merchants are present at one trade point.

Customers can have one or more Merchants.

In addition, Customer like a Trade Point has a geographical address. A customer address is the legal address of the customer office.

The domain data area that includes basic information about the POS-network devices and their parameters, device owners and operating parameters of the POS-network administration system, is contained in the relational database. PostgreSQL was chosen as the DBMS.

Figure 2 shows the database logical structure of the domain data layer, presented in the form of an ER diagram.

The diagram contains the main tables and columns describing the main POS-network structural elements and their key fields. The main POS-network structural elements include Physical Device (PD), Logical Device (LD), Trade Point, Merchant, Account, Customer, Address, City, Region, and Country. Each of the entities has its own key parameters:

- Physical Device: device identifier (Device.Id); device serial number (Device.SerialNumber); device model (DeviceModel.Name).
- Logical Device: device identifier (Device.Id); unique string terminal identifier Terminal ID (Device.TerminalId).
- Trade Point: unique string trade point identifier Trade Point ID (TradePoint.Id); trade point name (TradePoint.Name).
- Merchant: unique string merchant identifier Merchant ID (Merchant.Id); merchant name (Merchant.Name).
- Account: unique string account identifier Account ID (Account.Id); account name (Account.Name).
- Customer: unique string customer identifier Customer ID (Customer.Id); customer name (Customer.Name).

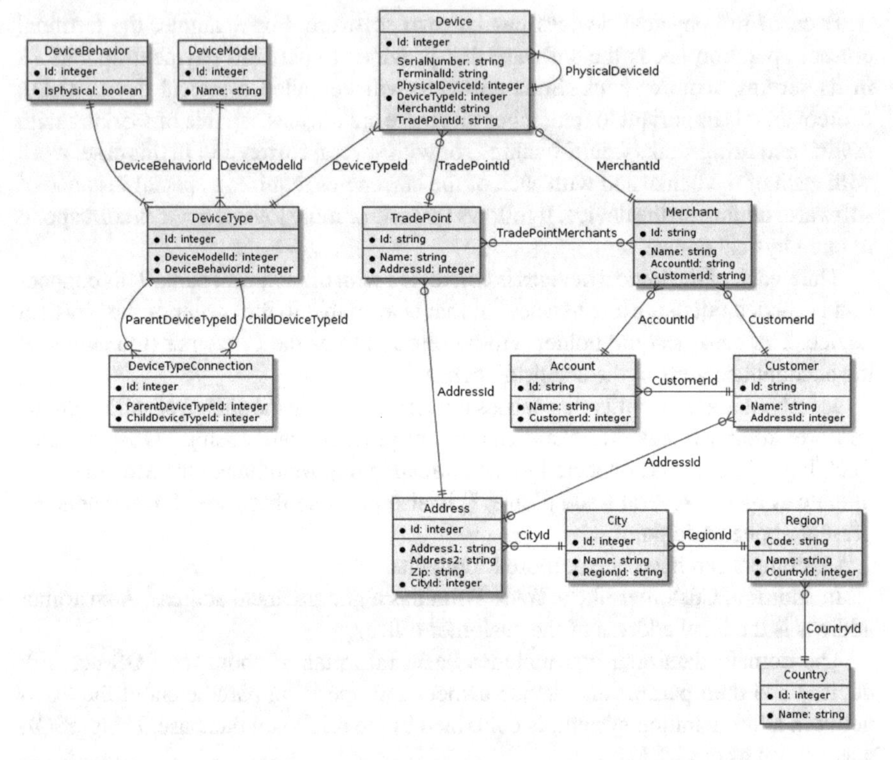

Fig. 2 ER-diagram of the POS-network administration system domain data layer

- Address: unique address identifier (Address.Id); the string, which contains the street name and house number (Address.Address1).
- City: unique identifier (City.Id); city name (City.Name).
- Region: region code (Region.Code); region name (Region.Name).
- Country: unique country identifier (Country.Id); country name (Country.Name).

As can be seen from the above description the POS-network structural elements have the following relations among themselves:

- Physical Device (one)—(many) Physical Device;
- Physical Device (one)—(many) Logical Device;
- Trade Point (one) —(many) Physical Device;
- Merchant (one)—(many) Logical Device;
- Trade Point (many)—(many) Merchant;
- Account (one)—(many) Merchant;
- Customer (one)—(many) Account;
- Address (one)—(many) Customer;
- Address (one)—(many) Customer;

- City (one)—(many) Address;
- Region (one)—(many) City;
- Country (one)—(many) Region.

In the context of this work, it is important to note that such a list of structural elements and the relations between them in their totality form a graph.

4 POS-Network Monitoring System Data Model

A graph model [12] of data presentation was chosen [6] for the most detailed and understandable to the POS-network monitoring system operator presentation since the POS-network structural elements and the relations between them have a graph structure. An example of a POS-network graphical representation is presented in Fig. 3.

Operations on the graph data model in the relational model are not natural. There is a special kind database for this purpose called graph database [13]. Graph databases are much more efficient for working with arbitrary graphs of unknown structure. However, graph databases are not always necessary in tasks where the structure of the graph is clear and predictable [14]. The graph structure is known and its processing does not require a recursive traversing in the presented case. In addition, the relational database is needed to perform the POS-network administration. Therefore the implementation of a graph database would require not only additional resources but also the availability of synchronization methods. Data storage in both relational and graph databases is redundant. Thereby it was decided to use the relational data model to perform the task.

It is also worth noting that the problem point when using the graph representation of the POS-network is the display of networks with a large number of nodes. Its construction requires a large number of computational resources and it is difficult to navigate in it. It was decided to use the following approaches to solve this problem:

1. Filtering POS-network nodes by their key parameters. In this case, not only the filtered elements should be displayed, but also all the nodes directly connected with them. A subgraph that satisfies specified filters should be returned as a result of the filtering.
2. The possibility of hiding particular type elements of POS-network.
3. The possibility of network nodes collapse.

It is much more convenient to operate not with separate nodes but with their hierarchies to implement this functionality. It is necessary to select the most "common" nodes at the top of the hierarchy and the most "private" at the down of the hierarchies to do this. In this case, each neighboring levels of the hierarchy should have one of the previously described relations between each other. Two basic hierarchies of node types (from "higher" to "lower") were defined as a result of this analysis:

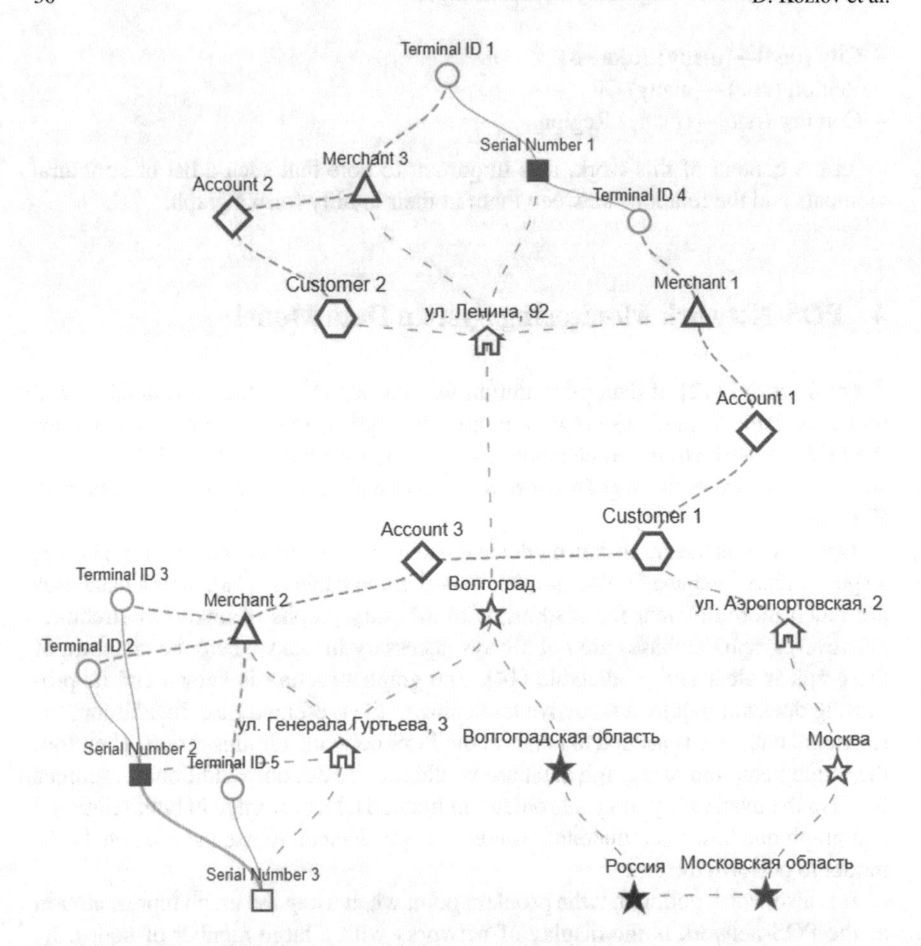

Fig. 3 Example of a POS-network graphical representation

1. Country—Region—City—Address—Customer—Account—Merchant—Logical Device—Physical Device.
2. Country—Region—City—Address—Trade Point—Physical Device—Logical Device.

Some basic rules for the POS-network display were denoted after definition the concept of hierarchy. These rules can help the monitoring system operator to find POS-devices and their relationships with other POS-network structural elements:

1. By default, have no more than 50 nodes on one screen and all the rest are collapsed.
2. The operator is not limited in the number of nodes on one screen by expanding the nodes independently.
3. Not only the nodes themselves but also nodes from their hierarchies fall into the result when filtering nodes by their attributes.

4. The operator is shown connections between the nodes located down the hierarchy and the node located up the hierarchy when hiding POS-network elements of a particular type.
5. Nodes located down the hierarchy are not displayed when the operator collapse node.

Information about all POS-network nodes and their relationships is stored in a relational database. However, they contain many fields that are not significant for representing the graph model of the POS network. Therefore, these fields need to be constantly sifted out when selecting data. These fields also need to be constantly discarded when sampling data. In addition, the nodes of each specific type are stored in their database table. It is very difficult to handle them as equivalent entities (nodes of the graph). Thereby it was decided to use the DBMS tool "views" to bring data about nodes into a single format. It was also decided to use the built-in PostgreSQL data type "jsonb" in views in order to avoid code redundancy for additional conversion to the JSON format. This helped convert the data about nodes of all types to a single format. For example, the SQL-script for creating a view for the nodes of the type "Logical Device" is shown below.

```
CREATE OR REPLACE VIEW public."JsonLogicalView" AS
  SELECT j.entity::jsonb || jsonb_build_object('id',
md5(j.entity::text)) AS "row"
  FROM (
    SELECT row_to_json(ld.*) AS entity
    FROM (
      SELECT
        'logical' AS type,
        d."Id" AS "deviceId",
        d."TerminalId" AS "terminalId",
        d."PhysicalDeviceId" AS "physicalDeviceId",
        d."MerchantId" AS "merchantId",
        1 AS "devicesCount"
      FROM "Device" d
      JOIN "DeviceType" dt ON dt."Id" = d."DeviceTypeId"
      JOIN "DeviceBehavior" db ON db."Id" =
dt."DeviceBehaviorId"
      WHERE db."IsLogical" = true AND d."Deleted" = false
    ) ld
  ) j;
```

The execution result of the above script is presented in Fig. 4.

The first 50 network nodes are displayed starting from the top of the hierarchy on the monitoring system main page. A general list of all network nodes has to be known to implement this functionality. To do this, an additional view "JsonDomainView" was created. The corresponding script is shown below.

1	{ "id": "09f5b365e66c318bd6bb21c0bf721d57", "type": "logical", "deviceId": 31147, "merchantId": "MERCHANT0000001", "terminalId": "TERM0001", "devicesCount": 1, "physicalDeviceId": 31145 }
2	{ "id": "9c61d9df10e19903d76e637d4b84f330", "type": "logical", "deviceId": 30171, "merchantId": "1013", "terminalId": "01/12/2017", "devicesCount": 1, "physicalDeviceId": 30170 }
3	{

Fig. 4 JsonLogicalView rows data

```
CREATE OR REPLACE VIEW public."JsonDomainView" AS
  SELECT "JsonCountryView"."row" FROM "JsonCountryView"
UNION ALL
  SELECT "JsonRegionView"."row" FROM "JsonRegionView"
UNION ALL
  SELECT "JsonCityView"."row" FROM "JsonCityView"
UNION ALL
  SELECT "JsonAddressView"."row" FROM "JsonAddressView"
UNION ALL
  SELECT "JsonTradePointView"."row"
  FROM "JsonTradePointView"
UNION ALL
  SELECT "JsonCustomerView"."row" FROM "JsonCustomerView"
UNION ALL
  SELECT "JsonAccountView"."row" FROM "JsonAccountView"
UNION ALL
  SELECT "JsonMerchantView"."row" FROM "JsonMerchantView"
UNION ALL
  SELECT "JsonPhysicalView"."row" FROM "JsonPhysicalView"
UNION ALL
  SELECT "JsonLogicalView"."row" FROM "JsonLogicalView";
```

The data about existing hierarchies are computable and should not be stored in separate tables in order to avoid excessive computational resources and to keep them up to date. All operations related to the acquisition of graph data structures and manipulations with them should be performed on the DBMS side because of the large number of possible nodes. Therefore, it was also decided to use the views tool to operate with hierarchies, rather than individual events. The script for creating such a representation is presented below.

```
CREATE OR REPLACE VIEW "JsonDomainHierarchyView" AS
SELECT
  logical."row" AS logical, physical."row" AS physical,
  child_physical."row" AS child_physical,
  child_logical."row" AS child_logical,
  merchant."row" AS merchant, account."row" AS account,
  customer."row" AS customer,
  tradepoint."row" AS "tradePoint",
  tradepointaddress."row" AS "tradePointAddress",
  tradepointcity."row" AS "tradePointCity",
  tradepointregion."row" AS "tradePointRegion",
  tradepointcountry."row" AS "tradePointCountry",
  customeraddress."row" AS "customerAddress",
  customercity."row" AS "customerCity",
  customerregion."row" AS "customerRegion",
  customercountry."row" AS "customerCountry",
  parent_physical."row" AS parent_physical,
  parent_logical."row" AS parent_logical
FROM "JsonPhysicalView" physical
JOIN "JsonLogicalView" logical ON
  physical."row"->>'deviceId'::text = logical."row"-
>>'physicalDeviceId'::text
-- ... --
LEFT JOIN "JsonLogicalView" parent_logical ON
  parent_logical."row"->>'physicalDeviceId'::text = par-
ent_physical."row"->>'deviceId'::text;
```

Cells of the "JsonDomainHierarchyView" view rows contain data about POS-network nodes in JSON format from the hierarchies of both described types. Figure 5 shows an example of a POS-network graph, which displayed on the main page of the developed monitoring system.

A rather simple SQL-script which used to obtain data for display the main page of the monitoring system is shown below.

```
select row_to_json(row.*) from (
  select j."row" as j from "JsonDomainView" as j
  limit 50
) row;
```

More complex queries that receive a set of filtered hierarchies from the "JsonDomainHierarchyView" view are used to get subnet satisfying one or another filter.

5 POS-Network Event Storage Data Model

As can be seen from Fig. 5, some nodes are highlighted in a special way. In this case, the highlighted nodes signal the operator that there is an error in the devices located lower in the hierarchy. This error is presented as an event with the type "error", which is stored in the events database. There is a document named "Events" for storing events of any type in the events database. Each event is stored as a JSON record. A complete list of event fields is presented in Table 1.

The described list of fields allows:

- Get information about the type of event and its main characteristics.
- Know by whom and when the event was generated.
- Correlate the event with each POS-network node from the hierarchy corresponding to the event by the node key field value.

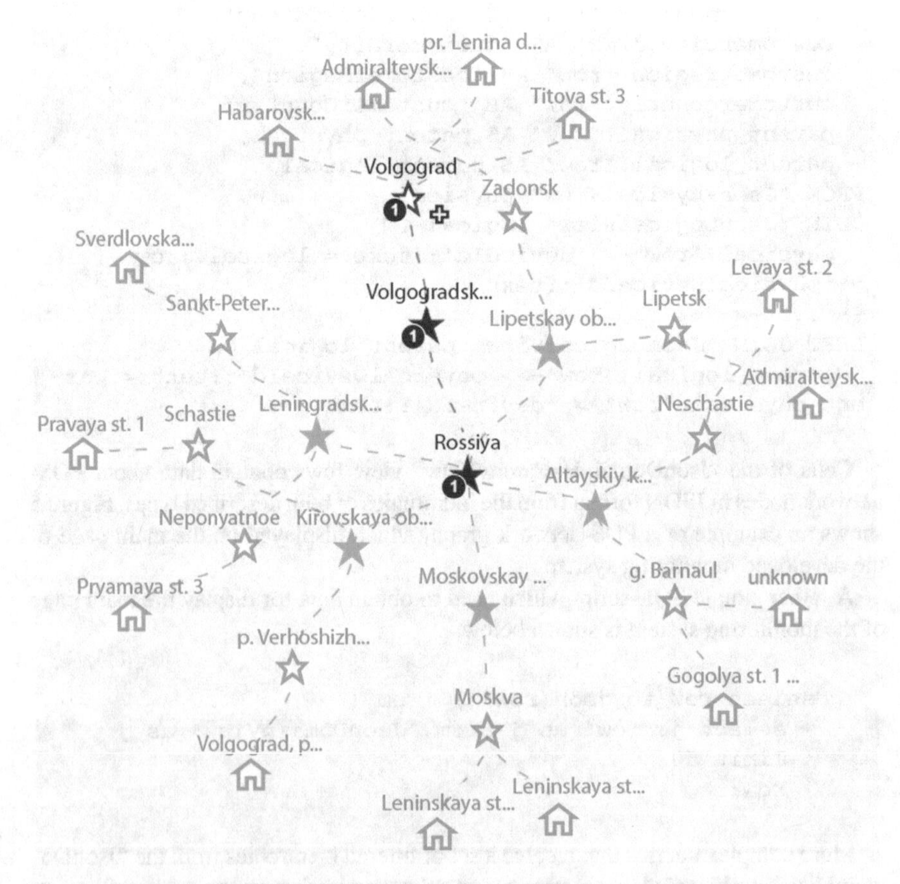

Fig. 5 An example of a POS-network graph on the monitoring system main page

Table 1 Description of all POS-network event fields

Field name	Field type	Description
id	string	The unique identifier of the event
type	string	Class of event: {'error', warning', 'info', 'success'}
subtype	string	The type of process to which the event belongs. Defines the business process in which the event was generated
acceptedAt	date time	Date and time of event creation
code	number	Event code/type. Corresponds to a specific localized message for presentation to the user. It allows classifying events according to their logical affiliation to a particular business process
emitter	string	The type of actor that generated the event
emitterId	string	The unique identifier of the actor that generated the event
content	object	Additional information about the event. For example, the progress of the download update kit
resolved	boolean	A sign that the event is "resolved" (it was hidden from display to the user and is no longer relevant)
terminalId	string	Logical device Terminal ID
serialNumber	string	Serial number of physical device
merchantId	string	Merchant identifier
accountId	string	Customer bank account identifier
customerId	string	Customer identifier
tradePointId	string	Trade point identifier
addressId	number	Address identifier
cityId	number	City identifier
regionId	string	Region code
countryId	number	Country identifier
customerAddressId	number	Customer address identifier
customerCityId	number	Customer city identifier
customerRegionId	string	Customer region code
customerCountryId	number	Customer country identifier

The following steps were taken to increase the speed of work with the records from the "Events" document since this document potentially stores a huge number of events:

- The indexes for the following event fields are set: "type", "subtype", "acceptedAt", "code", "emitter", "emitterId", "terminalId", "merchantId".
- The document has been partitioned by the "resolved" event field.

6 Conclusion

This chapter describes the storage data model of the POS-network monitoring system and the main algorithms for working with it. This, in its totality, will allow solving the problems described above.

As a result of the work, models and methods for storing and processing data on the structure of the POS-network and its elements were described. The described models and methods allow the operator of the monitoring system to display the POS-network structure as a whole or its subnets as graphs. In addition, it will allow filtering POS-network elements together with their relationships. The described method of displaying a POS-network in the process of monitoring the operation of POS-devices will make it possible to observe the status of the network as a whole, taking into account the parameters of each of its structural elements and the relationships between them. Ultimately, this will reduce the time to detect security threats for specific devices and the entire network as a whole. As a result, it will reduce the number of successful fraudulent attacks in the system and reduce the costs of leveling them. In addition, the proposed method will reduce the number of labor costs for analyzing the current state of the system by automating the process of collecting detailed information about each event.

Acknowledgements The reported study was funded by Russian Foundation for Basic Research according to the research project No. 18-37-20066_mol_a_ved.

References

1. Palshikar, G.K.: The hidden truth—frauds and their control: a critical application for business intelligence. Intell. Enterp. **5**(9), 46–51 (2002)
2. Kamaev, V.A., Finogeev, A.G., Finogeev, A.A., Parygin, D.S.: Attacks and intrusion detection in wireless sensor networks of industrial SCADA systems. J. Phys. Conf. Ser. **803**, 012063. http://iopscience.iop.org/papers/10.1088/1742-6596/803/1/012063/pdf. Last accessed 18 Mar 2019
3. Makarov V., Gaponenko V., Toropov B., Kupriyanov A.: Theoretical and applied aspects of orthogonal coding in computer networks technologies. In: Kravets A. (eds.) Big Data-Driven World: Legislation Issues and Control Technologies. Studies in Systems, Decision and Control, vol 181. Springer, Cham (2019)
4. Avdeyuk, O.A., Kozlov, D.V., Druzhinina, L.V., Tarasova, I.A.: Fraud prevention in the system of electronic payments on the basis of POS-networks security monitoring. In: Trapeznikov, V.A. (ed.) IEEE Tenth International Conference «Management of large-scale system development» (MLSD'2017). Moscow, Russia, 2–4 Oct 2017. Proceedings, 4p. Institute of Control Sciences of Russian Academy of Sciences, IEEE (Institute of Electrical and Electronics Engineers) (2017). https://doi.org/10.1109/mlsd.2017.8109597
5. Wingerath, W.: A Real-Time Database Survey: The Architecture of Meteor, RethinkDB, Parse & Firebase. https://medium.baqend.com/real-time-databases-explained-why-meteor-rethinkdb-parse-and-firebase-dont-scale-822ff87d2f87. Last accessed 30 Mar 2019
6. Kozlov, D., Druzhinina, L., Sadovnikova, N., Petrova, D. (eds.) Displaying the flow of the event of the POS-devices network for building an effective monitoring system. In: Proceedings of

2018 11th International Conference Management of Large-Scale System Development, MLSD 2018. Moscow (2018)
7. West, M.: Developing High-Quality Data Models. Morgan Kaufmann (2011)
8. Leggetter, P.: Real-Time Web Technologies Guide. https://www.leggetter.co.uk/real-time-web-technologies-guide/. Last accessed 15 Feb 2019
9. Fowler, M.: Event Sourcing: Capture all Changes to an Application State as a Sequence of Events, 12 Dec 2005
10. Buchmann, A.: Real-time database systems. In: Rivero, L.C., Doorn, J.H., Ferraggine, V.E. (eds.) Encyclopedia of Database Technologies and Applications. Idea Group (2005)
11. The open-source database for the realtime web. https://www.rethinkdb.com/. Last accessed 10 Mar 2019
12. Xu H., Hipel K.W., Kilgour D.M., Fang L.: Conflict models in graph form. In: Conflict Resolution Using the Graph Model: Strategic Interactions in Competition and Cooperation. Studies in Systems, Decision, and Control, vol 153. Springer, Cham (2018)
13. Robinson, I., Webber, J., Eifrem, E.: Graph databases. In: New Opportunities for Connected Data, 2nd edn., p. 238. O'Reilly Media, June 2015
14. Vicknair, C., Macias, M., Zhao, Z., Nan, X., Chen, Y., Wilkins, D: A comparison of a graph database and a relational database: a data provenance perspective. In: ACM Southeast Regional Conference. Published 2010. https://doi.org/10.1145/1900008.1900067

Data Mining Integration of Power Grid Companies Enterprise Asset Management

Oleg Protalinskiy, Nikita Savchenko and Anna Khanova

Abstract The issues of integration at the level of enterprise asset management systems data (EAM-systems), ontological modeling systems and data mining on defect identification in electrical equipment of Power Grid Companies are considered. Transformation of EAM-system data will be provided by special converters at the syntactic level, and at the semantic level—by the ontological model of defects and equipment of Power Grid Companies. In order to integrate EAM data with the Data Mining module, ETL procedures (Extract, Transform, Load) are used to load information about defects and equipment of a Power Grid Company. It is proposed to use artificial neural networks and decision trees for the production processes intellectualization of Power Grid Companies. Ontology and data mining models are integrated at the level of metadata concerning defects and equipment.

Keywords Data integration · Defect · Electrical equipment · Data mining · Ontology · Neural networks · Decision trees

1 Background

The lack of necessary investments in the power grid complex over the past 20 years has caused significant physical and technological obsolescence of electrical grid equipment. The total wear of electrical distribution systems has reached 70% [1]. Defects of the equipment of a Power Grid Company (PGC) can cause emergencies, any equipment with a defect is almost impossible to be used for its intended purpose or its effectiveness drops.

O. Protalinskiy
Moscow Power Engineering Institute, Krasnokazarmennaya Street, 14, Moscow 111250, Russia
e-mail: protalinskiy@gmail.com

N. Savchenko · A. Khanova (✉)
Astrakhan State Technical University, Tatishcheva Street 16, Astrakhan 414056, Russia
e-mail: akhanova@mail.ru

N. Savchenko
e-mail: savtch.nickita1997@yandex.ru

A. G. Kravets et al. (eds.), *Cyber-Physical Systems: Industry 4.0 Challenges*, Studies in Systems, Decision and Control 260,
https://doi.org/10.1007/978-3-030-32648-7_4

PGCs use Enterprise Asset Management Systems (EAM-Systems) of various developers. The key IT market players in this segment are the SAP, IBM, various 1C configurations [2]. It is possible to identify defects in PGC equipment based on data accumulated in EAM-Systems on the bases of data mining technologies. It is becoming an urgent task to integrate heterogeneous EAM-Systems into a single infrastructure for managing fixed assets of PGCs [3] in terms of the implementation of the Industry 4.0 concept, the overcoming of the trend of none-detecting defects in electrical equipment, reducing investment and operating costs, creating smart grids (Smart Grid). One of the currently most common approaches for integrating intelligent information systems is a data-level integration [4].

2 Data Power Grid Companies Enterprise Assets Management Integration

A preventive and planned maintenance system (PPM) is used in the Power Grid Companies in the process of Enterprise Asset Management. Planned maintenance is carried out in several stages: PPM scheduling, performing maintenance repairs and drawing up acts and reports based on the repairs performed. The preventive and planned maintenance schedule is drawn up on the basis of data received from the automatic equipment monitoring systems and the frequency of the preventive maintenance specified in the reference information for this equipment (Fig. 1a). Another information source about the equipment state is a routine inspection, which is also performed at the frequency indicated in the reference information, as well as in case of deviation detection in the monitoring data for specifying breakdowns.

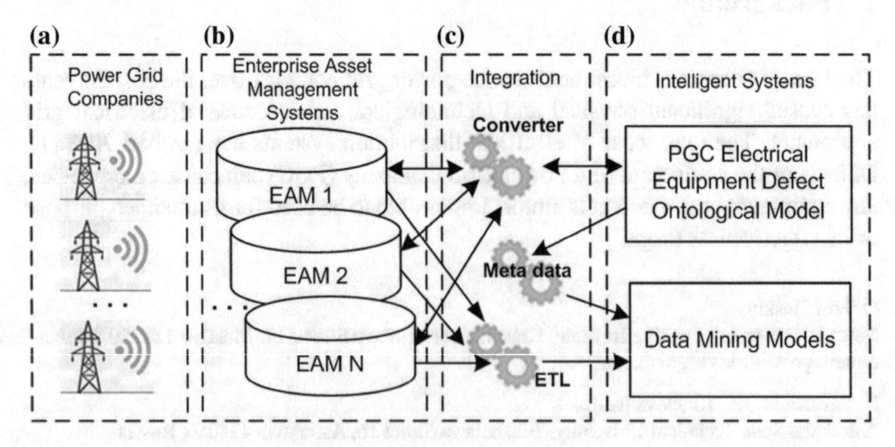

Fig. 1 PGC enterprise asset management systems integration

The resulting maintenance requests are ranked according to their criticality and are distributed by date, making up a maintenance schedule.

Spare parts are purchased for each maintenance request or, if possible, they are manufactured by the deadline specified in the maintenance schedule. On the specified date, all the activities related to the maintenance are performed—equipment turning off, the vacation of premises from the employees who are not involved in the maintenance, if necessary, and the maintenance is performed. After the maintenance, the equipment is tested and put into operation, and the removed parts are disposed of in accordance with the technical documentation instructions.

Due to the complexity, duration, and manpower effort of the equipment inspection process in Power Grid Companies, there arise situations where defects remain non-detected for a long time. If the defect is not detected and eliminated in time, and the equipment continues to be operated, the defect will begin to worsen, leading to an increase in the maintenance cost, accidents, emergency operation stoppage, injuries at work, and damage to other equipment assets. The introduction of intellectual Data Mining technology into the production cycle of asset management allows identifying defects in PGC equipment on the basis of data accumulated in EAM-Systems.

PGC EAM-Systems are focused on collecting, processing and storing the same types of information: login information, geodetic, object, textual information (Fig. 1b). Information support of different EAM-Systems is distinguished by semantic heterogeneity: the set of groups and brands of equipment, mismatch, and inconsistency of the equipment description levels, parameters, defects, measured characteristics and parameter dimensions of the equipment, as well as of the terminology used [5]. There are differences in the interpretation of these types of data by individual systems that is why solving the problem of semantic interpretation of data when transferring them from one control system to another one becomes a priority in order to ensure the data-level systems integration [4]. Ontologies are used to solve such problems in information systems. [6, 7]. In order to conduct research related to the equipment maintenance processes, the integration of the use of EAM-Systems data, models of ontologies and Data Mining is proposed (Fig. 1d).

To ensure the interaction of EAM systems and the ontological module, you can use a converter that will solve the syntactic data transformation problem, and the ontology will solve the semantic matching problem [4]. ETL-procedures (Extract, Transform, Load) [8] are used to integrate EAM-systems data with the Data Mining module, with which the information about equipment defects is also loaded. Equipment defects metadata level integration is proposed for the integration of the ontological model and the Data Mining models (Fig. 1c). The solution of these problems will allow the use of heterogeneous control systems for solving complicated complex problems not only in the field of PGC Enterprise Asset Management but also in various fields of activity.

3 Electrical Equipment Defect Ontological Model

PGC electrical equipment defect ontological model can be represented by a tuple [9]:

$$Q = \langle C, M, R \rangle,$$

where $C = \{c_i\}$—a set of concepts (concepts or classes when implementing an ontology in Ontostudio), building up an ontology Q, $i = \overline{1, I}$, that is $|C| = I$; $M_i = \{m_{1_i}, \ldots, m_{d_i}\}$—a set of attributes of the concept c_i (d—the number of attributes describing this concept); $R \subseteq C \times C$—direct inheritance relationship (a simple, symmetric, transitive or inverse one).

The ontological machine model of PGC electrical equipment defects developed in OntoStudio software includes fields containing the concept name, the attributes composition concept and generic connections concept:

$$C = \langle N, A, S, D \rangle,$$

where N—the concept name, A—the set of concept attributes, S—the set of parent concepts—class hierarchy (Classes) Defects and Electrical Equipment, D—the set of associated concepts (subclasses) that build up subgroups of equipment models with similar characteristics and encountered defects (Fig. 2). In addition to the above $\langle A \subseteq M \rangle$.

Each of the ontological classes of PGC electrical equipment defects has its own attributes—characteristics and entities (Instances). The concept attribute, in turn, is characterized by its name, type, and value:

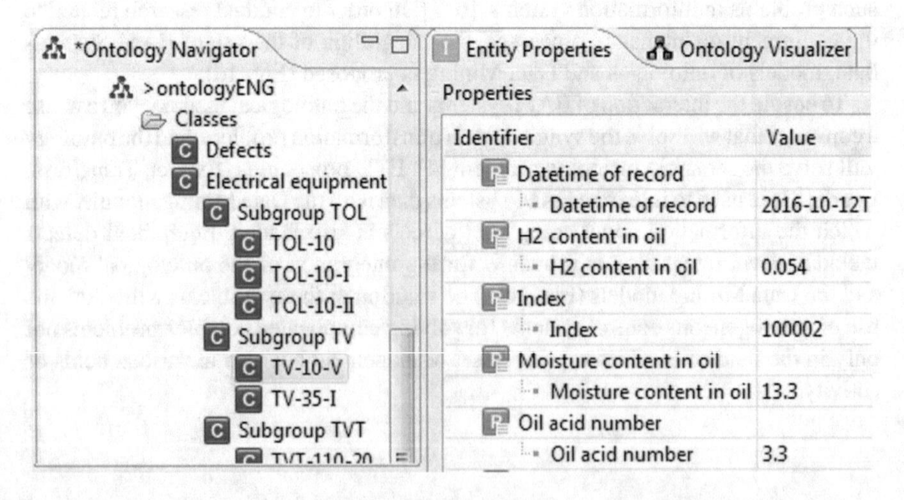

Fig. 2 Ontological classes tree of PGC equipment defects (using the example of transformers)

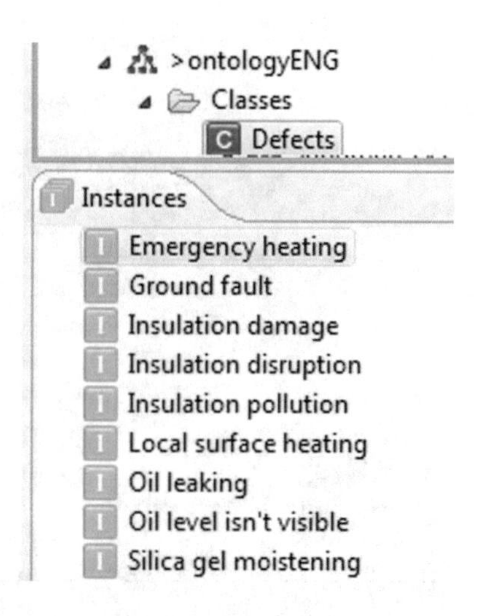

Fig. 3 Defects stored as entities

$$A = \langle N_A, T, V \rangle,$$

where N_A—attribute name, T—attribute type, V—attribute value. Ontological c_i concept attributes of PGC electrical equipment defects have a certain type of data T. Entities of the Electrical equipment class are separate records, for individual equipment items manufactured on a specific date. These records show the electrical equipment characteristics in a specific period of time. For the Defects class, in contrast to the Electrical equipment class, the entities are specific defects in electrical equipment (Fig. 3).

The entities of the Defects class are attached to a special field in the records of the Equipment class. On the basis of the obtained data on the equipment characteristics and the detected defects, a semantic search was made and the model was built like a tree, which shows what kind of defects and characteristics are belonging to the equipment. The ontological defects visualization in PGC equipment allows you to create a network of concepts, as well as clearly demonstrate the classes and the relationships between them (Fig. 4).

The ontology contains axioms (rules) that are valid in the modeled enterprise—defects and PGC equipment. Similar rules are expressed based on relationships R. A rule can include one or more relationships. For example:

IF the equipment N has attributes $A1$ <has values less than> some allowable limit $U1$ AND $A2$ <has values higher than> some limit $U2$, THEN N <has a defect> Z

In this example N—concept, $A1$ and $A2$—attributes, and <has a defect>—relation R, specifying for N connection with entity Z (Fig. 5). Different types of axioms are used to describe the integrity constraints imposed on the ontology, functional relationships, logical dependencies, as well as relationships between facts.

The use of ontologies in integration with EAM-systems technologies provides the possibility of obtaining a logical conclusion as a result of a query based on stored data, complex connections, and relationships modeling, eliminating semantic

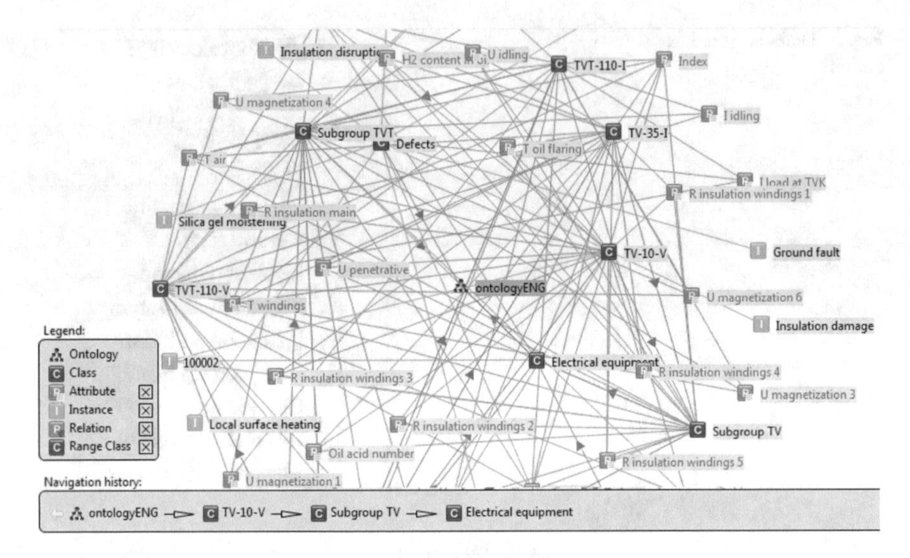

Fig. 4 The ontological defects visualization in PGC equipment

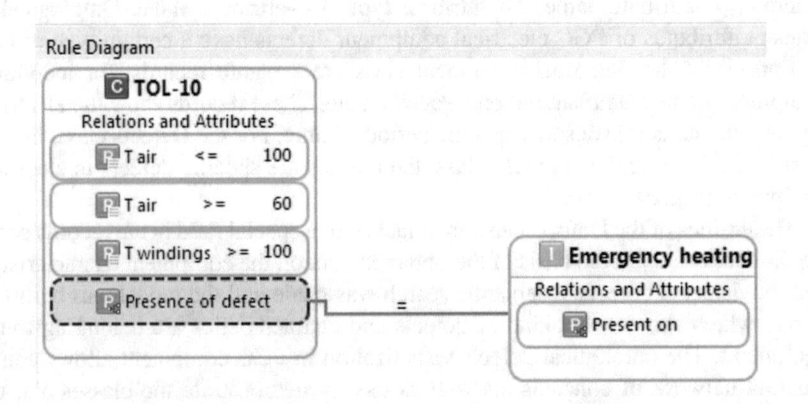

Fig. 5 An example of a rule in Ontostudio

heterogeneity and using consistent terminology that allows using data from different sources [10].

4 Electrical Equipment Defects Intellectual Data Analysis

The results of applying one or another Data Mining model are not always uniquely interpreted. In order to increase the efficiency of solving the problem of managing the PGC Enterprise Assets, we use an approach based on assembly models, which are a set of models used together [8]. The assembly model for the electrical equipment

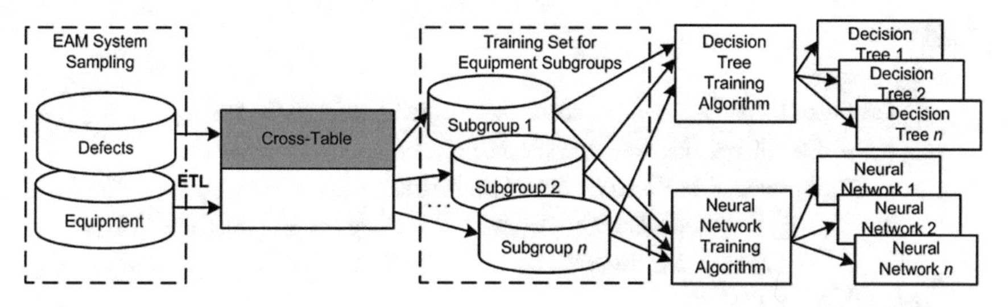

Fig. 6 The structure of the assembly models of PGC electrical equipment defects Data mining

defects analysis consists of neural networks [11] and decision trees [12] (Fig. 6). The unified data of the cross-table obtained on the basis of data on defects and equipment from the EAM-System is used as a training set. Based on the resampling procedure, several subsamples are extracted from the cross-table (training set), each of which is used to train one of the assembly models for a certain equipment subgroup. The original set is divided into a test and a training one in the ratio of 95–5%. Further, the models of decision trees and an artificial neural network are used parallelly and the same data set is "run" simultaneously through several models (Fig. 6).

As the analytical platform for the implementation of an assembly model of PGC electrical equipment defects data mining has been selected analytical platform Deductor [8]. Let us consider in detail the organization and implementation of the assembly models of PGC electrical equipment defects in data mining. Data from EAM-Systems come in the form of two arrays of data:

- equipment defect records, each of which includes: the type of equipment, inventory number, date of record and defect found;
- equipment parameters records, each of which includes: type of equipment, inventory number, date of record, described parameter and parameter value.

As a part of the ETL process, it is necessary to check for duplicates and conflicting records for arrays of data that are removed from the analyzed data and build up cross-tables so that records of different defects and parameters are stored in different columns. After that, the records of equipment parameters and defects are merged by dates of making records and equipment identification number. And then we get the table with the stored equipment parameters and defects in each row, currently present on this equipment item.

In order to ensure data-level integration of the ontological module and the Data Mining module, scenario processing of data in Deductor is also organized in two main areas: Defects and Electrical equipment (Fig. 7).

The analyzed equipment group includes several different types of equipment, for each of which a different set of parameters and encountered defects are build up. It was decided to divide the group into subgroups, within each of which the equipment has the same sets of parameters and common defects, and to carry out the analysis within these subgroups separately (Fig. 8). Decision trees and neural

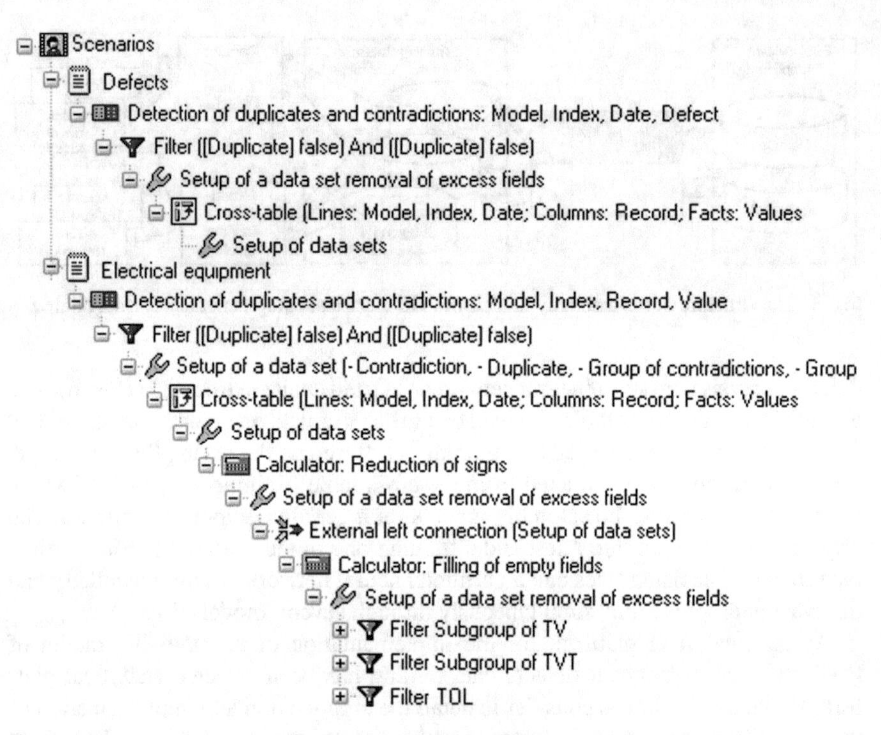

Fig. 7 The scenarios tree of PGC equipment defects data mining

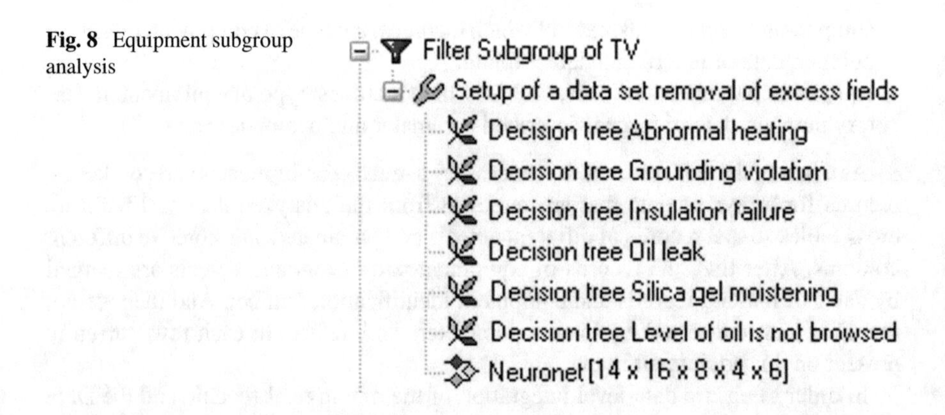

Fig. 8 Equipment subgroup analysis

networks actually play the role of experts and provide the information necessary for reasonable decision making [8].

As a result, a decision tree was build-up for each defect in each subgroup on the basis of the C4.5 algorithm, showing the relationship between the parameters and defects of the PGC equipment [12]. Each rule is read as follows:

IF <CONDITION> THEN <INVESTIGATION>

Condition	Investigation	Support	Reliability
IF		4275	3237
R of isolation of windings 5 < 105,17		4272	3237
Air T < 66,015		4269	3237
Circuit T < 55,6		1326	1296
Winding T < 54,566	0	1291	1270
Winding T >= 54,566		35	26
Circuit T >= 55,6		2943	1941
Air T >= 66,015	1	3	3
R of isolation of windings 5 >= 105,17	1	3	3

Fig. 9 Decision tree visualizer

Two numeric parameters are shown for each decision node: support and credibility. A graphical representation of the decision tree is built up in Decision Tree Visualizer (Fig. 9).

In order to analyze the connection of all parameters and all the defects occurring in a subgroup, a neural network is built up (Fig. 10) [13, 14]. The sigmoid activation function is used for the output layer neurons since it returns the probability of the relationship of an object to a specific class. Neural network learning algorithm Resilient Propagation (Rprop) was selected.

There were analyzed 192,000 records of parameters of the considered equipment group and 37,430 records of defects corresponding to these parameters. These records were grouped by equipment identification numbers and inspection dates; as a result, records were obtained containing in each row all parameters and defects that were registered on a specific equipment item on a particular day, including 14,970 and 12,739 records, respectively. These data were used to train neurals to identify specific

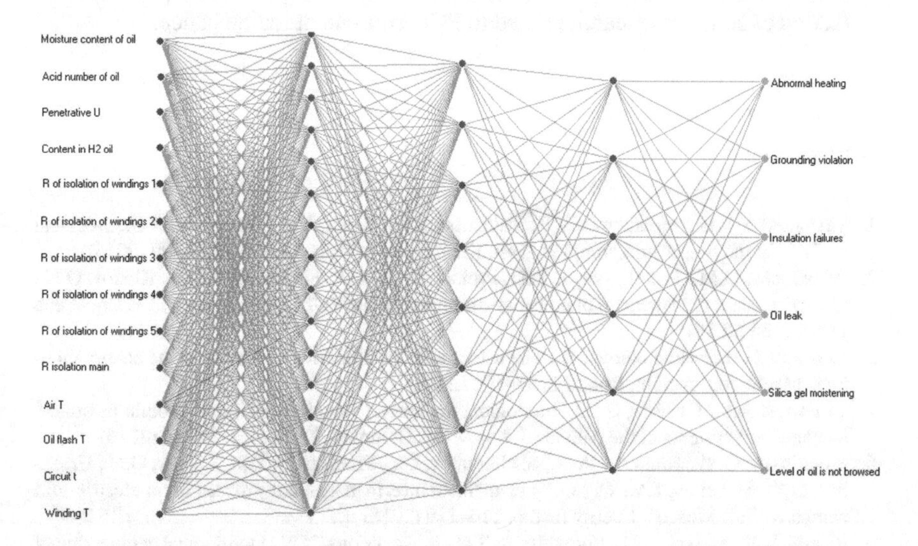

Fig. 10 Neural network graph visualizer

defects in equipment. These neural networks were able to correctly recognize 25,708 records from 27,000; thus, 95.2% of events were recognized.

The shared use of neural networks and decision trees is aimed at improving the understanding of the structure of the process of electric grid companies equipment maintenance and repair by providing the result obtained in the course of training the neural network in the form of a hierarchical, consistent structure of rules like "IF-THEN" [15].

5 Conclusion

The main scientific contribution of this article is to develop a conceptual solution in the field of Enterprise Asset Management of an Electric Grid Company based on the integration of EAM-Systems and the knowledge obtained from ontological models and data level mining.

The following results were obtained:

1. The integration mechanisms for the components of the proposed conceptual solution are detailed.
2. The models of Defects and Equipment of Power Grid Companies (using the transformers equipment example) were developed in ontological OntoStudio editor and the assembly models by equipment groups, including neural networks and decision trees, based on Deductor analytical platform.
3. The shared use of intellectual technologies will significantly increase the efficiency of PGCs, by using the accumulated experience, without repeatedly contacting experts for research related to PGC equipment maintenance.

References

1. Zakharenko, S.G., Malakhova, T.F., Zakharov, S.A., Brodt, V.A., Vershinin, R.S.: Accident analysis in the power grid complex. Bull. Kuzbass State Tech. Univ. **116**, 94–99 (2016)
2. Protalinsky, O.M., Protalinskaya, YuO, Protalinskaya, I.O., Shcherbatov, I.A., Kladov, O.N.: Enterprise asset management of power grid companies EAMOPTIMA. Autom. IT Energy Sect. **110**, 24–26 (2018)
3. Tsurikov, G.N., Shcherbatov, I.A.: The use of industrial Internet of Things at the energy facilities. Mechatron. Autom. Robot. **2**, 97–100 (2018)
4. Grigoriev, A.S., Uvaisov, S.U.: Ontological approach to the integration of mobile personnel management systems at the data level. Innov. Inf. Commun. Technol. **1**, 64–66 (2016)
5. Protalinsky, O.M., Khanova, A.A., Shcherbatov, I.A., Protalinsky, I.O., Kladov, O.N., Urazaliev, N.S., Stepanov, P.V.: Ontology of maintenance management process in an electric grid company. Bull Moscow Energy Inst. **6**, 110–119 (2018)
6. Massel, L.V., Massel, A.G., Vorozhtsova, T.N., Makagonova, N.N.: Ontological engineering of situational management in power sector. In the collection: Knowledge—Ontologies—Theories (ZONT-2015). RAS, SB. Sobolev Institute of Mathematics, pp. 36–43 (2015)

7. Tolk, A.: Ontology, Epistemology, and Teleology for Modeling and Simulation. 372p. Springer (2013)
8. Paklin, N.B., Oreshkov, V.I.: Business Analytics: From Data to Knowledge. 704p. SPb., St. Pete (2010)
9. Antonov, I.V., Voronov, M.V.: The method of building up the ontology of the enterprise. In: Bulletin of St. Petersburg State University of Technology and Design. Natural and Technical Sciences, vol 2, pp. 28–32 (2010)
10. Gonchar, A.D.: Comparative analysis of databases and knowledge bases (ontologies) applicable to the modeling of complex processes. Mod. Sci. Res. Innov. **37**, 26 (2014)
11. Efimov, P.V., Shcherbatov, I.A.: Algorithm for identifying obvious defects in process equipment in the power sector based on the neural network model. In: News of the South-West State University. Management, Computing Technology, Informatics. Medical Instrument, vol 27, pp 32–40 (2018)
12. Miftakhova, A.A.: Application of the decision tree method for solving classification and forecasting problems. Infocommun. Technol. **1**, 64–70 (2016)
13. Bobyr, M.V., Kulabuhov, S.A., Milostnaya, N.A.: Training neuro-fuzzy system based on the method of area difference. Artif. Intell. Decis. Making **4**, 15–26 (2016)
14. Diamantaras, K.: Artificial Neural Networks—ICANN 2010. 20th International Conference on Artificial Neural Networks, ICANN 2010, held in Thessaloniki, Greece, 15–18 Sept 2010, 543p. Springer (2010)
15. Evdokimov, I.A., Solodovnikov, V.I., Filipkov, S.V.: Using decision trees for data mining and extracting rules from neural networks. New Inf. Technol. Autom. Syst. **15**, 59–67 (2012)

On the Application of the Photogrammetric Method to the Diagnostics of Transport Infrastructure Objects

Pavel Elugachev and Boris Shumilov

Abstract The implementation of plans to create "smart cities" as one of the most important parts of the digital economy requires the priority development of transport infrastructure, ensuring the movement of people and goods within the city and surrounding areas. The safe operation and maximum throughput of this cyber-physical system are possible provided that a diagnostic technology is created for transport infrastructure objects, including those based on a video recording of road conditions. The algorithm of technical vision, which is proposed to be implemented as a program on mobile devices, for recognizing objects of the transport infrastructure and their defects using stereometry is investigated. The obtained data can be used when planning road repairs, in the analysis of road accidents, to process applications of road users, etc.

Keywords Roads as a cyber-physical system · Photogrammetry · Calibration

1 Introduction

Automation of diagnostics of the state of transport infrastructure facilities and the related need to develop modules for mathematical modeling, design and production of devices and diagnostic tools for inclusion in the management system of a transport cyber-physical system that provides physical accessibility of territories and cities is an urgent problem within the framework of "smart cities" as one of the most important areas of the digital economy [1]. First, the quality of the road surface, along with the category of the road, its visibility, road width, the location of the corresponding road signs, have a significant impact on road safety and traffic flow capacity. Together, they define the concept of "road conditions" on which the trajectory and speed of the car depend. According to experts, the impact of road conditions on the occurrence of

P. Elugachev · B. Shumilov (✉)
Tomsk State University of Architecture and Building, 2 Solyanaya, Tomsk 634003, Russia
e-mail: sbm05@yandex.ru

P. Elugachev
e-mail: elugachev@mail.ru

A. G. Kravets et al. (eds.), *Cyber-Physical Systems: Industry 4.0 Challenges*, Studies in Systems, Decision and Control 260,
https://doi.org/10.1007/978-3-030-32648-7_5

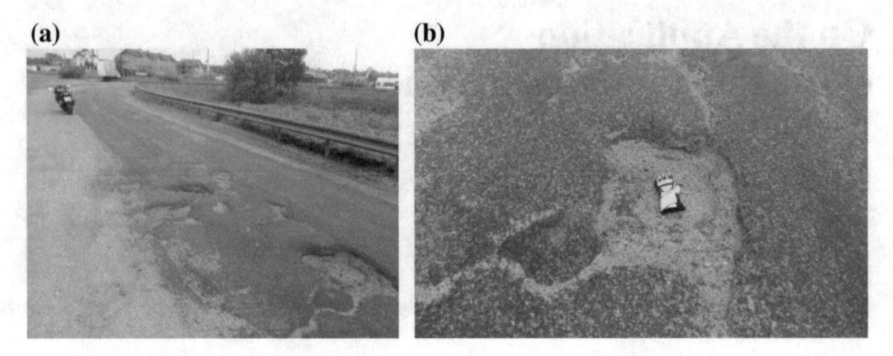

Fig. 1 Photos of potholes in the road surface (made by the teacher of TSUAB, V. Kalinichenko)

road traffic accidents is from 60 to 80% of cases [2, 3]. In this case, accidents are one of the most common causes of mortality in the world. More than 1.35 million people die every year, which is equivalent to the loss of about 3% of their gross domestic product by countries affected by this disaster [4]. Secondly, according to the results of the analysis of road accidents, there are the tasks of identifying those guilty of violating traffic rules and estimating the cost of repairing damaged motor vehicles (MV) [5]. Thirdly, the economic and social losses due to a decrease in the speed of the traffic flow and the corresponding throughput capacity of the highway, caused by the presence of cracks and holes in the road surface, tend to increase if they are not timely detected and repaired [6].

In most cases, road incident participants use mobile devices to record a situation (Fig. 1). In Fig. 1a, the motorcycle driver recorded a general traffic situation. In Fig. 1b, to understand the scale and size of the gouges, the driver used the glove as a scale marker. In this situation, the motorcycle owner will have to prove that the road surface does not meet the requirements for operating conditions acceptable for road safety reasons.

Based on the foregoing, the development of autonomous mobile devices capable of "on-site" to perform the tasks of collecting and analyzing traffic data, including photographs and video filming, in a form suitable for use according to the protocol [7] of the cyber-physical urban traffic management system, as well as for investigating and examining controversial situations, is an urgent task.

2 A Summary of the Theory of Stereo Measurements

2.1 A Linear Approach

The key component here is the technical vision system [8], based on the method of photogrammetric processing of stereoscopic images of a three-dimensional 3D object obtained from different angles [9–13].

The solution of the problem is reduced to the formation of geometric transformations according to graphic labels of the elements of the matrix (Fig. 2) and the restoration of the three-dimensional coordinates of the object points from the perspective projections in each image plane (Fig. 3).

Here $[x\ y\ z]$ is the vector of the geometric coordinates of the point P in three-dimensional space, x^* and y^* are the coordinates of the perspective projection p of the point P onto the picture plane of the photo image $z = 0$, z_c is the distance of the shooting point C from the origin.

Recall that with the introduction of a homogeneous system of geometric coordinates, an arbitrary point in three-dimensional space is represented by a four-dimensional vector: $[x'\ y'\ z'\ h] = [x\ y\ z\ 1]$ [T], where $[T]$ is the matrix of some transformation, and h is the scale factor. In turn, the linear perspective transformation is represented as a 4 × 4 matrix

$$[T'] = \begin{bmatrix} T'_{11} & T'_{12} & T'_{13} & T'_{14} \\ T'_{21} & T'_{22} & T'_{23} & T'_{24} \\ T'_{31} & T'_{32} & T'_{33} & T'_{34} \\ T'_{41} & T'_{42} & T'_{43} & T'_{44} \end{bmatrix}.$$

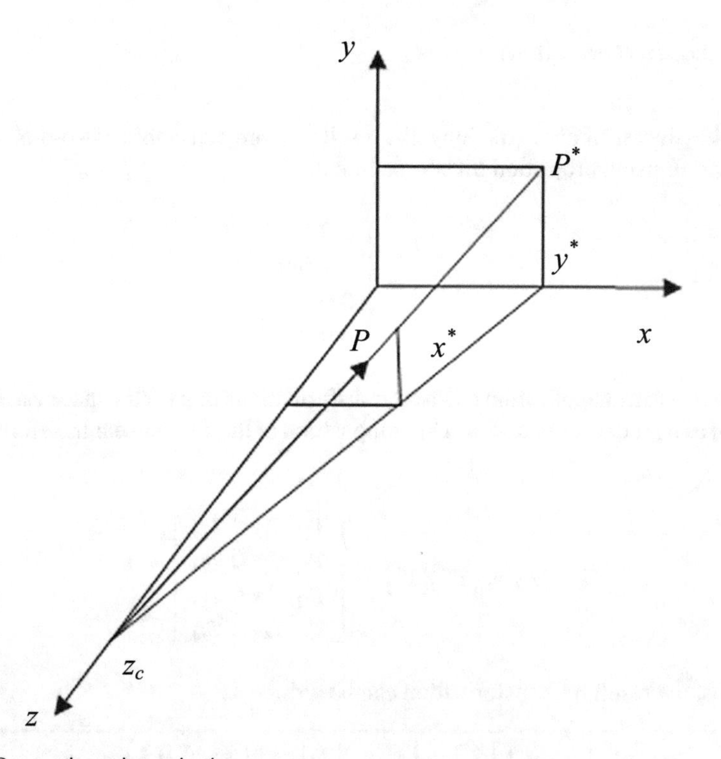

Fig. 2 Perspective point projection

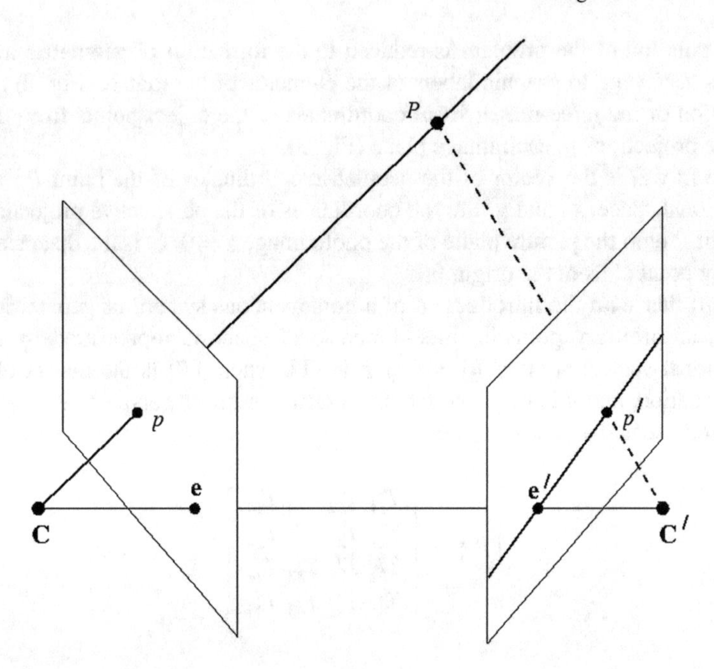

Fig. 3 Diagram of stereo vision

In the process of photographing, the results are projected onto a two-dimensional plane $z = 0$ using projection transformation

$$[T''] = \begin{bmatrix} 1\,0\,0\,0 \\ 0\,1\,0\,0 \\ 0\,0\,0\,0 \\ 0\,0\,0\,1 \end{bmatrix}.$$

The consistent application of linear transformations of a vector space can be considered as a product of matrices. The composition of these two linear transformations gives

$$[T] = [T''][T'] = \begin{bmatrix} T_{11}\ T_{12}\ 0\ T_{14} \\ T_{21}\ T_{22}\ 0\ T_{24} \\ T_{31}\ T_{32}\ 0\ T_{34} \\ T_{41}\ T_{42}\ 0\ T_{44} \end{bmatrix}.$$

Thus, the resulting transformation can be written as

$$\begin{bmatrix} x\ y\ z\ 1 \end{bmatrix}[T] = \begin{bmatrix} x'\ y'\ 0\ h \end{bmatrix} = h\begin{bmatrix} x^*\ y^*\ 0\ 1 \end{bmatrix}.$$

After eliminating the scale factor h, we obtain two scalar equations:

$$(Q_{11} - Q_{31}x^*)x + (Q_{12} - Q_{32}x^*)y + (Q_{13} - Q_{33}x^*)z + (b_1 - b_3x^*) = 0,$$
$$(Q_{21} - Q_{31}y^*)x + (Q_{22} - Q_{32}y^*)y + (Q_{23} - Q_{33}y^*)z + (b_2 - b_3y^*) = 0, \quad (1)$$

where Q_{ij} are the elements of the matrix of the projective transformation of size 3×3 obtained by deleting the last row and the third column from the matrix T and transposing, and $b = [b_1, b_2, b_3]$ are the three nonzero elements of the last row of the matrix T, which constitute the coordinates of the normalized vector of a shooting point.

Assuming that Q, b, x, y, z are known, these equations can be used to model the photographing process itself. If x^*, y^*, x, y, z are known, then (1) represents two equations with 12 unknown elements Q, b. Applying these equations to n given points in the space of objects and to their images on the perspective projection, we obtain a homogeneous system of $2n$ equations with 12 unknowns. Provided that the given points are not coplanar and $n \geq 6$, there is a non-trivial solution of the resulting system. For a numerical search for a solution, transfer the terms containing the coefficient b_3 to the right and set the value $b_3 = 1$. Thus, in order to find the solution Q, b, one gets an overdetermined system of equations whose matrix cannot be inverted, since it is not square. As is known from the theory of the least squares method, one can calculate the best-averaged solution by multiplying both sides of the matrix equation by the transposed matrix of the system. Then a system of 11 linear equations for 11 unknowns with a symmetric square matrix is obtained, to solve which you can apply the well-known square root method. Thus, a transformation is determined by known coordinates, which led to a given perspective projection, for example, a photograph (this is the task of calibrating the camera for an individual frame).

In the photogrammetry method, a stereo pair is used as an object image (Fig. 3), that is, two projections of the same three-dimensional object captured at different angles.

In this case, Q, b, x^*, y^* in Eq. (1) are assumed to be known. Then two equations are obtained for three unknown spatial coordinates x, y, z. This is an underdetermined system of equations, so it is impossible to solve it. However, if two perspective projections are known, say, two photographs taken at different angles, then these equations can be written for both projections:

$$(Q^1_{11} - Q^1_{31}x^{*1})x + (Q^1_{12} - Q^1_{32}x^{*1})y + (Q^1_{13} - Q^1_{33}x^{*1})z = x^{*1} - b^1_1,$$
$$(Q^1_{21} - Q^1_{31}y^{*1})x + (Q^1_{22} - Q^1_{32}y^{*1})y + (Q^1_{23} - Q^1_{33}y^{*1})z = y^{*1} - b^1_1,$$
$$(Q^2_{11} - Q^2_{31}x^{*2})x + (Q^2_{12} - Q^2_{32}x^{*2})y + (Q^2_{13} - Q^2_{33}x^{*2})z = x^{*2} - b^1_1,$$
$$(Q^2_{21} - Q^2_{31}y^{*2})x + (Q^2_{22} - Q^2_{32}y^{*2})y + (Q^2_{23} - Q^2_{33}y^{*2})z = y^{*2} - b^1_1. \quad (2)$$

As a result, four equations are obtained for three unknown spatial coordinates x, y, z, where superscripts 1 and 2 denote the first and second perspective projections.

Thus, an overdetermined system of equations is again obtained, and the least squares and square root methods can be used to find the coordinates of the point x, y, z.

2.2 Nonlinear Approach

Note that in the above approach, for each new frame, the shooting camera needs to be re-calibrated to get 11 new parameters with each camera movement in space. However, when shooting with the same camera or camcorder, not all of them are independent. Namely, five parameters are responsible for the so-called internal parameters of the camera (focal lengths, the coordinates of the main point and the distortion parameter), which do not change when it is moved. To define them, the matrix of the projective transformation Q leads to an upper triangular form, so that the result is the values of the angular and linear movements of the chamber R, t [14]

$$A = \begin{pmatrix} \alpha & \gamma & u_0 \\ 0 & \beta & v_0 \\ 0 & 0 & 1 \end{pmatrix}, \quad R = A^{-1} \cdot Q, \quad t = A^{-1} \cdot b. \tag{3}$$

In turn, the matrix R contains only three independent parameters—the angles of rotation of Euler. For their finding, it is important that it is orthogonal. The application of the orthogonality conditions makes the problem of determining the R, t parameters of the perspective transformation (external calibration) solvable, although nonlinear. Thus, it becomes possible to use $n \geq 3$ calibration points of an object in space to construct each next perspective transformation.

2.3 New "Single Point" External Calibration Method

In conditions when the camera of a modern smartphone is used for video recording, the spatial orientation angles of the smartphone's camera become available for measurement. Then, assuming that the internal parameters of the camera do not change, we formulate a new "single-point" method for external calibration of the camera of the smartphone. Let the projective transform matrix T be obtained for the first frame in some way and the angles of the smartphone's camera orientation for each frame of the stereo pair are fixed. Then you can get the rotation matrix [15] of the second frame relative to the first frame:

$$S = \begin{pmatrix} \cos(\vartheta)\cos(\varphi) & -\cos(\psi)\sin(\varphi) + \sin(\psi)\sin(\vartheta)\cos(\varphi) & \cos(\varphi)\cos(\psi)\sin(\vartheta) + \sin(\psi)\sin(\varphi) \\ \sin(\varphi)\cos(\vartheta) & \cos(\psi)\cos(\psi) + \sin(\psi)\sin(\vartheta)\sin(\varphi) & \sin(\vartheta)\cos(\psi)\sin(\varphi) - \cos(\varphi)\sin(\psi) \\ -\sin(\vartheta) & \sin(\psi)\cos(\vartheta) & \cos(\vartheta)\cos(\psi) \end{pmatrix}, \tag{4}$$

where the Euler angles ψ, θ, φ of the rotation matrix are equal to the differences of the corresponding measured angles. Then the matrix of the projective transformation Q^2 for the second frame is calculated using the formula $Q^2 = Q^1 \cdot S$ and the coordinates of the normalized vector of the shooting point for the second frame are found from the equations:

$$\begin{aligned} b_1^2 &= x^{*2} - (Q_{11}^2 - Q_{31}^2 x^{*2})x - (Q_{12}^2 - Q_{32}^2 x^{*2})y - (Q_{13}^2 - Q_{33}^2 x^{*2})z, \\ b_2^2 &= y^{*2} - (Q_{21}^2 - Q_{31}^2 y^{*2})x - (Q_{22}^2 - Q_{32}^2 y^{*2})y - (Q_{23}^2 - Q_{33}^2 y^{*2})z. \end{aligned} \quad (5)$$

Thus, the resulting solution allows you to perform external calibration of the smartphone camera at a single marker point $[x\ y\ z]$; in addition, the nonlinearity associated with the Euler angles disappears. Here, as in the previous example, you can use additional markers—for subsequent averaging.

3 3D-Modeling

One of the main tasks in identifying defects in transport infrastructure facilities, in particular, road pavement based on the results of mobile shooting is to build a three-dimensional mathematical model using a set of object interpolation points, followed by calculating the area and volume of the defect as a result of comparison with a typical (passport) object model [5]. The algorithm for constructing the interpolation points of an object from a set of pairs of corresponding points on stereo images generally consists of five (see below) stages.

Stage 1: The calculation of the elements of the projective matrix in accordance with the relations of the form (1).
Stage 2: Calculation of the internal parameters of the camera according to the projective matrix in accordance with the relations of the form (3).
Stage 3: Determining the reference frames of the video sequence forming a stereo pair.
Stage 4: Calculation of external parameters for each frame of a stereo pair in accordance with relations (4), (5).
Stage 5: Solving the inverse problem for each pair of conjugate points in accordance with relations of the form (2).

After that, knowing the real points in space, one can transfer to the construction of a three-dimensional mathematical model, using which it is possible to calculate the parameters necessary for diagnosing a defect in a transport infrastructure object, in particular, the road surface (area and volume).

4 Automation of the Process of Searching for Graphic Markers and Reference Points in Photographs

In this section, it is proposed to dwell on the problem of localization and identification of projections of interpolation points on images determined from the reference frames of a video sequence and which are stereoscopic in pair.

Selecting the boundaries of flat images is one of the important auxiliary tasks in detecting defects in transport infrastructure objects, in particular, road pavement damage. They contain comprehensive information on their form for further analysis [16–18]. As a preliminary conclusion, we note that the images of objects of transport infrastructure are characterized by the presence of angles formed by the intersections of the generatrices (curves or straight lines). Therefore, to analyze such images, it is advisable to use the so-called "corner" filters; in particular, the use of the Harris detector is popular [19].

More accurately reproduce the details and boundaries of flat images allows binarization (i.e., the transformation of the original image into an image, the elements of which take only two values) [20–24]. The difference of the algorithm [25] is the enhancement of contrast and the selection of weak contrast details on images in shades of gray. This is useful for further data processing. Consequently, the accuracy of calculating damage parameters (area and volume) increases.

5 Experimental Results

To illustrate the photogrammetric method, consider a real stereo pair corresponding to two photographs of a typical road cone on the background of damage in the road surface (Fig. 4). We measured the coordinates of the seven vertices of a three-dimensional object using a ruler. In addition, we recorded the coordinates of the corresponding points on the images in the graphical editor using the mouse. Taking

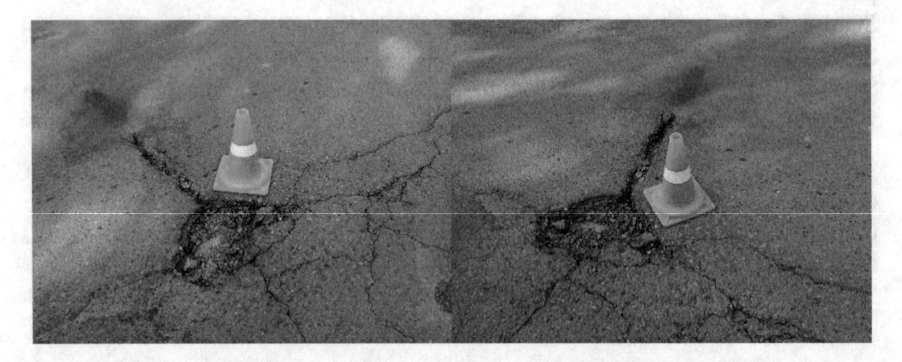

Fig. 4 Stereo pair: road cone on the background of damage to the road surface

into account the coordinates of the road surface point on the left and right images, the accuracy of the presented vision algorithm can be estimated. In our case, the distance from the top of the cone to the asphalt pavement, calculated according to the Pythagorean Theorem, was 31.975 cm, which is 0.078% different from the value of 32 cm in the datasheet.

Using the "three-point" method, provided that the internal parameters of the chamber were predetermined for all seven points, yielded a value of 21.174 cm, which gives a relative measurement error of 33.8%, and, according to the theory of errors, there is no correct number in the answer. However, adding in this context only one (that is, the fourth) point gave a value of 30.756 cm, which provides a relative measurement error of 3.8% and, consequently, at least one correct figure in the answer. The experiments performed allow us to recommend a device for video measurements on the site, such as a folding emergency stop sign, which has exactly four interpolation points' projections easily recognizable on flat images.

A more responsible task of calibrating the internal parameters of the camera can be successfully performed in office conditions at a much larger number of marker points. We emphasize that this requires special care when performing measurement work. For example, for the previous example, the internal parameters of the camera on the left and right images do not match (see Table 1).

In the next series of experiments, there are 10 calibration points, including the four corners of a label with a bar code (Fig. 5). The internal parameters of the camera on the left and right images are equal, respectively (Table 2).

Then, for the three-dimensional coordinates of the vertices of the dark tile next to the shaded area on the floor, a perfectly acceptable result is obtained: 9.757×3.738 instead of the true values of 10×4.

Another experiment was to test the ability to automatically search for graphic markers and reference points in photographs. Recall that these points must be found on each image in order to successfully solve the problem of photogrammetry. In particular, Fig. 6 shows the use of the Harris detector [19] for detecting conjugate points on the left and right frames of a stereo pair, which shows damage to the road surface.

Table 1 Internal parameters of the camera for Fig. 4

Photo	α	β	u_0	v_0	γ
Left	795.578	642.85	794.579	841.311	15.669
Right	401.112	296.649	824.506	667.454	13.096

Fig. 5 Stereo pair of two photographs of the packaging box (made by the Associate Professor of TSUAB, A. Titov)

Table 2 Internal parameters of the camera for Fig. 5

Photo	α	β	u_0	v_0	γ
Left	3599	3707	1340	2028	133.898
Right	3050	3233	981.91	2623	38.878

Fig. 6 Stereo pair from two photos of the real road surface (made by the Associate Professor of TSUAB, A. Titov)

Fig. 7 Photos of the car damaged as a result of the fall of the snow mass (made by the teacher of TSUAB, V. Kalinichenko)

6 Conclusion

The practical application of the algorithm for diagnosing the state of transport infrastructure facilities stated in the report is quite wide, including when fixing and determining the actual dimensions of MV damage based on the measured coordinates of body surface points on the left and right images (Fig. 7). Examination of nature and the list of damage to the MV provides for a detailed fixation of the damage to determine the possibility of their formation and involvement in the event under investigation. The results of the visual inspection of the damaged vehicle and the photographing are documented by the relevant act. Photographs and the inspection report of damaged MV are a mandatory attachment to the expert opinion or a report on the assessment of the cost of repairing the MV. Of course, the photogrammetric method is a weighty argument in solving controversial situations, and the determination of the actual size of damage to the elements of automatic trunk stations from photographs and video footage is an urgent task.

In conclusion, it must be said that mobile data collection on defects in transport infrastructure facilities, in particular, road pavement, may be much more efficient using the methods described in this report. It is also worth considering that due to the peculiarities of the image of the defects of the roadway (lack of clear boundaries, the presence of foreign objects, and the insignificance of some defects) it should be possible to "manually" mark the characteristic points on the images.

References

1. Kupriyanovsky, V., Namiot, D., Sinyagov, S.: Cyber-physical systems as a base for digital economy. Int. J. Open Inf. Technol. **4**(2) (2016). Homepage, http://injoit.org/index.php/j1/article/view/266/211. Last accessed 25 Mar 2019
2. Urfi, S.K., Amir, A., Khalil, S., Hoda, M.F.: Risk factors for road traffic accidents with head injury in Aligarh. Int. J. Med. Sci. Pub. Health **5**, 2103–2107 (2016)

3. Abdi, T., Hailu, B., Andualem, Adal T., Gelder, P.H.A.J.M., Hagenzieker, M., Carbon, C.-C.: Road crashes in Addis Ababa, ethiopia: empirical findings between the years 2010 and 2014. AFRREV **11**, 1–13 (2017)
4. Global status report on road safety (2018). Homepage, http://www.who.int/violence_injury_prevention/road_safety_status/2018/en/. Last accessed 25 Mar 2019
5. Morales, A., Sanchez-Aparicio, L.J., Gonzalez-Aguilera, D., Gutierrez, M.A., Lopez, A.I., Hernandez-Lopez, D., Rodriguez-Gonzalvez, P.: A new approach to energy calculation of road accidents against fixed small section elements based on close-range photogrammetry. Rem. Sens. **9**(1219), 1–18 (2017)
6. Bao, G.: Road distress analysis using 2D and 3D information. The University of Toledo, The University of Toledo Digital Repository Theses and Dissertations (2010)
7. Tsvetkov V.Y.: Control with the use of cyber-physical systems. Perspect. Sci. Educ. **3**(27), 55–60 (2017). Int. Sci. Electron. J. Homepage. http://psejournal.wordpress.com/archive17/17-03/. Last accessed 25 Mar 2019
8. Elugachev, P., Shumilov, B.: Development of the technical vision algorithm. In: MATEC Web of Conferences, vol 216, p. 04003. Polytransport Systems, 7p (2018)
9. Rogers, D.F., Adams, J.A.: Mathematical Elements for Computer Graphics. McGraw-Hill, New York (1990)
10. Tsai, R.: A versatile camera calibration technique for high-accuracy 3D machine vision metrology using off-the-shelf TV cameras and lenses. IEEE J. Robot. Autom. **3**, 323–344 (1987)
11. Slama, C. (ed.) Manual of Photogrammetry. American Society of Photogrammetry, Falls Church, VA (1980)
12. Davison, A.J., Reid, I.D., Molton, N.D., Stasse, O.: MonoSLAM: real-time single camera SLAM. IEEE Trans. Pattern Anal. Mach. Intell. **29**, 1052–1067 (2007)
13. Chiuso, A., Favaro, P., Jin, H., Soatto, S.: Structure from motion causally integrated over time. IEEE Trans. Pattern Anal. Mach. Intell. **24**, 523–535 (2002)
14. Hartley, R., Zisserman, A.: Multiple View Geometry in Computer Vision, 2nd edn. Cambridge University Press, Cambridge (2004)
15. Euler angles. Homepage. https://en.wikipedia.org/wiki/Euler_angles#Rotation_matrix. Last accessed 11 June 2018
16. Sonka, M., Hlavac, V., Boyle, R.: Image Processing. Analysis and Machine Vision. Thomson, Toronto (2008)
17. Harris affine region detector. Homepage. https://en.wikipedia.org/wiki/Harris_affine_region_detector. Last accessed 11 June 2018
18. Tuytelaars, T., Mikolajczyk, K.: Foundations and Trends in Computer Graphics and Vision, vol 3, pp. 177–280 (2007)
19. Harris, C., Stephens, M.: A combined corner and edge detector. In: Fourth Alvey Vision Conference, pp. 147–151. Manchester, UK (1988)
20. Niblack, W.: An Introduction to Digital Image Processing. Prentice-Hall, Englewood Cliffs, NJ, USA (1986)
21. Otsu, N.: A threshold selection method from gray-level histograms. IEEE Trans. Syst. Man Cybern. SMC **9**(1), 62–66 (1979)
22. Bayer, B.: An optimum method for two-level rendition of continuous tone pictures. In: IEEE International Conference on Communications, vol. 1, pp. 11–15 (1973)
23. Floyd, R.W.: An adaptive algorithm for spatial gray-scale. Proc. Soc. Inf. Disp. **17**(2), 75–78 (1976)
24. Wang, Z., Bovik, A.C.: Modern Image Quality Assessment. Morgan and Claypool Publishing Company, New York (2006)
25. Shumilov, B., Gerasimova, Y., Makarov, A.: On binarization of images at the pavement defects recognition. In: 2018 IEEE International Conference on Electrical Engineering and Photonics (EExPolytech), pp. 107–110. Saint Petersburg, Russia (2018)

Alternative Approach to Solving Computer Vision Tasks Using Graph Structures

Jiajian Li, Mark Makarychev and Aleksey Popov

Abstract An approach to recognizing objects on images is proposed, which uses graph structures and graph algorithms. The image being processed is converted into a grid graph, which is divided into image segments using Kruskal's algorithm and a Gaussian blur. Each resulting segment is characterized using descriptors, which are then grouped together to form the segment's fingerprint. In the knowledge base, which is also structured as a graph, groups of object fingerprints are linked via weighted edges, which indicate the degree of contextual association. During object recognition, neighboring segments and contextual associations are used to better predict what objects are presented in the input image.

Keywords Graphs · Object recognition · Computer vision · Segmentation · Image descriptors

1 Introduction

Singling out individual objects, analyzing the connections between them, and understanding information about the surrounding world—all of this is performed by the human visual system and constitutes an important ability of the human mind. Computers, which have become a crucial tool and aid to our society, will be able to do much more when they acquire the ability to see and interpret information about their surroundings. This makes computer vision a key field in the evolution of IT and robotics.

J. Li (✉) · M. Makarychev · A. Popov
Bauman Moscow State Technical University, 2-nd Baumanskaya st., 5/1, Moscow 105005, Russia
e-mail: dreki.li@mail.ru

M. Makarychev
e-mail: markmak77@gmail.com

A. Popov
e-mail: alexpopov@bmstu.ru

A. G. Kravets et al. (eds.), *Cyber-Physical Systems: Industry 4.0 Challenges*, Studies in Systems, Decision and Control 260,
https://doi.org/10.1007/978-3-030-32648-7_6

Computer vision is an interdisciplinary field that deals with how computers can be made to gain high-level understanding from digital images or videos [1]. From the perspective of engineering, it seeks to automate tasks that the human visual system is capable of doing.

Computer vision usually deals with the following tasks: recognition, detection, segmentation, restoring a 3D form using 2D images, motion assessment, and restoring images.

Computer vision, when dealing with the tasks of recognition, detection, and segmentation, is further complicated by the fact that there exist many factors that affect what information we receive as the input image. The viewing angle affects what we see and what faces of an object we see. Lighting affects shadows and exposition, which can hinder even the human visual system from receiving useful information from the image. Object representations can be deformed by perspective, accidental movement or shaking of the camera and/or object, and by other factors. Additionally, certain objects of the same class can have differing appearances (intra-class variance), which further encumbers the recognition of instances that have not been previously processed or saved.

However, certain attributes can play a key role in solving the tasks of object recognition. For example—texture can help achieve better results during image segmentation, while information about the context and content of the image (neighboring objects) can help improve assumption accuracy for segments during object recognition.

Today there exist two main methods of solving the tasks of computer vision: using neural networks and applying imaging science algorithms. Each of these methods has its own set of advantages, as well as flaws.

2 Flaws of Existing Approaches

One of the most popular methods of solving computer vision tasks is with the help of convolutional neural networks (CNNs). These solutions do indeed allow for creating computer vision systems that possess high identification accuracy. However, convolutional neural networks are not perfect and possess certain flaws.

The first flaw of CNNs is called translational invariance. This means that slight changes in an object's orientation or position can hinder activation of crucial neurons. As a result, the chain of neurons that used to be able to correctly recognize an object will no longer be activated, and the object will not be recognized. Data augmentation partially solves this problem, but it does not solve it completely.

The use of pooling layers in CNNs causes them to lose valuable information. One of the most important pieces of information that is lost during pooling is the relation between parts of a whole and the whole itself. A face detector, for example, needs to identify a combination of facial elements (a pair of eyes, nose, mouth, and face oval) before coming to the conclusion that the given image is an image of a face. A convolution neural network, however, will conclude that an image contains a face,

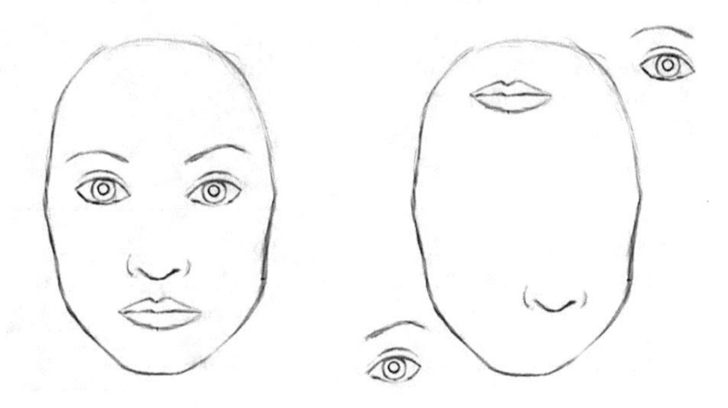

Fig. 1 The problem of using pooling layers in CNNs [2]

no matter where the elements are located in regards to each other. This problem is illustrated in the right half of Fig. 1.

Another flaw of neural networks is that they are not universal, as their amount of identifiable classes is limited by the dataset that was used during training. If new classes need to be added or the results received during testing are not satisfactory, then the network must be completely retrained. Creating an optimal training dataset that would provide acceptable test results is one of the most challenging and time consuming tasks when working with CNNs. If a training dataset is too specific, it will identify certain objects with a high degree of accuracy, but it will not be able to identify objects that do not completely match its strict criteria. On the other hand, if the dataset is not specific enough, it will identify all of the objects present, but very generally or with mistakes. Many CNN computer vision systems are trained using the ImageNet or COCO datasets, which relieve developers from the problems associated with creating their own dataset. However, even though the later of these two datasets contains more than 200 k marked images, it is capable of recognizing only 80 different classes of objects. The alternative approach that is proposed in this chapter successfully eliminates some of these flaws.

3 The Leonhard DISC Microprocessor

With the introduction of the Leonhard co-processor, based on a DISC (discrete mathematics instruction set computing) architecture, there came the idea of implementing computer vision using graph structures. Leonhard is an innovative co-processor that allows for energy-efficient and improved graph processing via specialized hardware. Implementing a graph-based computer vision algorithm and running it on such a processor could lead to quicker and more energy-efficient applications in the fields of robotics and computing in general.

The Leonhard microprocessor is a state-of-the-art computing system with multiple instruction streams and a single data stream architecture (MISD in accordance with Flynn's Taxonomy), which was developed at Bauman Moscow State Technical University's Department of Computer Systems and Networks by Professor Aleksey

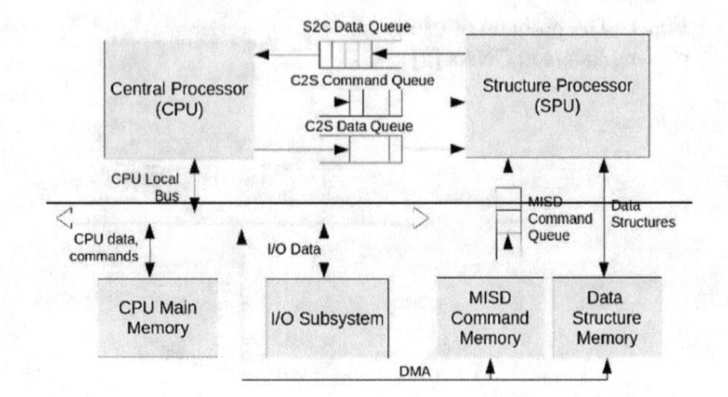

Fig. 2 Leonhard multiple instructions and single data stream computer architecture [3]

Popov. The Leonhard microprocessor was named in honor of Leonhard Euler and contains a discrete mathematics instructions set. Leonhard can work as a stand-alone microprocessor, or as a co-processor (SPU) for aiding a CPU in graph processing. Figure 2 shows the general structure of a system utilizing the Leonhard microprocessor.

Utilizing the Leonhard microprocessor in applications that manage graph structures allows for significant improvements in speed and energy-efficiency when processing, storing, and reading such structures. Having heard about this new technological advance, we decided to develop a new approach to solving the tasks of computer vision that could be optimized on such a processor. This new approach should solve most of the shortcomings that were discussed earlier and should operate using graph structures, so that the Leonhard processor can boost its performance in an energy-efficient way.

The approach that we developed turned out to be biologically inspired, as it attempts to mimic the way children study the surrounding world and learn to recognize objects. Newborn children lack semantic information about the objects of this world. During their learning process, children start to associate objects with their placement and with other surrounding objects. As a matter of fact, this happens not only during a person's childhood, but throughout their entire life.

Having introduced the tasks that computer vision seeks to solves, what improvements can be made to existing approaches, and how our approach differs—we can now dive deeper into how exactly this alternative approach operates and examine it stage by stage.

4 Data Acquisition

The first stage of any algorithm is receiving an initial set of input data and parameters. In computer vision, this means receiving the initial visual data. This is unprocessed visual data, which can range from simple pictures and diagrams, to complex monocular, stereo, or even equirectangular video frames. During this stage certain hardware problems can occur, such as perspective warp, or improper aperture focus, depth of field, and/or field of view. Once these problems are eliminated (if possible), the resulting image can be fed into a computer vision system for further processing and recognition.

Since many visual input data formats are based on static images (video frames, separate images of a stereo shot, etc.), our algorithm was developed to work with this exact "fundamental format".

5 Conversion of Images to Graph Structures

To properly represent an image format as a graph structure, the following steps must be taken:

- Formulate the rules, in accordance with which the image format's components will be represented as elements of the graph structure;
- Establish the types of these representations (one-to-one, single-valued, multi-valued) and the properties of the relations defined on the elements of the graph (binary relation, reflexive, symmetric);
- Set the method of transferring the properties and characteristics of various components of the image format into characteristics of the graph and its elements.

All of this is defined by the relations already present between the components of the image format, as well as its properties and the elements that are necessary and sufficient for solving the tasks at hand [4].

Using the above algorithm, we can convert an image into a graph structure. Pixels in an image are positioned in a grid-like fashion, therefor a grid graph structure is best fitting. Each pixel can be represented by a vertex in the grid graph. Information about the pixel (e.g. RGB values) can be stored within the vertex. Pixel-vertices are connected to neighboring pixel-vertices in the graph by edges—three for vertices at the corners of the image, five for the outline, and eight edges for the rest of the vertices in the graph. In addition to connecting neighboring vertices, edges can be weighted and store additional information (e.g. the difference in color between neighboring vertices). Use of weighted edges can further influence processing, improving efficiency and accuracy.

Once the input image has been preprocessed and structured as a grid graph, we can move on to the next stage of computer vision—image segmentation.

6 Image Segmentation

After analyzing existing algorithms that can be used to help solve the task of image segmentation on graph structures, the conclusion was made that the «Efficient Graph-Based Image Segmentation» algorithm, developed by Pedro F. Felzenszwalb (MIT) and Daniel P. Huttenlocher (Cornell University), is suited for solving the task at hand. It is based on Kruskal's algorithm and a Gaussian blur filter.

Kruskal's algorithm is a minimum spanning tree (MST) algorithm, which searches for edges of minimum weight, connecting any two trees in the forest. The vertices (v_i) of the graph represent the pixels of the image, and the edges $\left(e\left(v_i, v_j\right)\right)$—the connections between neighboring pixels. The color distance between neighboring pixels $\left(v_i, v_j\right)$ is defined by the weight of the edge $\left(w\left(e\left(v_i, v_j\right)\right)\right)$. MSTs need to be found in the graph, or, in other words, trees, whose sum of all edges, included in the tree, is minimal. Additionally—all vertices must be accessible, to avoid losing pixels or parts of the image.

At first, a set is created for each vertex ($G_1 \ldots G_n$, where n represents the amount of vertex-pixels), where each vertex is the only element in its set (i.e. the initial forest). The algorithm joins these sets to form a number of $G\prime$ sets, which represents the various segments of the image ($G'_1 \ldots G'_m$, where m represents the amount of segments). Additionally, the vertices in each G'_i set (($i = 1 \ldots m$) form a MST.

A Gaussian filter is used to smooth the image slightly before computing the edge weights, in order to compensate for digitization artifacts [5].

As an additional improvement to the «Efficient Graph-Based Image Segmentation» algorithm, a formula for automatically calculating the optimal "σ" value (Gaussian blur coefficient) was established. The mean and deviation of edge weights were analyzed on more than one hundred sample images, and an approximation function was found (Fig. 3).

The optimal "σ" value can be calculated using the mean and deviation of the weights that are stored in the graph's edges (i.e. the color distances between neighboring pixels). The formula for calculating the optimal "σ" coefficient is as follows:

$$\sigma = 0.2 \cdot 10^{-\left(\frac{deviation}{mean} - 1.75\right)^2} + 0.35 \quad (1)$$

The «Efficient Graph-Based Image Segmentation» algorithm breaks up the initial image into three monochrome images (color channels), smooths each image using a Gaussian blur, and performs image segmentation based on pairwise region comparison [5]. However, the distance between RGB values, although often effective

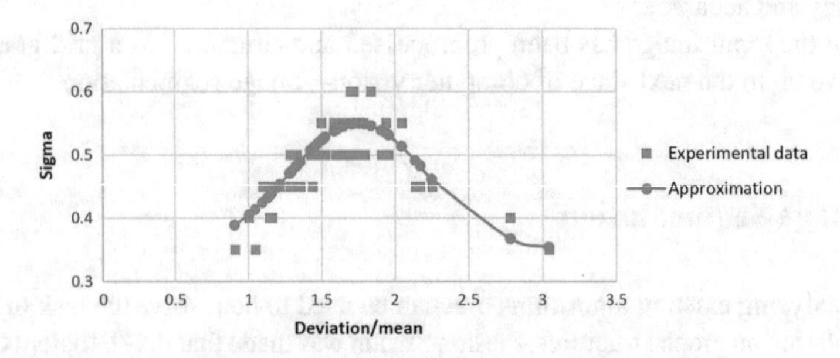

Fig. 3 Values of the selected coefficients and the approximation function

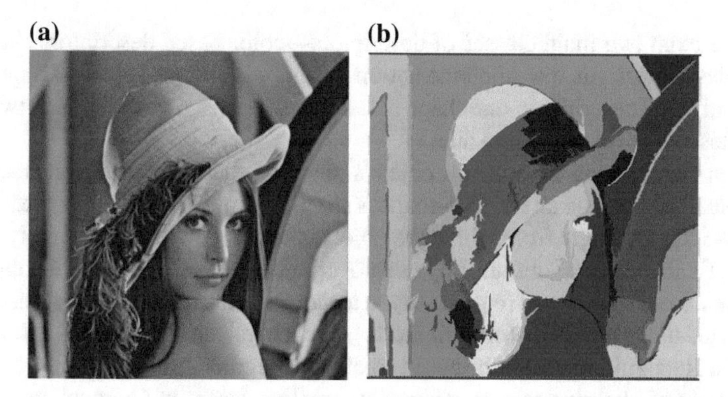

Fig. 4 Segmentation result using Kruskal's algorithm with a Gaussian blur. **a** initial image. **b** segmented image

enough, does fully correspond to how the human brain perceives distances between colors. Because of this, usage of the color model CIE LUV is recommended as a future improvement, in which color distances are more representative to those that are seen by the human visual system.

For testing this segmentation algorithm, the image of "Lenna", well known in the field of computer vision, was used. Once segmentation is complete, the algorithm colors the resulting $G'_1 \ldots G'_m$ sets (segments) using random colors, to visualize the result of the segmentation process. This result is shown in Fig. 4.

7 Segment Description

After breaking up our grid graph into individual segments and finding the boundaries between objects, we need to determine what characteristics can be digitized and used as segment descriptors. The goal of this step is to find such a set of descriptors that would characterize, as uniquely as possible, the given segment. A sub-goal is compressing these descriptors into a form that would allow for quick and efficient difference calculation and searching.

The operations of searching and comparing are much easier to fulfill with a single unit of information than with, for example, a sub-graph, consisting of various vertices and edges. Consequently, a set of descriptors must be found that would be able to, as uniquely as possible, characterize various segments and, at the same time, be compressible into a single unit of information. For simplicity, we decided to name such a set of descriptors "the segment's fingerprint". These segment fingerprints are used to populate our algorithm's knowledge base, where they are grouped in accordance with the object class they belong to. These classes, in turn, are linked with each other in accordance with their degree of association.

There exist two main classes of descriptors—color based descriptors and shape based descriptors. In our implementation of this algorithm, segment fingerprints consist of three descriptors: one shape based descriptor (region-based) and two color based descriptor (gradient and hue).

When recognizing objects, an object's shape is the most useful characteristic available. However, on an image—shape is simply a projection of a 3D object, which changes with the angle from which the object is being viewed. Additionally, part of an object's shape can be obscured by other overlapping objects, and certain objects do not have a fixed shape at all (e.g. the sky in a photo). In these cases, objects need to be recognized using primarily (or exclusively) other characteristics, such as color and/or gradient. Recognition cannot be based solely on a segment's color, so this descriptor serves more of a helping role, to increase assumption accuracy. Gradient, on the other hand, can be used to characterize texture, which can be almost as informative as an object's shape. Many objects have characteristic textures, depending on the material that they are made from or consist of.

7.1 Region-Based Shape Descriptor

The region-based shape descriptor can be used to efficiently describe the form of an object. This type of descriptor is capable of processing objects that consist of multiple outlines, each of which can include multiple holes in its shape. When using contour-based shape descriptors, the object being described must consist of one, completely filled, contour. The region-based shape descriptor, on the other hand, can be used to describe a wide variety of forms—from simple triangles and ovals, to more complex forms, whose holes carry as much information as the outer boundary does.

Since this descriptor is based on the characteristics of various image zones, it's robust to noise that may have been introduced during segmentation. The descriptor itself is also characterized by a small size and fast retrieval time. The descriptor's size is fixed to 17.5 byte, and the operations of retrieval and comparison are simple and have minimal computational complexity [6].

The region-based shape descriptor uses a set of ART (Angular Radial Transform) coefficients. ART is a 2D complex transformation, defined in polar coordinates:

$$F_{nm} = \langle V_{nm}(\rho, \theta), f(\rho, \theta) \rangle = \int_0^{2\pi} \int_0^1 V_{nm}^*(\rho, \theta), f(\rho, \theta)\rho, d\rho\, d\theta \quad (2)$$

Here, F_{nm} is an ART coefficient of order n and m; $f\ (\rho,\theta)$ is an image function in polar coordinates; and $V_{nm}(\rho,\theta)$ is the ART basis function, which can be divided along its angular and radial directions, i.e.:

$$V_{nm}(\rho, \theta) = A_m(\theta) R_n(\rho) \quad (3)$$

The angular and radial basis functions are defined as follows:

$$A_m(\theta) = \frac{1}{2\pi}\exp(jm\theta) \quad (4)$$

$$R_n(\rho) = \begin{cases} 1 & n = 0 \\ 2\cos(\pi n\rho) & n \neq 0 \end{cases} \quad (5)$$

To calculate an object's region-based shape descriptor (ArtDE[35]), the following five steps must be taken:

1. Basis function generation: since the basis functions can be divided, $V_{nm}(x, y)$ is directly calculated in Cartesian coordinates, to avoid converting $V_{nm}(,)$ from polar coordinates to Cartesian on a later step. For this, a set of ART basis functions is created as two four-dimensional arrays, for the real and imaginary parts, respectively.
2. Size normalization: the object's center of mass (centroid) is aligned with the lookup table's center. If the size of the object does not match that of the lookup table, linear interpolation is used to map the object onto the lookup table. Here, the size of the object is defined as twice the maximum distance from the centroid of the object.
3. ART transformation: the real and imaginary parts of the ART coefficients, ArtR[12] [3] and ArtI[12] [3], are calculated by summing up the multiplication of a pixel ("1" if it's part of the shape, otherwise—"0") to each corresponding position in the lookup table, in raster scan order. Here, twelve angular and three radial basis functions are used.
4. Area normalization: the magnitude of each ART coefficient is calculated (ArtM[m][n]) by extracting the square root from the sum of the squares of its real and imaginary parts. The resulting values are then divided by the first coefficient (ArtM[0][0]) for normalization.
5. Quantization: the magnitudes of the ART coefficients, excluding ArtM[0][0], are then quantized to 16 levels using the quantization table shown in Table 1.

The distance between two shape descriptors can be calculated using the following formula:

$$Distance(A, B) = \sum_{i=0}^{34} |ArtDE_A[i] - ArtDE_B[i]| \quad (6)$$

7.2 Gradient Descriptor

To represent the gradient of an object, we can use the segment's graph that we received from Kruskal's algorithm during segmentation. Since gradient is defined by

Table 1 ArtM to ArtDE quantization Table [5]

Range	ArtDE	Range	ArtDE
0.000000000 ≤ ArtM < 0.003585473	0000_2	0.038508176 ≤ ArtM < 0.045926586	1000_2
0.003585473 ≤ ArtM < 0.007418411	0001_2	0.045926586 ≤ ArtM < 0.054490513	1001_2
0.007418411 ≤ ArtM < 0.011535520	0010_2	0.054490513 ≤ ArtM < 0.064619488	1010_2
0.011535520 ≤ ArtM < 0.015982337	0011_2	0.064619488 ≤ ArtM < 0.077016351	1011_2
0.015982337 ≤ ArtM < 0.020816302	0100_2	0.077016351 ≤ ArtM < 0.092998687	1100_2
0.020816302 ≤ ArtM < 0.026111312	0101_2	0.092998687 ≤ ArtM < 0.115524524	1101_2
0.026111312 ≤ ArtM < 0.031964674	0110_2	0.115524524 ≤ ArtM < 0.154032694	1110_2
0.031964674 ≤ ArtM < 0.038508176	0111_2	0.154032694 ≤ ArtM	1111_2

the direction and intensity of change in brightness between two pixels, new weights must be set for the edges of the graph. The brightness of a pixel is defined as the sum of its RGB values, preliminarily multiplied by the brightness coefficient for each color. The coefficients are as follows: 0.212656 for red, 0.715158 for green, and 0.072186 for blue. Afterwards, the weight of the edge between neighboring pixels is calculated as the difference of brightness values. If the difference is negative, then the direction of the edge is inverted and the edge's weight is set to the absolute value of the difference. This results in a graph, whose edges are directed and weighted in accordance with the direction and magnitude of the gradient between neighboring vertices (pixels).

The gradient descriptor is initialized as a vector diagram of eight vectors (one for each possible edge direction in the grid graph), in which all gradient differences (edge weights) are summed in accordance with their direction (see Fig. 5b). These sums are then normalized (divided by the number of pixels in the segment), allowing the descriptor to remain unchanged when changes in scale or segment size are made. The resulting descriptor must then be rotated, so that the largest normalized sum faces up. This final step allows the descriptor to be robust to segment rotation and simplifies descriptor comparison and searching.

Figure 5 shows this process using an example segment's directed gradient graph, consisting of 6 vertices (5a). The directions of all gradients are summed by all eight directions and stored in the gradient descriptor (5b—for simplicity, all gradient intensities are set to "1"). Lastly, the result is rotated, so that the largest normalized sum always faces up (5c).

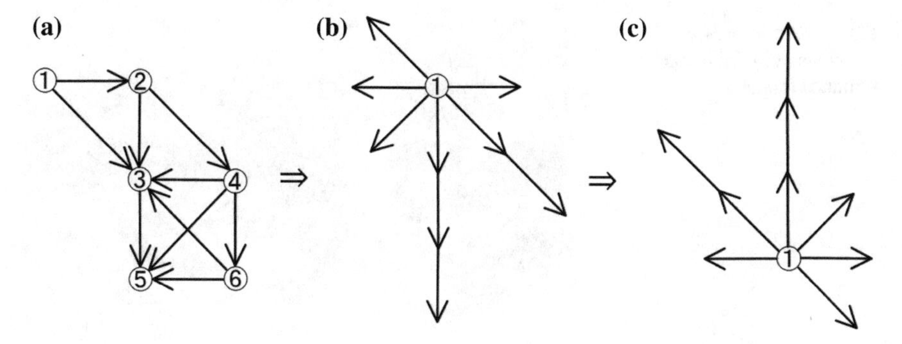

Fig. 5 Coding a segment's gradient using a directed graph of gradient directions

7.3 Hue Descriptor

A segment's hue descriptor can be calculated similarly to how its gradient descriptor was calculate. By separately summing the RGB values of the segment's pixels, and then dividing the resulting three sums by the total number of pixels in the segment, we receive the average RGB value for the given segment. This descriptor characterizes the "average color" of the segment, whose pixels were already grouped as similarly colored earlier on during segmentation.

7.4 Descriptor Examples and Comparison

To better explain the use of segment descriptors, let's examine the segment fingerprint that is generated by Lenna's "face" segment, as well as the one that is generated after rotating the initial image by 50°. It's important to point out that when rotating a raster image by a number of degrees that is not a multiple of 90°, the color values of the image's pixels are approximated, making them differ from the values of the original image. In other words—some sharpness is lost in the process of rotating a raster image, which can sometimes be noticed around object boundaries. When the grid graph for the rotated image is formed, the weight of its edges will differ from the ones in the initial image's graph. This means that the segments that are formed during image segmentation can differ from one another after image rotation (Fig. 6).

Table 2 shows both segments' shape descriptor values.

To conclude that two shapes are similar, their total shape descriptor difference must be less than 100, and the difference in each zone must be less than 5. In the example above, the total difference is 72, and the maximum difference in each zone is 4. This indicates that both segments are of the same shape, even though they are not identical, due to differences in segmentation. Table 3 demonstrates the values of these segments' gradient descriptors.

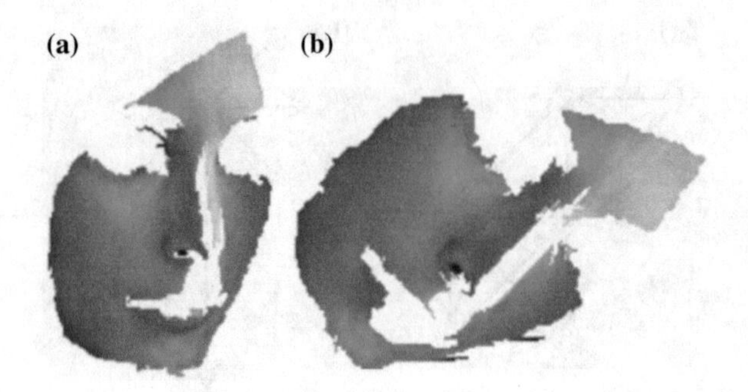

Fig. 6 Resulting face segments. **a** initial image. **b** rotated image

Since the gradient descriptor's values are normalized, all directions can be judged by a single maximum allowed difference, which we defined to be 0.3. As seen in Table 3—the maximum difference between the segment's descriptors is 0.112, which is less than the defined maximum difference.

The resulting hue descriptors are 222 121 120 and 221 121 120 for the initial and rotated segments respectively. From this, one can conclude that both segments are part of the same region (or located in similar regions).

8 Identification

Once we've generated a segment's fingerprint, we can search for similar fingerprints in our object knowledge base to form object class assumptions for the given segment. This knowledge base contains "smartly" averaged fingerprints from previous runs, allowing the algorithm to attempt to find a fingerprint that is similar to the one being processed. If no matches are found, or the object is new to the knowledge base (and, in turn, its segment's fingerprint), then the unidentified fingerprint can be added to the knowledge base by a teacher for future use.

Once we have formed object predictions for each segment fingerprint, we can use context-based recognition to improve the accuracy of these predictions and read the image as a whole, and not as separate, out of context objects.

9 Interpretation

To increase the accuracy of our solution, we structured our knowledge base of fingerprints as a graph, whose vertices are grouped and interconnected in accordance with their semantic features. Since objects of the same class can look very unalike, several fingerprints are stored for each object and grouped together. Afterwards, we can assign weights to the links (edges) that connect object classes (groups of segment

Table 2 Comparing shape descriptors

Zone#	01	02	03	04	05	06	07	08	09	10	11	12	13	14	15
0°	14	15	15	10	07	15	15	14	12	08	12	15	13	07	14
50°	15	15	15	12	08	15	15	13	15	09	13	15	12	10	13
Zone#	16	17	18	19	20	21	22	23	24	25	26	27	28	29	30
0°	11	07	12	10	06	12	07	03	12	09	06	02	03	05	05
50°	08	09	11	08	02	11	10	06	08	06	03	06	07	08	03
Zone#	31	32	33	34	35										
0°	07	08	05	03	02										
50°	04	05	08	05	06										

Table 3 Comparing gradient descriptors

Direction	T	TR	R	BR	B	BL	L	TL
0°	1.235	1.073	1.138	0.842	1.063	0.952	1.068	0.905
50°	1.215	0.961	1.094	0.778	1.032	0.816	0.959	0.883

fingerprints) in the following manner: if a pair of objects are frequently encountered together in various scenes, then larger weights must be assigned—otherwise, smaller or even "0" (null) weights are assigned. When attempting to interpret, what objects are present in a given scene, we simultaneously search for multiple object fingerprints, taking into account each object's surroundings.

To better explain, let's take a look at the following example: general computer vision systems can easily mistake a window curtain for a towel. In our solution, the edge between the group of "window" fingerprints and "curtain" fingerprints will be greater, than that between the "window" and "towel" groups. This allows the system to look for fingerprints that are linked to each other with greater weights, improving the accuracy of predictions for all objects in the scene and making the predictions scene-specific. Many CNN-based computer vision systems lack this functionality, which often leads to erroneous predictions [7].

10 Populating the Knowledge Base

If no knowledge base fingerprints match the segment fingerprint that is being processed, or if the segment was mistaken for an object of a different class, then a teacher can help improve the system through additional training and extension of the knowledge base. Once the teacher, who knows what real-world object a segment represents, adds the new segment fingerprint to a new or existing object class, the algorithm can make predictions as to whether or not objects of this class are represented on future images. However, for this to work as needed, the teacher must maintain integrity and avoid using synonyms when naming objects of the same class.

This computer vision system can be trained using already existing marked datasets, many of which are readily available online and open to the public. Using specially developed datasets helps quickly populate the knowledge base with a wide variety of object examples. The images contained in such datasets are often specially selected to be difficult to segment, which helps test the capabilities of a system's segmentation algorithm. However, training with marked datasets has its drawbacks [8]. The marks that are used to describe images in a dataset are extremely simplified, which leads to loss of object traits during object recognition. Because of this, we recommend training with the help of a teacher, as in this case, the algorithm's results can be made to be more detailed and accurate when describing the objects of a scene.

11 Conclusion

In this chapter, an alternative approach to solving computer vision tasks using graph structures was introduced. Prior to image segmentation, a Gaussian filter is used to smooth images and compensate for digitization artifacts. Use of the CIE LUV color model is suggested as a further improvement, which would allow calculating color distances in a way that more closely corresponds to the differences seen by the human visual system. Image segmentation is performed using pairwise region comparison.

Various descriptors were analyzed, and a standard set of descriptors for segment characterization was proposed and named the "segment's fingerprint". A graph-based knowledge base structure was proposed, that groups segment fingerprints into classes, and connects classes based on their degree of association. The advantages of such a structure were shown, and the process of improving assumption accuracy using neighboring assumptions was described.

Such an algorithm can be constantly recognizing objects and constantly learning, so there is no clear separation between the training and validation modes. When an object is not recognized, the algorithm can ask a teacher to determine the correct class of the object, and the teacher can correct the algorithm's decision to increase accuracy in future runs. As a result, the number of recognizable classes is not limited.

Additionally, since this alternative algorithm takes into account the relationships between objects, the output can include descriptions of simple actions, which were recognized as taking place on the scene. For example—an object that is simply hanging in the air can be characterized as falling or flying, and a rippling flag can indicate the presence of wind.

References

1. Liu, D., Xie, S.: Neural information processing. In: 24th International Conference, ICONIP 2017, Guangzhou, China, November 14–18, 2017, Proceedings, Part 2
2. Pechyonkin, M.: Understanding Hinton's Capsule Networks. Part I: Intuition. [Online]. Available: https://medium.com/ai%C2%B3-theory-practice-business/understanding-hintons-capsule-networks-part-i-intuition-b4b559d1159b. Accessed: 13 June 2018
3. Popov, A.: An introduction to the MISD technology. In: Proceedings of the 50th Hawaii International Conference on System Sciences, HICSS50, Hawaii, 3–7 January 2017, pp. 1003–1012 (2017)
4. Ovchinnikov, V.A.: Graphs in problems of analysis and synthesis of structures of complex systems. Bauman Moscow State Technical University, Moscow (2014). [In Russian]
5. Felzenszwalb, P.F., Huttenlocher, D.P.: Efficient graph-based image segmentation. Int. J. Comput. Vis., **59**(2), September 2004
6. Yamada, A., Pickering, M., Jeannin, S.: MPEG-7 visual part of experimentation model version 8.1// ISO/IEC JTC1/SC29/WG11/M6808. Pisa, Italy. January (2001)

7. Kravets, A.G., Lebedev, N., Legenchenko M.: Patents images retrieval and convolutional neural network training dataset quality improvement. In: Proceedings of the IV International research conference information technologies in Science, Management, Social sphere and medicine (ITSMSSM 2017) DEC 05–08, 2017, Tomsk, Russia
8. Korobkin, D.M., Fomenkov, S.A., Kravets, A.G.: Methods for extracting the descriptions of sci-tech effects and morphological features of technical systems from patents (2019). In: 2018 9th International Conference on Information, Intelligence, Systems and Applications, IISA 2018, art. no. 8633624

Efficient Computational Procedure for the Alternance Method of Optimizing the Temperature Regimes of Structures of Autonomous Objects

Mikhail Yu. Livshits, A. V. Nenashev and B. B. Borodulin

Abstract The method of increasing the efficiency of the computational procedure for determining an optimal control of the temperature field of load-bearing structures of autonomous objects is proposed. Optimization of temperature distributions using controlled heat sources ensures the reduction of the temperature component of the measurement information error, which comes from heat-releasing information measuring systems placed on the structure. As an example, a supporting structure in the form of a rectangular isotropic prism is analyzed. The computational procedure uses a finite element mathematical model of the optimization object in the ANSYS software environment.

Keywords Alternative optimization method · Optimal control · Computational procedure · Finite element model

1 Introduction

The problems of reducing the temperature component of the measurement information error, which comes from information measuring systems (IMS) placed on the supporting structure of an autonomous object (AO), have become aggravated not only and not so much due to the increase in the number of AO but, first and foremost, due to increased intensity of their work, the degree of saturation by heat generating equipment, and the extreme conditions of operation of the AO. This also increases the relevance of improving the weight and size characteristics of not only IMS, but supporting systems as well. The latter relates to the systems for ensuring the thermal

M. Yu. Livshits (✉) · A. V. Nenashev · B. B. Borodulin
Samara State Technical University, Samara, Russia
e-mail: usat@samgtu.ru

A. V. Nenashev
e-mail: alexvlnenashev@gmail.com

B. B. Borodulin
e-mail: borodulinbb@gmail.com

A. G. Kravets et al. (eds.), *Cyber-Physical Systems: Industry 4.0 Challenges*, Studies in Systems, Decision and Control 260,
https://doi.org/10.1007/978-3-030-32648-7_7

regime of AO in general, and the thermal stabilization of the temperature mode of operation of the IMS supporting structure (SS), in particular.

There is a proven system of thermal stabilization of the IMS temperature regime using controlled heat sources (CHS) as located on the surface of AO SS [1]. With their help, compensation for the irregularity of the SS temperature field, caused by heat emission of the IMS's operating in accordance with their own schedule, external factors, etc., is achieved. Irregularity of the temperature field causes thermal deformation of AO SS which, in turn, results in temperature errors, especially information from optical systems due to the shifting of the optical axis and the focal length. The system of automatic control of the heat sources which compensate for this unevenness is quite effective; however, it requires significant energy consumption and features considerable weight and dimensions due to the need for placing a sufficiently large number of CHS's on the structure [1]. Optimal control of the CHS's according to the criteria of maximum accuracy and speed will ensure extremely high uniformity of the temperature field in the critical cut of the AO SS at a given placement of the maximum permissible number of CHS's on AO SS, as limited by weight and size of CHS's.

2 Statement of the Problem

In order to solve this problem, the authors use the effective alternate method of optimization (AMO) [2]. In this case, the AMO computational procedure is performed on a finite-element mathematical model of the process in the ANSYS software environment in the offline mode, and the implementation of optimal control is carried out in real time using an onboard computer. The computational procedure (CP) of AMO [2, 3] is rather problematic because it reduces to solving, normally, a transcendental system of defining equations. The method of increasing the efficiency of the AMO CP is proposed.

Let us consider for definiteness the design of NC in the form of a rectangular prism made of an isotropic material (Fig. 1) with faces $\alpha = \overline{1, 6}$ which are exposed to heat from ambient environment by way of radiant heat exchange, and on which the heat generating IMS's are attached at points U1–U16. On mutually opposite faces $\alpha = 3$ and $\alpha = 4$, the rectangular regions S1 and S2 are highlighted, on which at points 1–6 CHS's are affixed in order to compensate for thermal deformations which occur in the responsible cut of SS due to the irregularity of the temperature field [1].

In [1, 4, 5], for the control object (CO) analyzed the problems of optimal, in terms of accuracy and speed, control of the temperature distribution in a prism was established. These problems were reformulated into a spatially one-dimensional non-smooth boundary problem of semi-infinite optimization, in order to ensure the minimum possible deviation from the given temperature T_z in the most critical cut of the structure in the segment $L = \{l_x : l_x \in [0, 1], l_y^*, l_z^*\}$ where (l_x, l_y, l_z) the relative Cartesian coordinates.

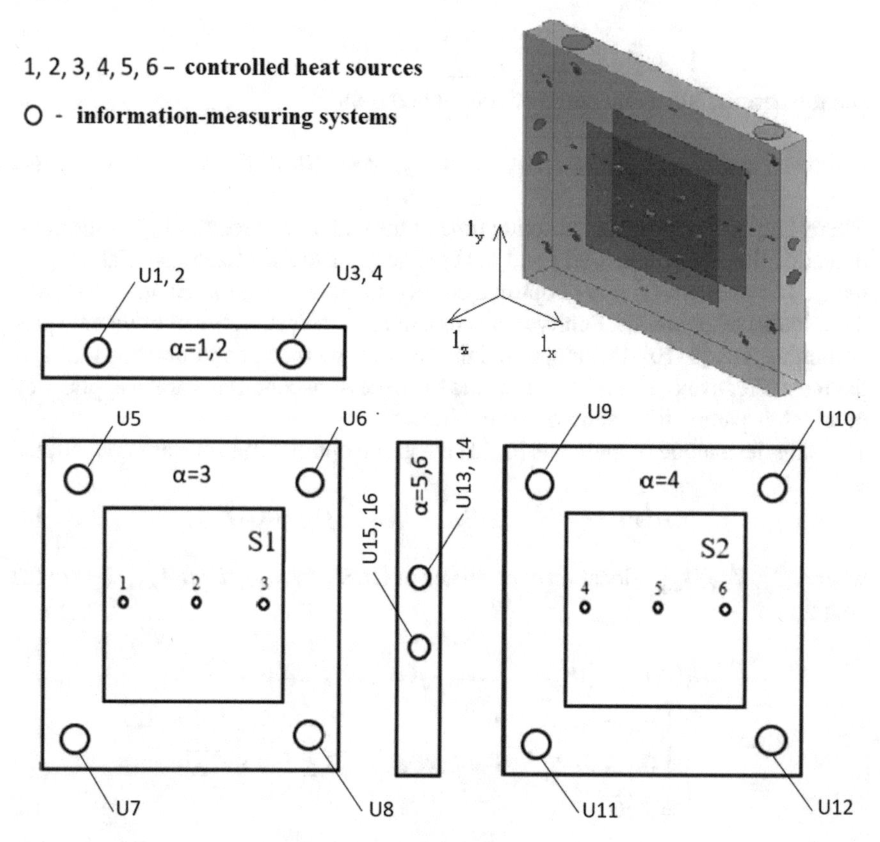

Fig. 1 Location of the heat sources on the supporting structure (example)

A solution of the problem is reduced to the determination of the optimal group change in the intensity of CHS's heat flux $q_j(\phi),\ j = \overline{1,6}$, which is an optimal control that provides the minimum value of functional:

$$J(\Theta,\ q_j(\phi),\ \phi) = \min_{q_j(\phi)} J\ ,$$
$$\min_{q_j(\phi)} J = \min_{q_j(\phi)} \left\| \Theta(l_x, l_y^*, l_z^*, \phi_k,\ q_j(\phi)) \right\|_{L_\infty[l_x \in L]}, \tag{1}$$

which represents the norm in space L_∞ of the deviations

$$\Theta(l_x, l_y^*, l_z^*, \phi_k,\ q_j(\phi)) = T(l_x, l_y^*, l_z^*, \phi_k,\ q_j(\phi)) - T_z \tag{2}$$

of temperature distribution $T(l_x, l_y^*, l_z^*, \phi_k,\ q_j(\phi))$ along the line $l_x \in [0, 1]$ in the critical cut $l_y^* \in l_y, l_y^* = const$; $l_z^* \in l_z, l_z^* = const$ of SS at a given point in time $\phi_k = \frac{\lambda \cdot \tau_k}{\cdot x_{\max}^2}$ in relative units, under the constraints of the control resource

$$0 \le q_j(\phi) \le q_{j,\,\max} \tag{3}$$

and the permissible temperature deviation in the SS

$$\left|\Theta(l_x, l_y, l_z, \phi)\right| \le \Theta_d,\ \forall\phi \in [0,\ \phi_k] \tag{4}$$

where: $\phi_k,\ \tau_k$—relative and absolute time of the end of the process, λ, C—the coefficient of thermal conductivity and heat capacity of the structure material, respectively. The problem (1)–(4) of optimal control can be solved, for example, by using the calculus of variations, Pontryagin's maximum principle, Bellman's dynamic programming, etc. [6–10]. In our opinion, the alternate method of optimization [2, 11] is the most effective one in this case,, while the necessary condition for the application of which is parametric nature of the control.

Using the method of moments [6, 11] the group optimal control was parametrized

$$u_{opt}(\phi) = u(\phi)\cdot q_j(l_{x,\,j},\ l_{y,j},\ l_{z,j}) = u(\Delta)\cdot q_j \tag{5}$$

where $l_{x,\,j},\ l_{y,j},\ l_{z,j}$—location coordinates of CHS's $q_j = q_j(l_{x,\,j},\ l_{y,j},\ l_{z,j})$ in the area S:

$$u(\Delta) = \left\{u_1(\Delta),\ \ldots,\ u_j(\Delta),\ldots,\ u_6(\Delta)\right\}$$

$$u_j(\Delta) = \begin{cases} q_{j,\,\max},\ \Delta = \Delta_\delta^{(i)},\ \delta = 1,3,5\ldots,\ \delta \le i,\ i = \overline{1,I} \\ 0,\ \Delta = \Delta_\delta^{(i)},\ \delta = 2,4,6\ldots,\ \delta \le i,\ i = \overline{1,I} \end{cases},$$

$$j = \overline{1,6} \tag{6}$$

$\Delta_\delta^{(i)}$—duration of the δ time interval $\phi = \frac{\lambda\cdot\tau}{\cdot x_{\max}^2}$, where the intensity of the heat flow of CHS retains one of the boundary values according to (3), i—a total number of the intervals of the constancy of optimal control in the i class of controls under consideration, I—the maximum possible number of constancy intervals of optimal control. Thus, the solution of the optimal control problem (1)–(4) is reduced to the problem of finding a finite number and values of the parameters $\Delta_\delta^{(\mathrm{i})}$, on which each vector-function $u(\Delta)$ component alternately takes it's maximum possible values, taking into account constraints (3). The duration $\Delta_\delta^{(\mathrm{i})}$ and the required number of intervals i are determined by the alternate method [2] from the system of transcendental equations [4]:

$$\Theta(l_k, \phi_k,\ u(\Delta))|_{l_k\in L_k} = \pm\varepsilon^{(\mathrm{i})},\quad \left.\frac{\partial\Theta(l_k,\ \phi_k,\ u(\Delta))}{\partial l_k}\right|_{l_k\in L_k} = 0, \tag{7}$$

$$k = \overline{1,R}\left(R = i,\ \varepsilon_{\min}^{(i-1)} > \varepsilon = \varepsilon_3 > \varepsilon_{\min}^{(i)};\ R = i+1,\ \varepsilon = \varepsilon_{\min}^{(i)}\right),$$
$$\omega = \overline{1,r} l_\omega \in \{l_k \in L_k, r = R-2, r = R-1, r = R\}$$

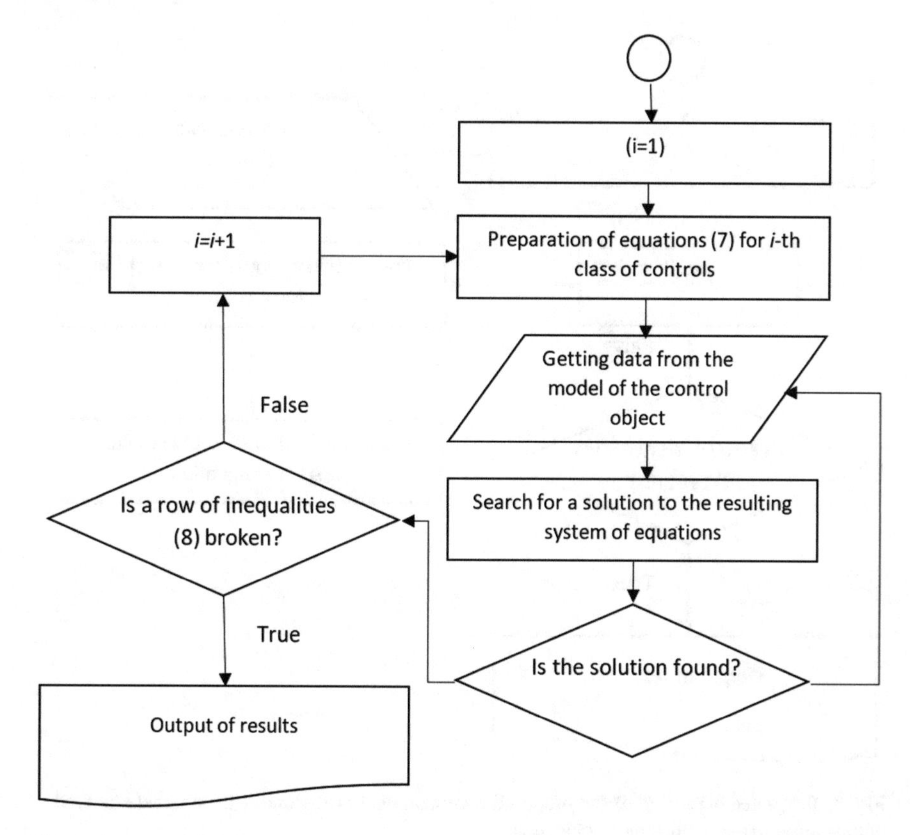

Fig. 2 Integrated algorithm of the computational procedure of the alternance method of optimization

where $\Theta(l_k, \phi_k, u(\Delta))|_{l_k \in L_k}$—numerical solution of the boundary value problem (1)–(4) at a finite point in time ϕ_k at the points $l_k(l_x^{(k)}, l_y^*, l_z^*) \in L_k(l_x^{(k)}, l_y^*, l_z^*, \phi_k) \subset \subset L(l_x, l_y^*, l_z^*)$ of the considered section line $L(l_x, l_y^*, l_z^*)$, which represent a countable set of points, where the relation (7) is satisfied. Depending on the optimal control problem, the following problems may be solved: The problem of speed with the ε_3—given deviation of the resultant temperature field from the required $T_z(R = i,\ \varepsilon = \varepsilon_3)$ or the problem of maximum accuracy $\left(R = i + 1,\ \varepsilon = \varepsilon_{\min}^{(i)}\right)$, with the minimum possible deviation in the i parametric class of optimal control—$\varepsilon_{\min}^{(i)}$. In this case, the optimal number $i = I$ of intervals of the constancy of the optimal control is not known in advance, and in accordance with the AMO procedure [2, 11–13] it is determined by a number of inequalities (Fig. 2):

$$\varepsilon_{\min}^{(1)} > \varepsilon_{\min}^{(2)} > \cdots > \varepsilon_{\min}^{(i)} \geq \varepsilon_3 \leq \varepsilon_{\min}^{(\xi=i+1)} > \cdots > \varepsilon_{\min}^{(n)} = \varepsilon_{\inf} = \inf J \quad (8)$$

where $I = \xi$—in case of solving the problem of speed, $I = n$—in case of reaching the maximum possible error—the exact lower bound of its assessment (1). In order to

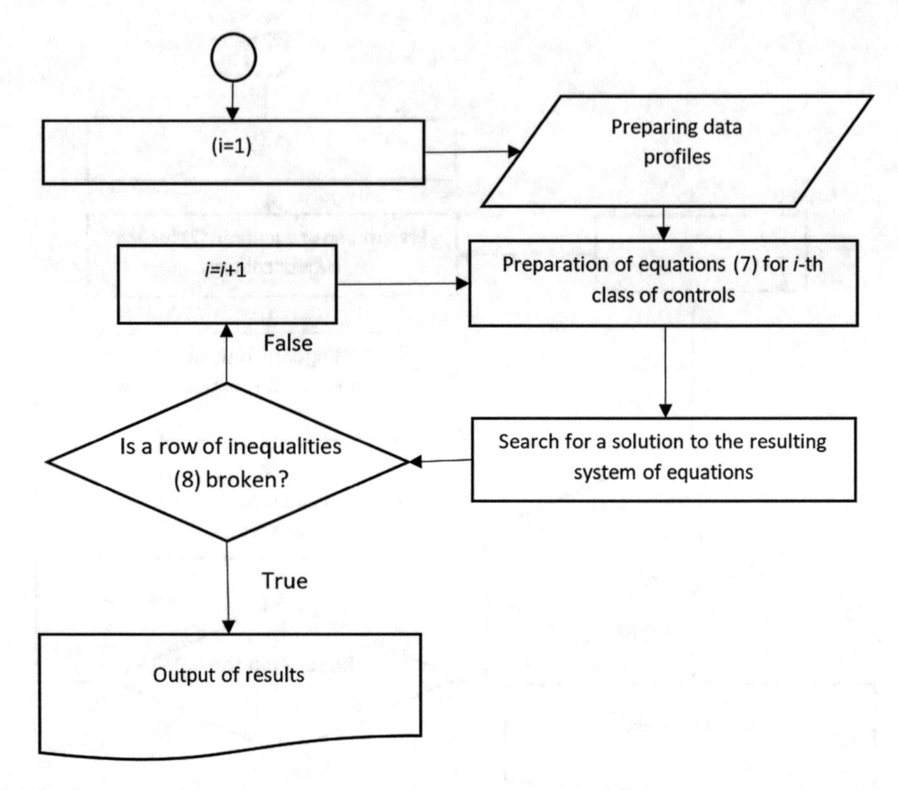

Fig. 3 Integrated algorithm of the proposed computational procedure of the alternate method of optimization in the multi-dimensional area

determine the parameters $\Delta_\delta^{(I)}$, the problems of maximum accuracy are sequentially solved for each $i \in Z^+$, for which the conditions (8) are fulfilled, for which I times (Fig. 3) the system (7) is compiled and solved, with a different number of variables and equations. Normally, in this case, its analytical solution is difficult to achieve, and numerical methods are used [14–20].

3 Computational Procedure

In order to solve the system (7), a specialized software package of AMO CP has been developed, and its application is effective under the conditions which are fair for the stated problem: $\phi_k = \sum_{\delta=1}^{i} \Delta_\delta^{(i)}$ and optimal control $u(\Delta^{(i)})$ can be represented as a simple Borelean vector function. Since CHS's commute simultaneously, the vector function (6) takes the following two possible values $u_{\text{max}} = \{q_{1,\text{max}}, \ldots, q_{6,\text{max}}\}$ and $u_0 = \{0, 0,0,0,0, 0\}$.

In accordance with the AMO CP algorithm [3] the points $M_{g,\,b} = (l_b, \phi_g)$:$\{\phi_g \in (0,\ \phi\text{max}],\ \phi_g = \phi_{g-1} + \phi\text{max}/N_\phi,\ \phi_k \le \phi\text{max},\ g = \overline{1,\ N_\phi};\ l_b \in L,\ b == \overline{1, N_{l_x}},\ N_{l_x} = L/\Gamma_{set,\,x}\}$ are prepared, which are located in the cut $L(l_x, l_y^*, l_z^*)$, the spatial coordinates l_b of which coincide with the nodes of the finite element grid of the numerical solution, which corresponds to the direct boundary value problem of heat conduction in the ANSYS package. Where $\Gamma_{set,\,x}$—the linear size of a finite element in the ANSYS package along the axis x. For each l_b two profiles are determined on the numerical finite element model of this boundary value problem:

$$\Theta_b^{(u\text{max})} = \left\{\Theta_{b,1}^{(u\text{max})}, \ldots, \Theta_{b,\,N_\phi}^{(u\text{max})}\right\},\ \Theta_{b,\,g}^{(u\text{max})} = \Theta\left(M_{g,\,b},\ u\text{max}\right),$$

$$\Theta_b^{(0)} = \left\{\Theta_{b,1}^{(0)}, \ldots, \Theta_{b,\,N_\phi-1}^{(0)}\right\},\ \Theta_{b,\,g}^{(0)} = \Theta\left(M_{g,\,b},\ u_0\right),$$

thereupon for any $\Delta^{(i)} = \left\{\Delta_1^{(i)}, \ldots, \Delta_\delta^{(i)}, \ldots, \Delta_i^{(i)}\right\}$ in the points $M_{g,\,b}$ using the fit method it is possible to calculate:

$$\begin{gathered}
\Theta(M_{g,b}, u(\Delta^{(i)})) = \Theta^*_{g,b}(\Delta^{(i)}), \\
\Theta^*_{g,b}(\Delta^{(i)}) = \begin{cases} \Theta_{b,g}^{(u_{\max})}, \delta = 1, 3, 5\ldots \\ \Theta_{b,y}^{(u_{\max})} + \Theta_{b,g-y}^{(0)} - \Theta_{b,0}, \delta = 2, 4, 6\ldots \end{cases},\quad \delta = \overline{1, i}, \\
\Theta_{b,y}^{(u_{\max})} = \Theta\left(l_b, \phi_y = \sum_{p=1}^{\delta-1} \Delta_p^{(i)}, u_{\max}\right)
\end{gathered} \tag{9}$$

where $\Theta_{b,\,0}$—the value of deviation (2) of the temperature from T_z at the point with the spatial coordinates $\left(l_b, l_y^*, l_z^*\right)$ at the moment of time $\phi = 0$, $b = \overline{1, N_{l_x}}$, $g = \overline{1,\ N_\phi}$. No further calculations of the configuration of the CO model are required. The initial temperature distribution $\Theta(M_{g,\,b},\ 0)$, the points $M_{g,\,b}$, the profiles $\Theta_b^{(q\text{max})}$ and $\Theta_b^{(0)}$ are fed to the input (Fig. 3) of the AMO CP software package. In order to solve the problem of optimal control in the software package for each i class of controls, the functional of discrepancies of equations in the system (7) is automatically composed:

$$\begin{gathered}
\Theta(l_k, \phi_k,\ u(\Delta^{(i)}))\big|_{l_k \in L_k} \pm \varepsilon = f_k(l_k, \phi_k,\ u(\Delta^{(i)}), \varepsilon), \\
\left.\frac{\partial\Theta(l_\omega, \phi_k,\ u(\Delta^{(i)}))}{\partial l_\omega}\right|_{l_\omega \in L_k} = f_\omega(l_k, \phi_k,\ u(\Delta^{(i)}), \varepsilon)
\end{gathered} \tag{10}$$

$$\begin{gathered}
k = \overline{1, R}\left(R = i,\ \varepsilon_{\min}^{(i-1)} > \varepsilon = \varepsilon_3 > \varepsilon_{\min}^{(i)};\ R = i + 1,\ \varepsilon = \varepsilon_{\min}^{(i)}\right), \\
\omega = \overline{1, r} l_\omega \in \{l_k \in L_k, r = R - 2, r = R - 1, r = R\}
\end{gathered}$$

in the form of the penalty function

$$\begin{aligned} & I\left(l_k, \phi_k,\ u\left(\Delta^{(i)}\right), \varepsilon\right) \\ & \quad = I_{big}\left(l_k, \phi_k,\ u\left(\Delta^{(i)}\right), \varepsilon\right) + I_{small}\left(l_k, \phi_k,\ u\left(\Delta^{(i)}\right), \varepsilon\right) \\ & \quad = I^{(i)}\left(l_k, \phi_k,\ u\left(\Delta^{(i)}\right),\ i, \varepsilon\right), \end{aligned} \tag{11}$$

where

$$I_{big}\left(l_k, \phi_k,\ u\left(\Delta^{(i)}\right), \varepsilon\right) = [f_{abs}]^2;\ I_{small}\left(l_k, \phi_k,\ u\left(\Delta^{(i)}\right), \varepsilon\right) = \sqrt{f_{abs}};$$

$$f_{abs} = \sum_{k=1}^{R}\left|f_k\left(l_k, \phi_k,\ u\left(\Delta^{(i)}\right), \varepsilon\right)\right| + \sum_{\omega=1}^{r}\left|f_\omega\left(l_\omega, \phi_k,\ u\left(\Delta^{(i)}\right), \varepsilon\right)\right|$$

and thereupon the problem of mathematical programming is solved:

$$I^{(i)}\left(l_k, \phi_k,\ u\left(\Delta^{(i)}\right),\ i, \varepsilon\right) \to \inf_{\bar{R}_c} I^{(i)}\left(l_k, \phi_k,\ u\left(\Delta^{(i)}\right),\ i, \varepsilon\right) \tag{12}$$

$$\bar{R}_c = \left(l_k, \phi_k,\ u\left(\Delta^{(i)}\right),\ i, \varepsilon\right), = \dim R_c = 2(i+1)$$

In order to solve the problem (12) the S points are prepared

$$\begin{aligned} X_s^{(i)} = \left(l_{k,s}, \phi_{k,s}, u\left(\Delta^{(i)}\right), \varepsilon\right) : \big\{ & l_{k,s} \in l_b \in L, b = \overline{1, N_{l_x}}, \\ & \phi_{k,s} \in \phi_g \in (0, \phi_{\max}], g = \overline{1, N_\phi}, s = \overline{1, S}\big\} \end{aligned}$$

$$\Delta_\delta^{(i)} \in \phi_g \in (0, \phi_{\max}], \quad \delta = \overline{1, i}, \quad \phi_{k,s} = \sum_{\delta=1}^{i} \Delta_\delta^{(i)}$$

in which the values of the function $I_s^{(i)}\left(X_s^{(i)}\right)$ are determined in accordance with (9), (11).

Using the values obtained of the $I_s^{(i)}\left(X_S^{(i)}\right)$ function (11) and the set of points $X_S^{(i)}$ the problem (12) is solved in accordance with the method of Ψ—transformation as proposed by Chichinadze [3, 14]. This method reduces to the transformation of the minimized multidimensional function (11) into a one-dimensional, continuous, and monotony decreasing numerically defined metric $\Psi(\varsigma)$, the zero of which corresponds to the value of $I_\Psi^{(i)}\left(X_\Psi^{(i)}\right)$ which lies in the region most approximated to $I_{\inf}^{(i)} = 0$. The solution of the optimal control problem as achieved by the AMO CP for the CO under consideration (Fig. 4) corresponds to the results published in [1].

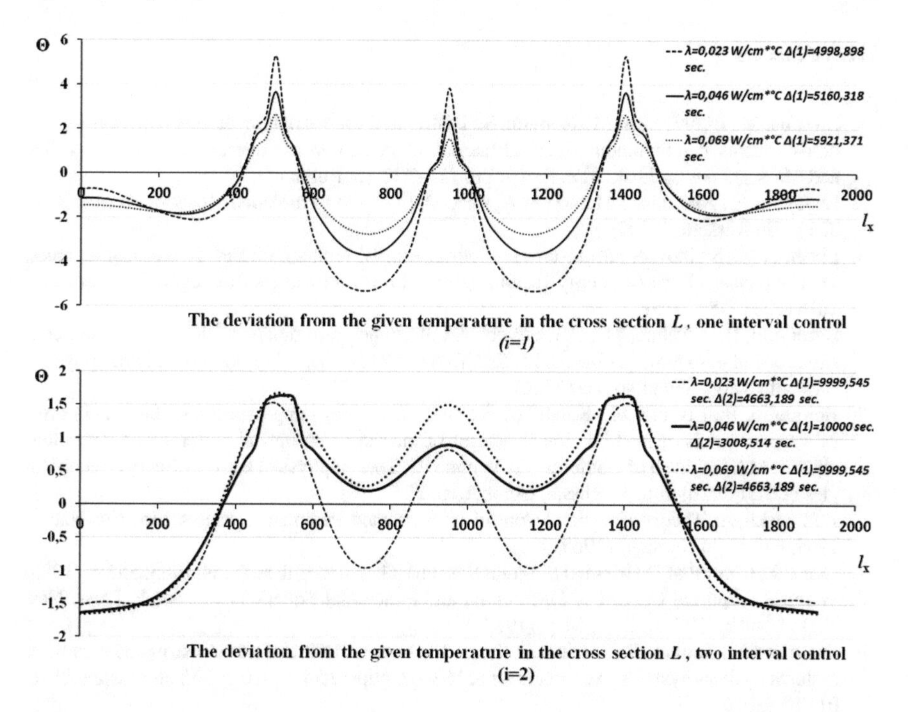

Fig. 4 Deviation from the given temperature on the cut line $L(l_x, l_y^*, l_z^*)$, $l_y^* = 0,5$, $l_z^* = 0,5$, $\rho = 1440\,\text{kg/m}^3$, $c = 1480$ J/Kg K, $N_{l_x} = 40$, $N_\phi = 40$ with the control which is optimal in terms of accuracy

4 Conclusion

Solution of the problem (1)–(4) as achieved without the use of AMO CP required 20 or more calculations of the configuration of the numerical finite element model of the CO within the ANSYS package. At the same time, the computational complexity of solving the problem (1)–(4) turns out to be a random value that varies over a wide range.

The proposed computational procedure requires only two calculations of the configuration of the CO model in the ANSYS package, in order to prepare a sufficient set of data to solve the problem (1)–(3) of the optimal control in accordance with AMO CP [1].

The results obtained enabled us to state the following:

1. Application of the proposed AMO CP has reduced the computational complexity of solving the problem (1)–(4) by the factor of 10 (and more);
2. The computational complexity of solving the problem (1)–(4) has become a determinate constant value—thanks to the use of AMO CP.

Acknowledgements The work was supported by the Russian Foundation for Basic Research projects No. 17-08-00593.

References

1. Livshits, M., Derevyanov, M., Kopytin, S.: Distributed control of temperature regimes of structural elements of autonomous objects. Materials of the XIV Minsk International Forum on Heat and Mass Transfer, Minsk, V. 1. Part 1. pp. 719–722. [in Russian] (2012)
2. Rapoport, E.: Alternance Method in Applied Optimization Problems. Moscow. Nauka, 2000, 335 p. [in Russian] (2000)
3. Livshitc, M., Sizikov, A.: Multi-criteria optimization of refinery. In: EPJ Web of Conferences. Thermophysical Basis of Energy Technologies 2015. vol. 110. https://doi.org/10.1051/epjconf/201611001035 (2016)
4. Borodulin, B., Livshits, M.: Optimal control of temperature modes of the instrumental constructions of autonomous objects. In: EPJ Web of Conferences. Volume 110, Thermophysical Basis of Energy Technologies (2016)
5. Borodulin, B., Livshits M., Korshikov S.: Optimization of temperature distributions in critical cross-sections of load-bearing structures of measurement optical systems of autonomous objects. MATEC Web of Conferences. Volume 92, Thermophysical Basis of Energy Technologies (TBET-2016) Tomsk, Russia, October 26–28 (2016)
6. Butkovskii, A.: Theory of Optimal Control of Distributed-Parameter Systems. Moscow: Nauka, 1965, 474 p. [in Russian] (1965)
7. Lions, J.: Control of Distributed Singular Systems. Gauthier-Villars, Paris, 1985. 552 p. (1985)
8. Warga, J.: Optimal Control of Differential and Functional Equations. Academic Press, New York, London, 1972. xiii + 531 p. (1972)
9. Di Loreto, M., Damak, S., Eberard, D., Brun, X.: Approximation of linear distributed parameter systems by delay systems. Automatica, pp. 162–68. https://doi.org/10.1016/j.automatica.2016.01.065 (2016)
10. Felgenhauer, U., Jongen, H.Th., Twilt, F., Weber, G.: Semi-infinite optimization: structure and stability of the feasible set. J. Optim. Theory Appl., **3**, 529–452. (1992)
11. Rapoport, E.Y.: Optimal Control of Distributed-Parameter Systems. Moscow: Vysshaya Shkola, 2009, 677 p. [in Russian] (2009)
12. Pleshivtseva, Y., Rapoport, E.: Parametric optimization of systems with distributed parameters in problems with mixed constraints on the final states of the object of control. J. Comput. Syst. Sci. Int. **57**, 723 (2018). https://doi.org/10.1134/S1064230718050118
13. Rapoport, E., Pleshivtseva, Y.: Optimal control of nonlinear objects of engineering thermophysics. Optoelectron. Instrument. Proc. **48**, 429 (2012). https://doi.org/10.3103/S8756699012050019
14. Chichinadze, V.: Solving non-convex nonlinear optimization problems. Nauka, Moscow 1983, 256 p. [in Russian] (1983)
15. Gill, F., Murray, W., Wright, M.: Practical optimization. Academic Press, New York, 1981. 509 pp. (1981)
16. Li, R., Liu, W., Ma, H., Tang, T.: Adaptive finite-element approximation for distributed elliptic optimal control problems. SIAM J. Contr. Optim., **4**, 1244–1265 (2003)
17. Murat, F., Tartar, L.: On the control of the coefficients in partial equations. SIAM J. Contr. Optim., **4**, 1244–1265 (2003)
18. Lian, T., Fan, Z., Li, G.: Lagrange optimal controls and time optimal controls for composite fractional relaxation systems. Adv Differ Equ. **1**, 233. https://doi.org/10.1186/s13662-017-1299-7. (2017)
19. Felgenhauer, U.: Structural properties and approximation of optimal controls. Nonlinear Anal. **3**, 1869–1880 (2001)
20. Buttazzo, G., Kogut, P.: Weak optimal controls in coefficients for linear elliptic problems. Rev. Mat. Complut. **24**, 83–94 (2018)

Cyber-Physical Systems Intelligent Control

Parametric Synthesis Method of PID Controllers for High-Order Control Systems

Andrey Prokopev, Zhasurbek Nabizhanov, Vladimir Ivanchura and Rurik Emelyanov

Abstract A method for designing a model of PID controllers for high-order systems based on the modal method for linear systems is proposed taking into account the assignment of the roots of the characteristic polynomial of the corrected automatic feedback control system. From the stability condition of the system, the coefficients of the characteristic polynomial must be positive. It is provided to set the test conditions of the positive values of the coefficients of the PID regulators. The calculation of the controller parameters is performed according to transition characteristics of the system. Functionality test of method has provided by the example of synthesis of PID controllers parameters of the control system of third order object.

Keywords Control · Parametric synthesis · PID controller · A polynomial of the system · The desired polynomial · Transient response

1 Introduction

The spread of information modeling technologies for a construction project (the BIM-technology concept) contributes to the development of directions for the implementation of automated control systems (ACS) in industrial-civil and road construction. It is demonstrated in scientific works on the design of automatic control systems for road and building machines, the development of the theory of building cybernetics [1, 2].

A. Prokopev (✉) · Z. Nabizhanov · V. Ivanchura · R. Emelyanov
Siberian Federal University, 79 Svobodny, Krasnoyarsk 660041, Russia
e-mail: prok1@yandex.ru

Z. Nabizhanov
e-mail: jasur150691@yandex.ru

V. Ivanchura
e-mail: ivan43ura@yandex.ru

R. Emelyanov
e-mail: ert-44@yandex.ru

A. G. Kravets et al. (eds.), *Cyber-Physical Systems: Industry 4.0 Challenges*, Studies in Systems, Decision and Control 260,
https://doi.org/10.1007/978-3-030-32648-7_8

Fig. 1 Control system structure

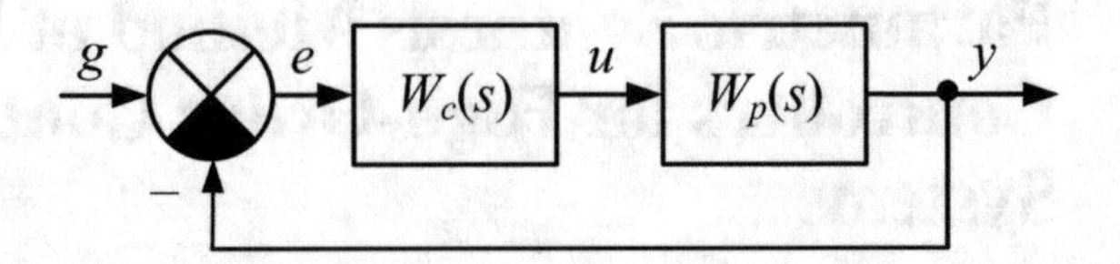

From scientific sources it is possible to single out methods of the synthesis of PID regulators [3–11]: empiric tuning; Ziegler-Nichols methods [3] and derived methods [4]; algebraic methods; methods of modal synthesis; methods of frequency domain synthesis; methods of optimal synthesis; method of the optimal transfer function (PF) of a closed system (technical and symmetric optima). The main trends in the development of methods for the synthesis of PI and PID regulators are discussed in Aidan O'Dwyer 2006 [9] (443 methods for the synthesis of PID regulators) and in Aidan O'Dwyer 2009 [10] (1731 methods for the synthesis of PID regulators).

Among the methods of synthesis of PID-regulators developed by scientists, is the modal method [5, 11]. Mathematical models of working processes of road and construction machines, depending on the assumptions accepted, lead to models of the second and higher order [12–19].

Previously, the authors obtained expressions for determining the PID controller parameters for the control object of the second order [2].

The development of designing regulators methods for objects above the second order is an actual scientific and practical task.

2 Research Objective

We consider a linear automatic control system with feedback, with a transfer function (TF) of an object higher than the second order (Fig. 1).

In Fig. 1, the following symbols are used: e—an error signal equal to the difference between the drive signal g and the y—regulated signal; u—control signal.

The initial data for the analytical synthesis of the ACS PID controller of the parameters are the TF model of the control object higher than the second order, the TF model of the regulator and the given system quality indicators: overshoot and control time.

The transfer function $W_p(s)$ of the high-order control object and the PID controller $W_c(s)$ is given as follows:

$$W_p(s) = \frac{Y(s)}{X(s)} = \frac{b_m \cdot s^m + b_{m-1} \cdot s^{m-1} + \ldots + b_0}{a_n \cdot s^n + a_{n-1} \cdot s^{n-1} + \ldots + a_0} = \frac{N(s)}{D(s)}, \text{ under } m < n, \quad (1)$$

$$W_c(s) = K_p + \frac{K_i}{s} + K_d \cdot s = \frac{K_d \cdot s^2 + K_p \cdot s + K_i}{s}, \quad (2)$$

where s—Laplace operator; K_p, K_i, K_d coefficients of, respectively, proportionality, integration, and differentiation.

The problem of determining the coefficients (parameters) K_p, K_i, K_d of the PID controller model of a closed control system that provide specified quality indicators when operating a control object with a high order TF (higher than the second one) W_p(s) for the variant of specifying the number and type of roots of the characteristic polynomial of this system is solved.

The algorithm of the PID model design method (approach) includes the following (see below) steps.

Step 1: Present the transfer function of the control object $W_p(s)$ (1) to the following form (in the notation in the language of the MathCAD program).

$$W_c(s) = K_p + \frac{K_i}{s} + K_d \cdot s = \frac{K_d \cdot s^2 + K_p \cdot s + K_i}{s}, \tag{3}$$

under given conditions $a_0 = 1$.

Step 2: Determine the transfer function $K(s)$ of the closed aided system, Fig. 1, according to the formulas (2) and (3).

$$K(s) = \frac{W_c(s) \cdot W_p(s)}{1 + W_c(s) \cdot W_p(s)} \rightarrow \frac{K_0 \cdot K_d \cdot s^2 + K_0 \cdot K_p \cdot s + K_0 \cdot K_i}{a_0 \cdot s^4 + a_1 \cdot s^3 + (a_2 + K_0 \cdot K_d) \cdot s^2 + (a_3 + K_0 \cdot K_p) \cdot s + K_0 \cdot K_i}. \tag{4}$$

By the denominator of the transfer function (4) of the closed-loop system, it can be seen that the PID-controller coefficients affect the last three terms of the characteristic polynomial of this system. The remaining coefficients of this polynomial do not depend on the parameters of the PID controller.

Therefore, it is possible to draw a conclusion influencing further studies on the use of the proposed method. Depending on the n-order of the control object model, the coefficient (s) of the characteristic polynomial of the corrected system, which is described by the TF model of the form (3), does not depend on the parameters of the PID controller.

$n = 3$–a_1;
$n = 4$–a_1, a_2;
$n = 5$–a_1, a_2, a_3;
$n = 6$–a_1, a_2, a_3, a_4;
$n = 7$–a_1, a_2, a_3, a_4, a_5.

Step 3: Specify the number and type of roots of this polynomial (or poles of the system being corrected).

Variants for specifying the number and type of the roots of the characteristic polynomial of a corrected closed system can be different. This condition determines

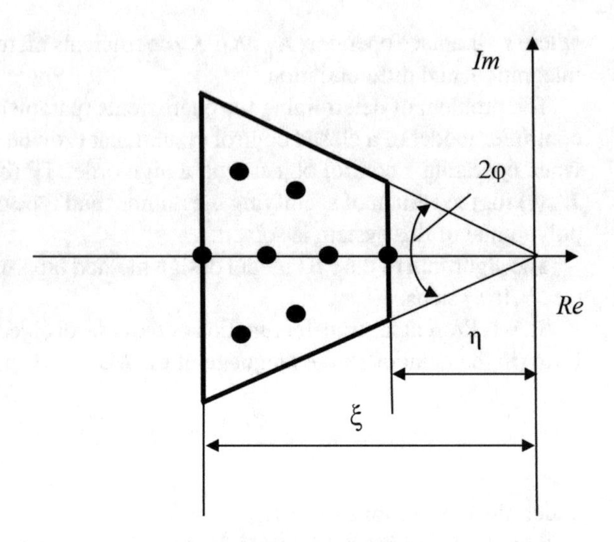

Fig. 2 Root quality data

the research task of the proposed method (approach) for designing a PID controller model of an object control system higher than the second order.

It is necessary to keep the sequence of the poles assignment. First, the real parts of these poles of the projected system characteristic polynomial are determined, since these values uniquely determine the constant coefficient of the polynomial a_1. Further

The roots of the polynomial are given taking into account the root quality data.

In Fig. 2 the following symbols are used: η—degree of stability, φ—angle, its maximum value is used to determine the root quality of the degree of oscillatory system $\mu = tg\varphi_{\max}$, ξ—distance from the imaginary axis to the most remote root (roots).

For the exploratory study, the following variant of the task was adopted: all the roots of the characteristic polynomial of an adjustable closed system are the complex conjugate

$$s_1 = -\alpha_1 + j \cdot \beta_1;\ s_2 = -\alpha_1 - j \cdot \beta_1;\ s_3 = -\alpha_2 + j \cdot \beta_2;\ s_4 = -\alpha_2 - j \cdot \beta_2.$$

Studies have shown that the coefficient a_1 of the transfer function (4) will be determined only by the real parts of these roots. For the third-order object model ($n = 3$), the characteristic polynomial of the closed aided system will be of the order of one more, i.e. $n_p = n + 1$. So, if all the roots of this polynomial are complex conjugate, then

$$a_1 = (2 \cdot \alpha_1 + 2 \cdot \alpha_2). \tag{5}$$

Formula (5) will determine the choice of the values of these real parts. For example, we specify α_1, but α_2 is determined from expression (5), i.e.

$$\alpha_2 := \frac{a_1 - 2 \cdot \alpha_1}{2}.$$

That is, a limit is set $2 \cdot \alpha_1$ should be less than a_1, which is due to the stability of the system.

Step 4: To determine the coefficients of the desired polynomial of the closed aided system with the found values of the characteristic polynomial roots of this system. These coefficients depend on the imaginary part of the complex-conjugate root β_i, where $i = 1, 2$.

From the structure type of the characteristic polynomial of the closed aided system, it follows that the last three coefficients depend on the parameters of the PID controller, i.e.

$$D(s) := a_0 \cdot s^4 + a_1 \cdot s^3 + (a_2 + K_0 \cdot K_d) \cdot s^2 + (a_3 + K_0 \cdot K_p) \cdot s + K_0 \cdot K_i. \tag{6}$$

To obtain an expression for the desired characteristic polynomial $D_p(\mathrm{s})$ of the aided system

$$D_p(s) := a_{0p} \cdot s^4 + a_{1p} \cdot s^3 + a_{2p} \cdot s^2 + a_{3p} \cdot s + a_{4p}. \tag{7}$$

In the polynomial (7), the coefficients a_{2p}, a_{3p}, a_{4p}, are determined by the following expressions

$$a_{2p} := a_2 + K_0 \cdot K_d;\ a_{3p} := a_3 + K_0 \cdot K_p;\ a_{4p} := K_0 \cdot K_i. \tag{8}$$

Characteristic polynomial development of a closed aided system expressed through complex poles (the notation in the language of the MathCAD program):

$$s_1 := -\alpha_1 + 1j \cdot \beta_1;\ s_2 := -\alpha_1 - 1j \cdot \beta_1;$$

$$\begin{aligned}(s - s_1) \cdot (s - s_2) &\to (s + \alpha_1 - \beta_1 \cdot 1j) \cdot (s + \alpha_1 + \beta_1 \cdot 1j)\ \text{simplify} \\ &\to s^2 + 2 \cdot \alpha_1 \cdot s + \alpha_1^2 + \beta_1^2.\end{aligned}$$

Then, using the function of collecting of terms (in the language of the MathCAD program), the characteristic polynomial of the closed aided system expressed in terms of complex poles will be received

$$\begin{aligned}&(s^2 + 2 \cdot \alpha_1 \cdot s + \alpha_1^2 + \beta_1^2) \cdot (s^2 + 2 \cdot \alpha_2 \cdot s + \alpha_2^2 + \beta_2^2)\ \text{collect},\ s \\ &\quad \to s^4 + (2 \cdot \alpha_1 + 2 \cdot \alpha_2) \cdot s^3 + \left(\alpha_1^2 + 4 \cdot \alpha_1 \cdot \alpha_2 + \alpha_2^2 + \beta_1^2 + \beta_2^2\right) \cdot s^2 \\ &\qquad + \left[2 \cdot \alpha_2 \cdot \left(\alpha_1^2 + \beta_1^2\right) + 2 \cdot \alpha_1 \cdot \left(\alpha_2^2 + \beta_2^2\right)\right] \cdot s + \left(\alpha_1^2 + \beta_1^2\right) \cdot \left(\alpha_2^2 + \beta_2^2\right).\end{aligned} \tag{9}$$

Introduce the notation taking into account the structure of the characteristic polynomials (6), (7), (9), expressed in terms of complex poles

$$a_1 := (2 \cdot \alpha_1 + 2 \cdot \alpha_2); \tag{10}$$

$$a_2 + K_0 \cdot K_d := \left(\alpha_1^2 + 4 \cdot \alpha_1 \cdot \alpha_2 + \alpha_2^2 + \beta_1^2 + \beta_2^2\right);$$
$$a_{2p} := \left(\alpha_1^2 + 4 \cdot \alpha_1 \cdot \alpha_2 + \alpha_2^2 + \beta_1^2 + \beta_2^2\right); \tag{11}$$

$$a_3 + K_0 \cdot K_p := \left[2 \cdot \alpha_2 \cdot \left(\alpha_1^2 + \beta_1^2\right) + 2 \cdot \alpha_1 \cdot \left(\alpha_2^2 + \beta_2^2\right)\right];$$
$$a_{3p} := \left[2 \cdot \alpha_2 \cdot \left(\alpha_1^2 + \beta_1^2\right) + 2 \cdot \alpha_1 \cdot \left(\alpha_2^2 + \beta_2^2\right)\right]; \tag{12}$$

$$K_0 \cdot K_i := \left(\alpha_1^2 + \beta_1^2\right) \cdot \left(\alpha_2^2 + \beta_2^2\right);\ a_{4p} := \left(\alpha_1^2 + \beta_1^2\right) \cdot \left(\alpha_2^2 + \beta_2^2\right). \tag{13}$$

It follows from the condition of stability of the system that the coefficients of the characteristic polynomial (7) must be positive. And the coefficient a_{1p} of this polynomial does not depend on the parameters of the regulator and its value can not be changed, because it is determined by calculations.

The following condition must be met

$$a_{1p}(\alpha_1, \alpha_2) = a_1.$$

The real part α_1 and imaginary parts β_1, β_2 of the complex roots of the characteristic polynomial of the corrected closed system are expertly specified.

The real part α_2 of the roots of the system under investigation is determined

$$\alpha_2 := \frac{(a_1 - 2 \cdot \alpha_1)}{2}.$$

To implement the proposed method of synthesis, conditions that define the requirement for positive parameters values of the PID controller model are set, i.e.

$$a_{2p} > a_2 + K_0 \cdot K_d;\ a_{3p} > a_3 + K_0 \cdot K_p. \tag{14}$$

Conditions (14) that expressed through the complex poles

$$\left(\alpha_1^2 + 4 \cdot \alpha_1 \cdot \alpha_2 + \alpha_2^2 + \beta_1^2 + \beta_2^2\right) > a_2; \tag{15}$$

$$\left[2 \cdot \alpha_2 \cdot \left(\alpha_1^2 + \beta_1^2\right) + 2 \cdot \alpha_1 \cdot \left(\alpha_2^2 + \beta_2^2\right)\right] > a_3. \tag{16}$$

Step 5: To calculate the coefficients of the PID controller model.

The coefficients of the PID controller are determined from the expressions (8):

$$a_{2p} > a_2 + K_0 \cdot K_d;\ a_{3p} > a_3 + K_0 \cdot K_p. \tag{17}$$

Expressions (17) for determination of coefficients of the PID controller model obtained through complex poles have the following form

$$K_d := \frac{(\alpha_1^2 + 4 \cdot \alpha_1 \cdot \alpha_2 + \alpha_2^2 + \beta_1^2 + \beta_2^2) - a_2}{K_0}; \tag{18}$$

$$\left[2 \cdot \alpha_2 \cdot (\alpha_1^2 + \beta_1^2) + 2 \cdot \alpha_1 \cdot (\alpha_2^2 + \beta_2^2)\right] > a_3; \tag{19}$$

$$K_i := \frac{(\alpha_1^2 + \beta_1^2) \cdot (\alpha_2^2 + \beta_2^2)}{K_0}. \tag{20}$$

Step 6: Check the results of coefficients calculating the PID controller model for the step response of the ACS.

The found roots values of the characteristic polynomial allow determining the numerical values of their coefficients.

3 Example of Synthesizing the Parameters of a PID Controller Model

As an example, let us consider the calculation of optimal parameters of the PID controller model of a system with a third-order object. Formulas are represented using the notation accepted in the MathCAD program.

A third-order control object $W_p(s)$ is defined, which is formed by a serial connection of the TF $W_1(s)$ and $W_2(s)$

$$W_1(s) := \frac{K_1}{T_1 \cdot s + 1};\ W_2(s) := \frac{K_2}{T_2^2 \cdot s^2 + 2 \cdot \xi \cdot T_2 \cdot s + 1};\ W_p(s) := W_1(s) \cdot W_2(s),$$

where coefficients are $K_1 := 3$, $K_2 := 4$.

Define the TF of the control object in the MathCAD environment using the command collect.

$$W_1(s) \cdot W_2(s) \text{ collect}, s \rightarrow \frac{K_3}{T_1 \cdot T_2^2 \cdot s^3 + (T_2^2 + 2 \cdot T_1 \cdot T_2 \cdot \xi) \cdot s^2 + (T_1 + 2 \cdot T_2 \cdot \xi) \cdot s + 1}. \tag{21}$$

Further, consider a third-order control object $W_p(s)$ and a controller $W_c(s)$ whose TFs have the form of (3), (2), and in a series connection form an open system with a transfer function $W(s)$

$$W(s) := W_p(s) \cdot W_c(s) \text{ collect}, s \rightarrow \frac{K_0 \cdot K_d \cdot s^2 + K_0 \cdot K_p \cdot s + K_0 \cdot K_i}{a_0 \cdot s^4 + a_1 \cdot s^3 + a_2 \cdot s^2 + a_3 \cdot s}. \tag{22}$$

A closed aided control system with a single feedback is determined

$$W(s) := W_p(s) \cdot W_c(s) \text{ collect}, s \rightarrow \frac{K_0 \cdot K_d \cdot s^2 + K_0 \cdot K_p \cdot s + K_0 \cdot K_i}{a_0 \cdot s^4 + a_1 \cdot s^3 + a_2 \cdot s^2 + a_3 \cdot s}. \tag{23}$$

For further calculations, the control object's TF, the time constants T_1, T_2, and the damping coefficient are defined ξ : $T_1 := 0.5$, $T_2 := 1.0$, $\xi := 0.5$.

We define the coefficients of the TF (23) taking into account the model of the TF (21) of the control object

$$K_0 := \frac{12}{T_1 \cdot T_2^2};\ a_1 := \frac{(T_2^2 + 2 \cdot T_1 \cdot T_2 \cdot \xi)}{T_1 \cdot T_2^2};\ a_2 := \frac{(T_1 + 2 \cdot T_2 \cdot \xi)}{T_1 \cdot T_2^2};\ a_3 := \frac{1}{T_1 \cdot T_2^2}.$$

And obtain $K_0 = 24$; $a_1 = 3$; $a_2 = 3$; $a_3 = 2$.

The ratio $a_0 := 1$ of the transfer function $K(s)$ is set.

Analysis of the control object. We investigate the stability of a closed system with a third-order control object by a direct method for determining the roots of its characteristic polynomial and the graph of the step response.

To obtain the TF of a closed system with the control object from the original data of the example

$$W_{p1}(s) := \frac{5.049 \cdot 10^{-29} \cdot s + 24}{s^3 + 3 \cdot s^2 + 3 \cdot s + 26}.$$

To define the roots of the characteristic polynomial of the control object using the command "polyroots"

$$V_0 := \begin{pmatrix} a_3 + K_0 \\ a_2 \\ a_1 \\ 1 \end{pmatrix};\ \begin{pmatrix} s_{30} \\ s_{20} \\ s_{10} \end{pmatrix} := \text{polyroots}(V_0) = \begin{pmatrix} -3.924 \\ 0.462 - 2.532 \cdot j \\ 0.462 + 2.532 \cdot j \end{pmatrix},$$

where V_0—the vector of the polynomial coefficients, starting with a smaller degree; s_{10}, s_{20}, s_{30}—polynomial roots. Positive values of the real parts of the roots are obtained—the system is unstable.

We construct a step response of a closed system with a control object in the MathCAD program environment. The expression for the step response $h(t)$ obtained using the function "invlaplace".

$$\begin{aligned} h(t) : = W_{p1}(s) \cdot 1/s \,\text{invlaplace},\ s \rightarrow 0.2384 \cdot e^{-3.924 \cdot t} \\ + 0.000005722 \cdot \cos(0.866 \cdot t) \cdot e^{-0.5 \cdot t} + 0.000004186 \cdot \sin(0.866 \cdot t) \cdot e^{-0.5 \cdot t} \\ - 0.6846 \cdot \cos(2.532 \cdot t) \cdot e^{-0.462 \cdot t} - 0.2446 \cdot \sin(2.532 \cdot t) \cdot e^{-0.462 \cdot t} + 0.9231. \end{aligned}$$

The graph of the step response of the investigated system is shown in Fig. 3.

Conclusion on the analysis of the control object: The closed system is unstable, therefore requires correction.

The characteristic polynomial of a corrected closed aided system expressed through complex poles is determined by expression (9).

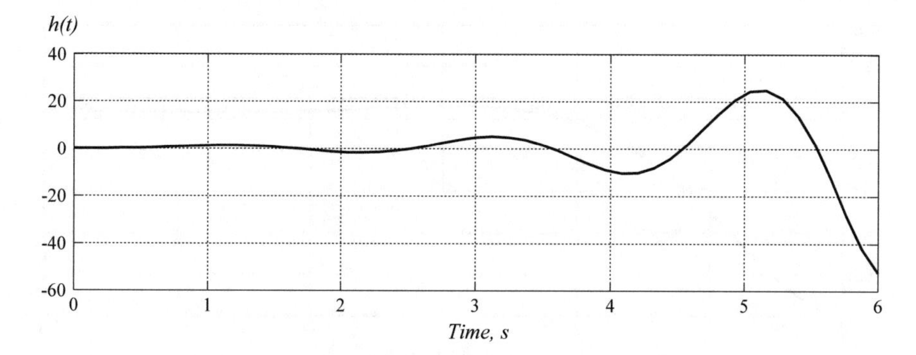

Fig. 3 Step response of the control object with feedback

Then the notation (10)–(13) is applied, taking into account the structure of the characteristic polynomials (6), (7), (9) expressed through complex poles.

Limits (15) and (16) values of the poles of the corrected system are established, caused by the requirement of positive values of the PID regulator coefficients.

Parameters of complex conjugate poles are set

$$s_1 = -\alpha_1 + j \cdot \beta_1;\ s_2 = -\alpha_1 - j \cdot \beta_1;\ s_3 = -\alpha_2 + j \cdot \beta_2;\ s_4 = -\alpha_2 - j \cdot \beta_2;$$

$$\alpha_1 := 0.8;\ \alpha_2 := \frac{(a_1 - 2 \cdot \alpha_1)}{2};\ a_1 = 3;\ \alpha_2 = 0.7;\ \beta_1 := 1.0;\ \beta_2 := 0.8.$$

The coefficients of the PID controller model are calculated from expressions (18)–(20):

$$K_p := \frac{\left[2 \cdot \alpha_2 \cdot \left(\alpha_1^2 + \beta_1^2\right) + 2 \cdot \alpha_1 \cdot \left(\alpha_2^2 + \beta_2^2\right)\right] - a_3}{K_0};\ K_p = 0.088;$$

$$K_i := \frac{\left(\alpha_1^2 + \beta_1^2\right) \cdot \left(\alpha_2^2 + \beta_2^2\right)}{K_0};\ K_i = 0.077;$$

$$K_d := \frac{\left(\alpha_1^2 + 4 \cdot \alpha_1 \cdot \alpha_2 + \alpha_2^2 + \beta_1^2 + \beta_2^2\right) - a_2}{K_0};\ K_d = 0.084.$$

Verification of the stability of the system and the correctness of the set values of the TF poles of the corrected closed system is performed by the direct method

$$V_1 = \begin{pmatrix} K_0 \cdot K_i \\ a_3 + K_0 \cdot Kp \\ a_2 + K_0 \cdot K_d \\ a1 \\ 1 \end{pmatrix};\ \begin{pmatrix} S_1 \\ S_2 \\ S_3 \\ S_4 \end{pmatrix} := \text{polyroots}(V_1) = \begin{pmatrix} -0.8 + j \\ -0.8 - j \\ -0.7 - j\,0.8 \\ -0.7 + j\,0.8 \end{pmatrix}.$$

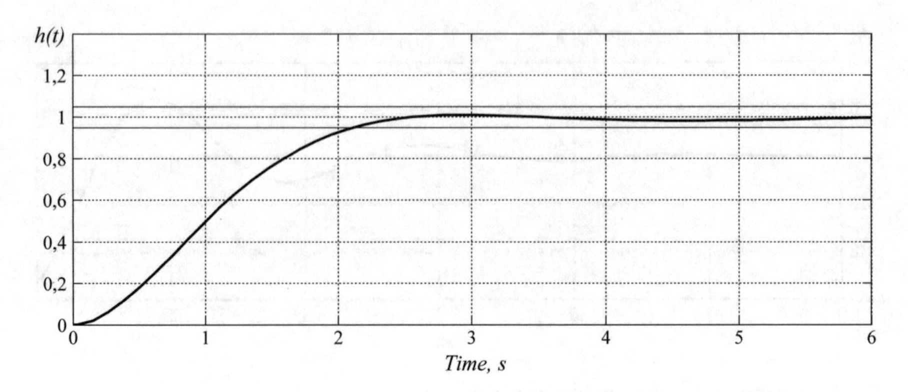

Fig. 4 Step response of a feedback control system

Conclusion: the adjusted system is stable. The values of the poles of the TF of a closed system correspond to given values.

Determination of the values of the coefficients of the TF (23) of the corrected closed control system

$$a_1 = 3;\ K_0 \cdot K_d = 2.01;\ K_0 \cdot K_p = 2.104;\ K_0 \cdot K_i = 1.853;$$
$$a_2 + K_0 \cdot K_d = 5.01;\ a_3 + K_0 \cdot K_p = 4.104.$$

Determination of the step response using the invlaplace command of the MathCAD environment:

$$h(t) := K(s) \cdot 1/s\,\text{invlaplace},\, s \rightarrow 1.398 \cdot \cos(0.8 \cdot t) \cdot e^{-0.7 \cdot t} - 0.92 \cdot \sin(0.8 \cdot t) \cdot e^{-0.7 \cdot t} - 2.398 \cdot \cos(0.999 \cdot t) \cdot e^{-0.7999 \cdot t} - 0.204 \cdot \sin(0.999 \cdot t) \cdot e^{-0.7999 \cdot t} + 1.0.$$

Construction of the step response $h(t)$ of the transfer function model (23) describing the corrected closed control system. The result of the construction of the step response $h(t)$ in the MATLAB program environment is shown in Fig. 4.

The expected control time is determined from the approximate formula [20–22]

$$t_p \approx 3/\eta,$$

where η is the parameter called the degree of stability, is the distance from the imaginary axis to the nearest root, that is, the minimum value of its real part $\eta{=}|\alpha|_{\min}$.

For the considered example the expected the time of the transient process $t_{p0} = 3/\min(\alpha_1, \alpha_2) = 4.28\,s$, where $\min(\alpha_1, \alpha_2) = 0.7$. The time of the transition process was $t_p = 2.1s$, the overshoot was $\sigma\% < 1\%$.

4 Conclusion

A method for designing the parameters of a PID controller model is proposed, this method provides the quality indicators of the automatic control system for speed, overshoot, and the type of the transient process for a given arrangement of poles, in systems with a high-order object (higher than the second).

The advantage of the proposed algorithm over methods based on restrictions on stability stocks is that the poles of the system are directly related to direct estimates of the quality of transient processes, such as control time and overshoot.

From the considered example for the third-order object, it can be seen that the implementation of the method provides positive parameters of the PID controller model and given estimates of the quality of the transient response—regulation time and overshoot.

In the parametric synthesis of the PID controller model, the following conditions must be taken into account by the proposed method:

it is necessary to set the actual poles values α_1, α_2 of the corrected system using the coefficient a_1 of the corrected system $a_1 = (2 \cdot \alpha_1 + 2 \cdot \alpha_2)$ and limitations determined by the coefficients a_2, a_3: $\left(\alpha_1^2 + 4 \cdot \alpha_1 \cdot \alpha_2 + \alpha_2^2 + \beta_1^2 + \beta_2^2\right) > a_2, \left[2 \cdot \alpha_2 \cdot \left(\alpha_1^2 + \beta_1^2\right) + 2 \cdot \alpha_1 \cdot \left(\alpha_2^2 + \beta_2^2\right)\right] > a_3$ that determine the values of imaginary parts β_1 and β_2 poles;
there is no spontaneous setting of the values of the coefficients a_2, a_3, which is possible only for the order of the system being corrected $n = 2$; the minimum time of the transient is limited.

References

1. Prokopev, A.P., Ivanchura, V.I., Emelianov, R.T., Palchikov, P.A.: Implementation of the concept of automation and intellectualization of management of road construction processes. Vestnik MGSU. 2018, **13**(1) (112), 61–70 (2018)
2. Prokopev, A.P., Ivanchura, V.I., Emelyanov, R.T.: Synthesis PID controller for objects second order with regard to the location poles. Zhurnal SFU. Tech. Technol. **9**(1), 50–60. https://doi.org/10.17516/1999-494x-2016-9-1-50-60 (2016)
3. Ziegler, J.G., Nichols, N.B.: Optimum settings for automatic controllers. Trans. Am. Soc. Mech. Eng. **64**, 759–768 (1942)
4. Astrom, K.J., Hagglund, T.: Advanced PID Control. Research Triangle Park, North Carolina: The Instrumentation, Systems, and Automation Society, 354 p (2006)
5. Vadutov, O.S.: Design of PID controller for delayed systems using optimization technique under pole assignment constraints. News of Tomsk Polytechnic University. Inf. Technol. **325**(5), 16–22 (2014)
6. Filips, Ch., Kharbor, R.: Feedback control systems. Moscow, Basic knowledge laboratory Publ., 2001, 616 p (2001)
7. Voronov, A.A.: Fundamentals of the Theory of Automatic Control. Automatic control of continuous linear systems. Energiya Publ., Moscow, 309, p (1980)
8. Lukas, V.A.: Automatic Control Theory. Nedra Publ., Moscow, 416 p (1990)

9. O'Dwyer, A.: Handbook of PI and PID Controller Tuning Rules. Imperial College Press, London, 564 p (2006)
10. O'Dwyer, A.: Handbook of PI and PID controller tuning rules. Imperial College Press, London, 623 p (2009)
11. Zamyatin, D.V., Lovchikov, A.N.: The time optimal systems synthesis methodic. Vestnik SibGAU, **4**, 28–30 (2005)
12. Besekersky, V.A., Popov, E.P.: Theory of automatic control systems. Moscow, Publishing House "Nauka", 768 p (1975)
13. Zamyatin, D.V., Lovchikov, A.N.: Correctional unit parameters obtaining for time optimal four order system. Vestnik SibGAU, **4**(11), 18–20 (2006)
14. Zamyatin, D.V., Lovchikov, A.N.: Synthesis of time optimal systems of high order. Vestnik SibGAU, **2**(48), 24–28 (2013)
15. Ivanchura, V.I., Prokopiev, A.P.: Optimization of a tracker system of automatic control. Vestnik SibGAU, **5**(38), 44–49 (2011)
16. Prokopev, A.P., Ivanchura, V.I., Emelyanov, R.T.: Features of the synthesis of controller for nonlinear system control. Mathematical methods in engineering and technology. MMTT-27: XXVII international scientific conference: proceedings. Section 2. The Ministry of education and science of the Russian Federation, Tambov state technical. University [etc.]. Tambov: [b. I.], 2014 (2014)
17. Prokopev A.P., Ivanchura V.I.: Features of the synthesis of controller electro-hydraulic control system XII All-Russia meeting on control problems VCPU-2014. Moscow, June 16–19, 2014: Proceedings. Moscow, Institute of control them. V.A. Trapeznikov Academy of Sciences, 2014, pp. 307–317 (2014)
18. Prokopev, A.P., Ivanchura, V.I., Emelyanov R.T.: Identification of nonlinear control systems with the PID control, SICPRO'15, Moscow, pp. 387–397; http://www.sicpro.org/sicpro15/code/r15_08.htm (2015)
19. Prokopiev, A.P., Ivanchura, V.I., Emelianov, R.T., Scurihin, L.V.: The technique of synthesis of regulators for objects of the second order. Vestnik SibGAU, **17**(3), 618–624 (2016)
20. Efimov, S.V., Zamyatin, S.V., Gayvoronskiy, S.A.: Synthesis of the PID controller with respect to the location of zeros and poles of the system of automatic control. News of Tomsk Polytechnic University, **317**(5), 102–107 (2010)
21. Efimov, S.V., Gayvoronskiy, S.A., Zamyatin, S.V.: Root tasks of analysis and synthesis and synthesis of automatic control systems. News Tomsk. Polytech. Univ. **316**(5), 16–20 (2010)
22. Uderman, E.G.: Root Locus Method in the Theory of Automatic Systems. Gosenergoizdat Publ., Moscow-Leningrad 112 p (1963)

Limiting Values of the Stability Margins in the Parametric Synthesis of PID-Controllers

Goerun Ayazyan and Elena Tausheva

Abstract The problem of parametric synthesis of PID controllers of integer and fractional orders is solved. PID-controllers are an integral part of many cyber-physical systems. Two well-known methods of synthesis are considered—for a relative stability margin and a maximum value of the sensitivity function. Algorithms have been developed for calculating the limiting values of the differential gain of the controller, at which the boundary of the region of a given stability margin has a cusp. In the case of using the criterion for low-frequency disturbance rejection (LFDR), the limiting values of the relative stability margin and the maximum value of the sensitivity function are determined.

Keywords PID-controller · Fractional order · Relative stability margin · Maximum value of the sensitivity function · Cyber-physical systems · Maple

1 Introduction

PID-controllers make up the majority of controllers used in industry; they play an important role in cyber-physical systems also.

The procedure of parametric synthesis of PID-controller is complicated by the presence of three tuning parameters. These are the gain k_p, integral time T_i and derivative time T_d, or equivalent parameters $k_1 = k$, $k_0 = k_1/T_i$, $k_2 = k_1 \cdot T_d$. Often the coefficient T_d or k_2 is assumed constant or determined by the formulas $T_d = \alpha T_i$ and $k_2 = \alpha k_1^2/k_0$. The parameter α is chosen, for example, in a range $\alpha = 0.15 \ldots 0.6$, and often empirically. In [1] the value of α is assumed equal to $\alpha = 0.25$. The [2–12] gives the values of Ti and T_d for various methods of the PID-controllers synthesis. In [13], based on the criterion for low-frequency disturbance rejection (LFDR) [14], formulas are given that allow one to uniquely determine k_2 as

G. Ayazyan · E. Tausheva (✉)
Ufa State Petroleum Technological University, Ufa, Russia
e-mail: TaushevaEV@mail.ru

G. Ayazyan
e-mail: AyazyanGK@rambler.ru

A. G. Kravets et al. (eds.), *Cyber-Physical Systems: Industry 4.0 Challenges*, Studies in Systems, Decision and Control 260,
https://doi.org/10.1007/978-3-030-32648-7_9

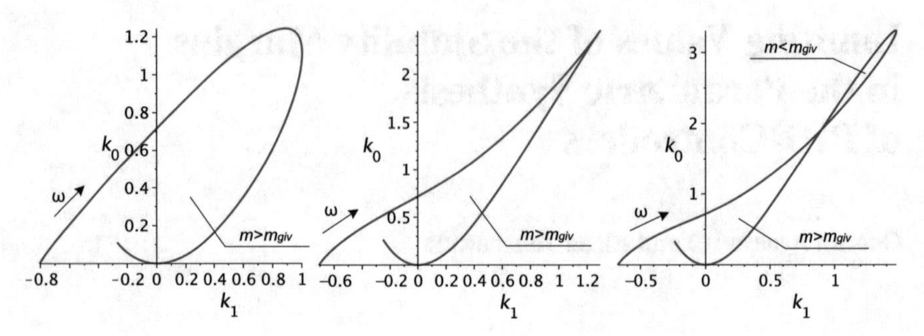

Fig. 1 D-partition for various α

a function of k_1, k_0 and the plant model parameters. On this basis, an algorithm for the parametric synthesis of a PID-controller with an ideal differentiator for a given relative stability margin has been developed $m \geq m_{giv}$. Where $m = \max|\text{Re}s_i/\text{Im}s_i|$, and $s_i, i = 1, 2, \ldots, n$ are the roots of the characteristic equation of a closed-loop system. In [15], the method was used to synthesize an ideal PID-controller, with constraints on the maximum value of the sensitivity function $M_s = \max_\omega |S(j\omega)|$ or the complementary sensitivity function $M_p = \max_\omega |T(j\omega)|$. In [16, 17], Volgin and Safronova showed that when calculating the PID-controller settings for a given relative stability margin (degree of oscillation) $m \geq m_{giv}$ and $k_2 = \alpha k_1^2/k_0$, the curve type of the D-partition (a boundary of a given stability margin) and the control quality depends on the value of parameter α. The D-partition curve is constructed in coordinates $(k_1 = k_1(\omega), k_0 = k_0(\omega))$, Fig. 1, as a function of the frequency ω [13, 16]. With increasing α, self-intersection of the D-partition curve is possible, which narrows the area of a given stability margin. And finally, there is some limit value $\alpha = \alpha_{lim}$, when a cusp is formed. For typical plants and relative stability margin values of $m = 0.221$ and $m = 0.366$, the values of $\alpha = \alpha_{lim}$ were determined in [16, 17] and an algorithm for solving the problem was proposed.

In [18] there is an example of a system with a PI-controller and a maximum limit of M_s, when there is also a self-intersection of the boundary curve. Below we show that for the same example with the PID-controller when selecting k_2 by the criterion of the LFDR, there is also a self-intersection of the boundary curve. In this chapter we propose, a new simple enough algorithm for determining $\alpha = \alpha_{lim}$ for PID-controllers of integer and fractional order corresponding to the cusp when calculating control systems for a given relative stability margin and maximum of the sensitivity functions. For the case when the coefficient k_2 is determined by the LFDR criterion, the problem of determining the limit values of $m = m_{lim}$ and $M_s = M_{s\_lim}$ or $M_p = M_{p\_lim}$ corresponding to the cusp is solved.

2 Problem Statement and Solution Algorithm

A control system with one input and one output is considered. The controller transfer function is

$$W_p = k_1\left(1 + \frac{1}{T_i s^\delta} + T_d s^\beta\right) = k_1 + \frac{k_0}{s^\delta} + k_2 s^\beta \tag{1}$$

where k_0, k_1, k_2 are gain coefficients, the controller settings; $T_i = k_1/k_0$, $T_\text{d} = k_2/k_1$ are integral time and derivative time respectively; if δ and β are equal to one, the controller is integer order or if they are non-negative fractional numbers other than one the controller is fractional order.

We consider plant models with transfer functions of the form

$$W_y(s) = \frac{1}{s^r}W_{y1}(s) = \frac{1}{s^r}\frac{b_0 + b_1 s \ldots + b_m s^m}{a_0 + a_1 s \ldots + a_n s^n}e^{-s\tau} \tag{2}$$

where $W_{y1}(s)$ is a proportional part of the plant model; $|W_{y1}(0)| = |b_0/a_0| < \infty$, $r = 0,\ 1,\ a_i,\ b_i$ are constant coefficients and $a_n,\ b_0 > 0,\ m \le n$; τ is a time delay.

The controller settings can be determined as the minimum of the integral criterion, subject to a constraint on the given relative stability margin m, or on the maximum of sensitivity function $M_s = \max_\omega |S(j\omega)|$ and complementary sensitivity function $M_p = \max_\omega |T(j\omega)|$.

The task of the settings' calculation is divided into two stages: construction in the plane of the controller settings k_1, k_0 of the region boundaries of the specified stability margin; determining the point at the border or inside the region that minimizes a given optimality criterion [1, 13].

The derivative term coefficients k_2 or T_i of the controller are chosen in two ways

$$k_2 = \alpha k_1^2/k_0 \text{ or } T_d = \alpha T_i\text{---the classicway} \tag{3}$$

and from the condition of low-frequency disturbance rejection (LFDR)

$$\begin{aligned} k_0 &= \max k_0, \\ k_2 &= \alpha(\alpha_1 + k_1)^2/k_0 + \alpha_2 k_0 + \alpha_3, \end{aligned} \tag{4}$$

where $\alpha = 1/2$, $\alpha_1 = 1/\mu_0$, $\alpha_2 = \left(\mu_1^2 - 2\cdot\mu_0\cdot\mu_2\right)/2\cdot\mu_0^2$, $\alpha_3 = \mu_1/\mu_0^2$,

$\mu_k = \frac{1}{k!}\frac{d^k}{ds^k}W_{y1}(s)\big|_{s=0}$, $k = 0, 1, 2$—moments of the transfer function W_{y1}.

The first constraint in (4) is equivalent to a minimum of the integral criterion IE [1]. The formulas for $\alpha_i,\ i \in 1, 3$, which is very important, depend on the properties of the plant model and the point of application of the disturbance. For the disturbance acting on plant input $\alpha_2 = 0$, for astatic plant $\alpha_1 = 0$ and $\alpha_3 = -1/\mu_0$ [13]. If $\alpha_1 = \alpha_2 = \alpha_3 = 0$ and $\alpha = const$ (is constant) we obtain the well known expression $k_2 = \alpha k_1^2/k_0$, $\alpha > 0$. In the following, we confine ourselves to the task of the rejection of the disturbance acting on plant input. It is important that when using LFDR, the choice of k_2 depends on the properties of the plant.

When using the criterion of LFDR, the problem of choosing k_2 is automatically solved. The reason for the appearance of self-intersection is the excessive large values of m and, as a rule, excessively understated values of M_s and M_p. During the PID-controller synthesis according to the method of dominant poles, poles giving large oscillations may appear [1].

Thus, the task of the limiting values determination of α and the limiting values of the stability margin criterions $m = m_{lim}$ and $M_s = M_{s\_lim}$, $M_s = M_{p\_lim}$ is of practical importance. The solution of the listed tasks is given below.

2.1 The Calculation Algorithm for a Given Relative Stability Margin, the Case of α = const

Consider a system with a PID-controller of an integer order.

We can find the equation of the region boundary $m \geq m_{giv}$ (D-partition curve) from the characteristic equation of the closed-loop system by the substitution $s = -m\omega + j\omega$

$$W_p(s) \cdot W_y(s) + 1 = \left(k_1 + k_0 / s + k_2 s\right) W_y(s) + 1 = 0 \tag{5}$$

where here and further $m = m_{giv}$, $\omega \geq 0$, $j = \sqrt{-1}$.

As a result, we obtain the equation

$$\left(k_1 + \frac{k_0}{-m\omega + j\omega} + k_2(-m\omega + j\omega)\right)(V_1(m, \omega) + jV_2(m, \omega)) + 1 = 0 \tag{6}$$

where $W_y(-m\omega + j\omega) = V_1(m, \omega) + jV_2(m, \omega)$ is the extended frequency response of the plant.

Extracting in (6) the real and imaginary parts and turn to an inverse frequency response of the plant, we get

$$\begin{aligned} F_1 &= k_0 - m\omega k_1 + \omega^2(m^2 - 1)k_2 - m\omega\tilde{V}_1 - \omega\tilde{V}_2 = 0 \\ F_2 &= \omega k_1 - 2m\omega^2 k_2 + \omega\tilde{V}_1 - m\omega\tilde{V}_2 = 0 \end{aligned} \tag{7}$$

where $W_y^{-1}(s)_{s=-m\omega+j\omega} = \tilde{V}_1(m, \omega) + j\tilde{V}_2(m, \omega) = \tilde{V}_1 + j\tilde{V}_2$ is the inverse frequency response of the plant.

The solution of a system of type (7) is examined in detail in [13] and when $k_2 = \alpha k_1^2 / k_0$ is

$$k_0 = (-B - \sqrt{B^2 - 4AC})/(2A),$$
$$k_1 = 2mk_0/(\omega(m^2+1)\,(\omega(m^2+1)) - \tilde{V}_1 - m\tilde{V}_2, \tag{8}$$

where $A = \alpha(4 \cdot \alpha \cdot m^2 - 1 - m^2)/(m^2+1)^2$, $B = \alpha \cdot \omega \cdot (-4\alpha \cdot m\tilde{V}_1 + (1 + m^2 - 4\alpha \cdot m^2) \cdot \tilde{V}_2)/(m^2+1)$, $C = \alpha\omega^2(\tilde{V}_1 + m\tilde{V}_2)^2$.

Given ω, we find from (8), then k_0 and construct a curve of k_1 D-partition.

Now we consider the problem of determining the limiting value of α. At the cusp point of the D-partition, the equalities $dk_0/d\omega = dk_1/d\omega = 0$ are valid [16, 19, 20]. Differentiating Eq. (7) by ω and considering k_0 and k_1 as functions of ω, we get

$$F_3 = \frac{\partial F_1}{\partial k_0}\frac{dk_0}{d\omega} + \frac{\partial F_1}{\partial k_1}\frac{dk_1}{d\omega} + \frac{dF_1}{d\omega} = 0$$
$$F_4 = \frac{\partial F_2}{\partial k_0}\frac{dk_0}{d\omega} + \frac{\partial F_2}{\partial k_1}\frac{dk_1}{d\omega} + \frac{dF_2}{d\omega} = 0 \tag{9}$$

Since $dk_0/d\omega = dk_1/d\omega = 0$ Eq. (9) are simplified. Attaching to them the Eq. (7) we get four equations to calculate the critical parameters (k_0, k_1, k_2, ω)

$$F_1 = k_0 - m\omega k_1 + \omega^2(m^2-1)k_2 - m\omega\tilde{V}_1 - \omega\tilde{V}_2 = 0$$
$$F_2 = \omega k_1 - 2m\omega^2 k_2 + \omega\tilde{V}_1 - m\omega\tilde{V}_2 = 0$$
$$F_1 = dF_1/d\omega = -mk_1 + 2\omega(m^2-1)k_2 - m(\tilde{V}_1 + \omega\tilde{V}_1^1) - \tilde{V}_2 - \omega\tilde{V}_2^1 = 0$$
$$F_2 = dF_2/d\omega = \omega k_1 - 2m\omega^2 k_2 + \tilde{V}_1 - m\tilde{V}_2 + \omega(\tilde{V}_1^1 - m\tilde{V}_2^1) = 0 \tag{10}$$

where $W_y^{-1}(s)_{s=-m\omega+j\omega} = \tilde{V}_1(m,\omega) + j\tilde{V}_2(m,\omega) = \tilde{V}_1 + j\tilde{V}_2$, $\tilde{V}_1^1 = d\tilde{V}_1/d\omega$, $\tilde{V}_2^1 = d\tilde{V}_2/d\omega$.

Since the system (10) is linear with respect to k_1, k_0, k_2, it is easier to determine the value of k_2, and then to find α by the formula (3). The rank of its coefficient matrix is three, since the third-order minor is non-zero if $\omega \neq 0$

$$\Delta_k = \begin{vmatrix} 1 & -m\omega & \omega^2(m^2-1) \\ 0 & \omega & -2m\omega^2 \\ 0 & -m & 2\omega(m^2-1) \end{vmatrix} = -2\omega^2$$

And the system has a solution if the determinant Δ of its extended matrix is zero

$$\Delta = \begin{vmatrix} 1 & m\omega & \omega^2(m^2-1) & -\omega(m\tilde{V}_1 + \tilde{V}_2) \\ 0 & \omega & -2m\omega^2 & \omega(\tilde{V}_1 - m\tilde{V}_2) \\ 0 & -m & 2\omega(m^2-1) & -m(\tilde{V}_1 + \omega\tilde{V}_1^1) - \tilde{V}_2 - \omega\tilde{V}_2^1 \\ 0 & \omega & -4m\omega & \tilde{V}_1 - m\tilde{V}_2 + \omega(\tilde{V}_1^1 - m\tilde{V}_2^1) \end{vmatrix} \tag{11}$$

By calculating Δ, after simple transformations we get

$$\Delta = -2\omega^2(m^2+1)(\omega\tilde{V}_1^1 + m\tilde{V}_2) = 0. \tag{12}$$

The frequency ω is found as the root of Eq. (12). For plant of the 3rd order without delay and some plant of the 4th order, the frequency is explicitly expressed as a function of m and plant parameters.

The expressions for the controller coefficients are found by solving the first three Eq. (10), for example, by the Kramer method, and α_{lim} from (3)

$$k_0 = -0.5\omega(m^2+1)\Big((m^2-1)\tilde{V}_2 + \omega(m\tilde{V}_1^1 + \tilde{V}_2^1)\Big)$$

$$k_1 = -(\tilde{V}_1 + m^3\tilde{V} + \omega m(m\tilde{V}_1^1 + \tilde{V}_2^1)^2)$$

$$k_2 = -0.5((m^2+1)\tilde{V}_2 + \omega(m\tilde{V}_1^1 + \tilde{V}_2^1))/2\omega$$

$$\alpha_{npe\partial} = \frac{(m^2+1)\Big(\omega(m\tilde{V}_1^1 + \tilde{V}_2^1) + (m^2+1)\tilde{V}_2\Big)\Big((m^2-1)\tilde{V}_2 + \omega(m\tilde{V}_1^1 + \tilde{V}_2^1)\Big)}{4\Big(\tilde{V}_1 + m^3\tilde{V}_2 + m\omega(m\tilde{V}_1^1 + \tilde{V}_2^1)\Big)^2}. \tag{13}$$

Consider a control system with a fractional-order PID controller.

The frequency response of the controller of fractional order after the substitution of $s = -m\omega + j\omega$ takes the form

$$W_p(m,\omega) = k_1 + k_0\omega^{-\delta}A + k_2\omega^{\beta}B + I\big(k_0\omega^{-\delta}C + k_2\omega^{\beta}D\big), \tag{14}$$

where $A = (m^2+1)^{-\delta/2}cos(0.5\pi\delta + \delta\cdot arctan(m))$,

$$B = (m^2+1)^{-\beta/2}\cos(0.5\pi\delta + \beta\cdot\arctan(m)),$$
$$C = (m^2+1)^{-\delta/2}\sin(0.5\pi\delta + \delta\cdot\arctan(m)),$$
$$D = (m^2+1)^{-\beta/2}\sin(0.5\pi\delta + \beta\cdot\arctan(m)).$$

Equation (10) and the extended determinant of the system take the following form

$$F_1 = k_1 + k_0\omega^{-\delta}A + k_2\omega^{\beta}B + \tilde{V}_1 = 0,$$
$$F_2 = k_0\omega^{-\delta}C + k_2\omega^{\beta}D + \tilde{V}_2 = 0,$$
$$F_3 = -k_0\omega^{-\delta}\delta A + k_2\omega^{\beta}\beta B + \omega\tilde{V}_1^1 = 0,$$
$$F_4 = -k_0\omega^{-\delta}\delta C + k_2\omega^{\beta}\beta D + \omega\tilde{V}_2^1 = 0 \tag{15}$$

$$\Delta = -C\beta B\tilde{V}_2^1\omega + C\beta\tilde{V}_1^1 D\omega + \delta AD\beta\tilde{V}_2 - \delta AD\tilde{V}_2^1\omega + \delta CD\tilde{V}_1^1\omega - \delta C\beta B\tilde{V}_2 \tag{16}$$

The controller settings are found by solving the first three Eq. (15)

$$k_0 = (D\tilde{V}_1^1\omega - \beta B\tilde{V}_2)\omega^{\delta}/\Delta x, k_2 = -(C\tilde{V}_1^1\omega + V_2\delta A)\omega^{-\beta}/\Delta x,$$
$$k_1 = (-\tilde{V}_1(CB\beta + AD\delta) + \tilde{V}_2 AB(\beta + \delta) + \omega\tilde{V}_1^1(BC - AD))/\Delta x,$$
$$\Delta \mathrm{x} = (\delta AD + \beta CB) \tag{17}$$

The value of α is found by the formula $\alpha_{lim} = k_0 k_2/k_1^2$, ω as a root of (16).

2.2 The Calculation for a Given Maximum of Sensitivity Function, the Case of $\alpha = Const$

The boundary of the $M \leq M_{giv}$ region in the settings plane is determined by solving the system of equations [15]

$$\begin{cases} A_C(\omega, k_0, k_1, \alpha) = M_{giv} \\ dA_C(\omega, k_0, k_1, \alpha)\big/ d\omega = 0 \end{cases} \tag{18}$$

where M_{giv} is the specified value M_s or M_p, $A_C(\omega, k_0, k_1, k_2)$ is the magnitude frequency response of the closed-loop system.

Equation (18) is a necessary condition for the maximum of the frequency response, as well as the condition of tangency the Nyquist curve of the open-loop system with the "sensitivity circle" [1, 15]. Figure 2 shows the curve of the specified stability margin $M = M_{giv}$ (Fig. 2b) and the magnitude frequency response of the closed-loop system (Fig. 2a), when $\alpha > \alpha_{lim}$. The frequency response has two resonant peaks of the same amplitude at two different frequencies. At the cusp point (Fig. 2c), when $\alpha = \alpha_{lim}$ the peaks merge, the smoothness of the frequency response at the resonance point increases.

For this case the conditions for the presence of a cusp we can get by adding to system (18) two equations [19, 20]

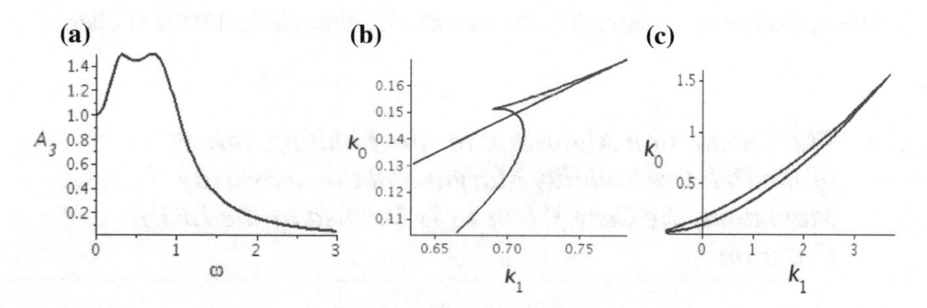

Fig. 2 The curves to the calculation for a given sensitivity maximum

$$
\begin{aligned}
&A_C(\omega, k_0, k_1, k_2) = M_{giv} \\
&dA_C(\omega, k_0, k_1, k_2)/d\omega = 0 \\
&d^2 A_C(\omega, k_0, k_1, k_2)/d\omega^2 = 0 \\
&d^3 A_C(\omega, k_0, k_1, k_2)/d\omega^3 = 0
\end{aligned} \tag{19}
$$

We have four equations for four unknowns ω, k_0, k_1, k_2.

Here, as in the previous case, for simplicity instead of α we consider k_2. The value of α_{lim} is found by the formula $\alpha_{lim} = k_0 k_2/k_1^2$.

After substitution expressions for the frequency characteristics of the plant and the controller into (19), and transformations, Eq. (19) take the form

$$
\begin{aligned}
&S_1 = Ak_0^2 - 2k_0k_2 + B\cdot k_0 + k_1^2 + C\cdot k_1 + Dk_2^2 + E\cdot k_2 + F = 0 \\
&S_2 = dS_1/d\omega = A_1k_0^2 + B_1\cdot k_0 + C_1\cdot k_1 + D_1k_2^2 + E_1\cdot k_2 + F_1 = 0 \\
&S_3 = d^2S_1/d\omega^2 = A_2\cdot k_0^2 + B_2\cdot k_0 + C_2\cdot k_1 + D_2\cdot k_2^2 + E_2\cdot k_2 + F_2 = 0 \\
&S_4 = d^3S_1/d\omega^3 = A_3\cdot k_0^2 + B_3\cdot k_0 + C_3\cdot k_1 + E_3\cdot k_2 + F_3 = 0
\end{aligned} \tag{20}
$$

where $A = 1/\omega^2$, $B = -2a\tilde{V}_2/\omega$, $C = 2a\tilde{V}_1$, $D = \omega^2$,

$$
\begin{aligned}
&E = 2a\omega\cdot\tilde{V}_2,\ F = (a^2 - R^2)(\tilde{V}_1^2 + \tilde{V}_2^2),\ A_k = d^kA/d\omega^k, B_k = d^kB/d\omega^k, \\
&C_k = d^kC/d\omega^k, D_k = d^kD/d\omega^k, E_k = d^kE/d\omega^k, F_k = d^kF/d\omega^k, k = 1, 2, 3,
\end{aligned}
$$

$W_y^{-1}(j\omega) = \tilde{V}_1(\omega) + j\tilde{V}_2(\omega)$ is the inverse frequency response of the plant,

$a = M_p^2/(M_p^2 - 1)$, $R = M_p/(M_p^2 - 1)$ and $a^2 - R^2 = a$, if $M_{giv} = M_p$ and $a = 1$, $R = 1/M_s$, if $M_{giv} = M_s$.

System (20) is not linear with respect to ω, k_0, k_1, k_2. To solve it, it is desirable to set a range of initial values of variables. For this, the boundary of the region of a given stability margin in the plane (k_1, k_0) is calculated for several values of α. To construct it, it is enough to have Eq. (18) or the first two equations of system (20) after the substitution $\alpha = k_0k_2/k_1^2$. The method of solving these equations is considered in [15].

The technique easily extends to the case of a fractional order PID-controller.

2.3 The Calculation Algorithm for the Limiting Values of the Relative Stability Margin and the Sensitivity Maximum, the Case When k_2 Is Defined by the LFDR Criterion

In this case, the problem of choosing k_2 is solved automatically. However, there remains the problem of determining the limiting values of system stability margin. As already noted, with an inaccurate choice of a given value relative stability margin m or sensitivity maximum M_s, M_p there are cusps or self-intersections of the boundary curves. The algorithm for calculating limiting values is based on Eqs. (10), (15) и (20).

In general, the corresponding equations are written as follows.

$$
\begin{aligned}
S_1(k_1, k_0, X, \omega, \beta, \delta) &= 0 \\
S_2(k_1, k_0, X, \omega, \beta, \delta) &= 0 \\
S_3(k_1, k_0, X, \omega, \beta, \delta) &= 0 \\
S_4(k_1, k_0, X, \omega, \beta, \delta) &= 0
\end{aligned} \tag{21}
$$

where β and δ are known, and X required limiting values m_{lim} or M_s, M_p.

The calculation algorithm of a liming relative stability margin value for a fractional order controller. Substituting into Eq. (15) the expression for k_2 from (4), we get

$$
\begin{aligned}
S_1 &= 0.5\omega^{\beta} Bk_1^2 + k_0k_1 + \alpha_1\omega^{\beta} Bk_1 + \left(\alpha_2\omega^{\beta} B + \omega^{-\delta} A\right)k_0^2 \\
&\quad + (\tilde{V}_1 + \alpha_3\omega^{\beta} B)k_0 + 0.5\alpha_1^2\omega^{\beta} B = 0 \\
S_2 &= 0.5\omega^{\beta} Dk_1^2 + \alpha_1\omega^{\beta} Dk_1 + \left(\alpha_2\omega^{\beta} D + \omega^{-\delta} C\right)k_0^2 \\
&\quad + \left(\tilde{V}_2 + \alpha_3\omega^{\beta} D\right)k_0 + 0.5\alpha_1^2\omega^{\beta} D = 0 \\
S_3 &= 0.5\omega^{\beta-1}\beta Bk_1^2 + \alpha_1\omega^{\beta-1}\beta Bk_1 + \left(\alpha_2\omega^{\beta-1}\beta B - \omega^{-(1+\delta)}\delta A\right)k_0^2 \\
&\quad + \left(V_1^1 + \alpha_3\omega^{\beta-1}\beta B\right)k_0 + 0.5\alpha_1^2\omega^{\beta-1}\beta B = 0 \\
S_4 &= 0.5\omega^{\beta-1}\beta Dk_1^2 + \alpha_1\omega^{\beta-1}\beta Dk_1 + \left(\alpha_2\omega^{\beta-1}\beta D - \omega^{-(1+\delta)}\delta C\right)k_0^2 \\
&\quad + \left(V_2^1 + \alpha_3\omega^{\beta-1}\beta D\right)k_0 + 0.5\alpha_1^2\omega^{\beta-1}\beta D = 0
\end{aligned} \tag{22}
$$

where A, B, C, D are defined by formulas (14) and depend on the relative stability margin m, $W_y^{-1}(s)_{s=-m\omega+j\omega} = \tilde{V}_1(m, \omega) + j\tilde{V}_2(m, \omega) = \tilde{V}_1 + j\tilde{V}_2$, $\tilde{V}_1^1 = d\tilde{V}_1/d\omega$, $\tilde{V}_2^1 = d\tilde{V}_2/d\omega$, $\tilde{V}_1$, $\tilde{V}_2$, their derivatives $\tilde{V}_1^1$, $\tilde{V}_2^1$ also depends on m.

To estimate the initial approximations, we can construct a series of curves of equal stability margin for several values of m in the (k_1, k_0) plane.

The values of k_0 and k_1 are found as functions of ω by solving the first two Eq. (22) using the elimination method described in [13] and a similar solution to Eq. (8).

The calculation algorithm of a liming sensitivity maximum value M_s and M_p. Consider a system with a PID-controller of an integer order. As a basis, we take Eq. (20). Note the feature of the formulas for the coefficients in (20). If the task of determining $M_s = M_{s\_lim}$ is solved, in formulas (20) $a = 1$, and M_s enters only the expressions for F, .., F_3 in the form of a factor $X = 1 - 1/M_s^2$. Having defined X we will find $M_s = M_{s\_lim}$. If $M_p = M_{p\_lim}$ is searched, we need to take into account $a^2 - R^2 = a$. Then, calculating $a = M_p^2/(M_p^2 - 1)$, we find M_p.

Substituting k_2 from (4) into Eq. (20), we obtain a system of four equations for the unknown (k_0, k_1, ω, X), and then $M_s = M_{s\_lim}$ or $M_p = M_{p\_lim}$.

To determine the initial values of variables, a family of curves of equal stability margin is constructed by the Eq. (18).

In the case of a fractional order PID-controller, the algorithm for solving the problem does not change, only the form of the coefficients of the equations becomes more complicated.

3 Examples

The considered problems were solved using programs developed in the environment of the package of symbolic calculations Maple.

1. A relative stability margin m is given. Determination of α_{lim}.

1.1. The plant is $W^{-1}(s) = s(Ts+1)^2$, the PID-controller has an integer order

$$\Delta = -8T\left(m^2+1\right)(2Tm\omega - 1) = 0 \Rightarrow \omega_{\lim} = 1\big/(2Tm),\ k_0 = (1+1\big/m^2)^2\big/(16T^2),$$
$$k_1 = (1+1\big/m^2)\big/(2T),\ k_2 = (1+1\big/m^2)\big/2,\ \alpha_{\lim} = (1+1\big/m^2)\big/8$$

1.2. The plant is $W^{-1}(s) = (Ts+1)^3$, the PID-controller has an integer order

$$\Delta = -4\omega^4(m^2+1)T^2(4Tm\omega - 3),\ \omega_{\lim} = 3\big/(4Tm)$$
$$k_0 = 81(1+1\big/m^2)^2\big/(256T),\ k_1 = (11+27\big/m^2)\big/16,\ k_2 = 3T(1+3\big/m^2)\big/8$$
$$\alpha_{lim} = 243(m^2+3)(m^2+1)^2\big/(8m^2(11m^2+27)^2)$$

1.3. The plant is $W^{-1}(s) = s(Ts+1)^3$, the PID-controller has a fractional order,

$$m = 0.25, = 2,\ \delta = 0.85,\ \beta = 1.1$$
$$\omega = 0.8758 \cdot T^{-1} = 0,4379,\ k_0 = 1.2393 \cdot T^{-1.85} = 0.3438$$
$$k_1 = 1.7217 \cdot T^{-1} = 0.8608,\ k_2 = 3.3337 \cdot T^{0.1} = 30.5730,\ \alpha_{lim} = 1.6575.$$

2. A maximum value of the sensitivity function is given M_p. Determination of α_{lim}.

The plant is $W_y^{-1}(s) = s(Ts+1)^3$, the PID-controller has an integer order $M_P = 1.5$.

$$\alpha_{lim} = 0.3099,\ k_0 = 0.1108,\ k_1 = 0.6550,\ k_2 = 1.1998,\ \omega_{lim} = 0.4534$$

For $\alpha = 0.45$ there is a self-intersection point with coordinates $k_0 = 0.1479$, $k_1 = 0.7038$, $k_2 = 1.5069$, $\omega_1 = 0.33$, $\omega_2 = 0.72$,

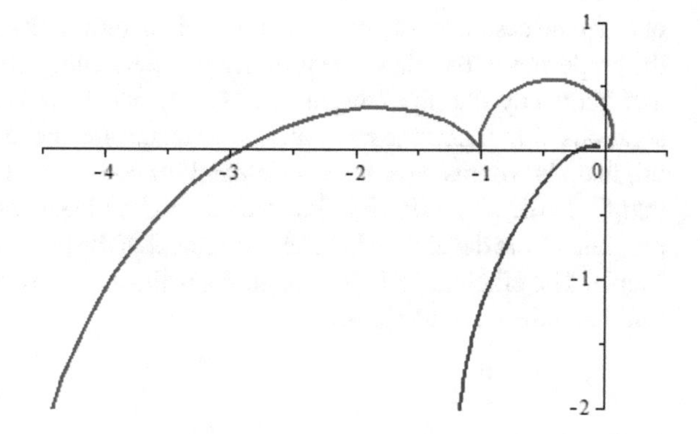

Fig. 3 Normal and extended open-loop frequency response

3. Determination of m_{lim}, k_2 is determined by the LFDR (Figs. 1, 2).
   The plant is $W_y^{-1}(s) = (2s + 1)(s + 1)(0.5s + 1)^2$, the PID-controller has a fractional order, $\delta = 0.85$, $\beta = 1.1$. Results: $m_{lim} = 0.7017$, $k_0 = 1.5644$, $k_1 = 3.7144$, $\omega = 0.9895$. The roots of the characteristic equation of a closed-loop system $s_{1,2} = -0.6943 \pm j\,0.9895$, $s_{3,4} = -0.6943 \pm j\,0.9895$.
   Figure 3 shows the normal (blue) and extended open-loop frequency response for $m = m_{lim}$. The extended Nyquist curve passes through a critical point with a cusp, which corresponds to the limiting stability margin.
4. Determination of the Limiting Value of M_S, k_2 is Determined by the LFDR.
   The plant transfer function is $W_y^{-1}(s) = (s + 1)^3 e^{5s}$, $\delta = 1$, $\beta = 1$. The limiting value is $M_s = 1.4558$, $k_0 = 0.1101$, $k_1 = 0.3898$, $\omega = 0.3275$. When $M_s = 1.4$ there is a self-intersection point.

4 Conclusion

The chapter discusses some features of the problem of parametric synthesis of PID controllers of integer and fractional orders. It is shown that the value of α in the formulas $T_d = \alpha T_i$ or $k_2 = \alpha k_1^2/k_0$, where the value of α can vary within wide limits, significantly depends on the nature of the region of a given stability margin (root or frequency) in the plane (k_1, k_0). An excessive increase in α causes self-crossing of a boundary curve and leads to a narrowing of the given stability margin region, in particular to a decrease in the maximum possible value of the system gain coefficient k_1. The integrated criterion IE as is well-known depends on the value of k_1 [1]. In work algorithms of calculation of the limiting value α are presented to and of the corresponding values ω, k_0, k_1, k_2 when the boundary curve has a cusp. In case of a pole stability margin the limiting frequency is the root of the nonlinear equation, and k_0, k_1, k_2 is the solution to the system of linear equations. A closed-loop system has two pairs of complex conjugate roots of the form $s_{1,2} = -m_{giv}\omega \pm j\omega$. It is important that α_{lim} depends on parameters of the plant and limits the range of possible values of α. In the case when k_2 or T_i are uniquely determined by the LFDR and depend on the properties of the plant, the problem of determining the limiting values of the pole and frequency stability margins m, M_s, M_p is solved. Limiting values are found as solutions of four nonlinear equations. To determine the initial values of the variables, original algorithms have been developed for constructing curves of a given stability margin in the plane (k_1, k_0). For the calculation based on the proposed algorithms, programs were developed in the environment of the package of symbolic calculations Maple. The efficiency of the algorithms is illustrated by examples of the parameters limiting values calculations.

References

1. Åström, K.J., Hägglund, T.: Advanced PID control, p. 460. ISA-The Instrumentation, Systems, and Automation Society, Research Triangle Park, NC (2006)
2. Ang, K.H., Chong, G., Li, Y.: PID control system analysis, design, and technology. IEEE Trans. Control Syst. Technol. **13**, 559–576 (2005)
3. Wallén, A., Åström, K.J., Hägglun, T.: Loop-shaping design of PID controllers with constant Ti/Td ratio. Asian J. Control. **4**, 403–409 (2008)
4. Åström, K.J., Hägglund, T.: The future of PID control. IFAC Proc. Vol. **33**, 19–30 (2000)
5. Tan, W., Liu, J., Chen, T., Marquez, H.J.: Comparison of some well-known PID tuning formulas. Comput. Chem. Eng. **30**, 1416–1423 (2006)
6. Tang, W., Wang, Q.G., Lu, X., Zhang, Z.: Why Ti = 4Td for PID controller tuning. Robotics and Vision 2006 9th International Conference on Control, pp. 1–2. Automation. IEEE, Singapore (2006)
7. O'Dwyer A.: An overview of tuning rules for the PI and PID continuous-time control of time-delayed Single-Input, Single-Output (SISO) Processes. In: Vilanova R., Visioli A. (eds.) PID Control in the Third Millennium: Lessons Learned and New Approaches Springer London, pp. 3–44. London (2012)
8. Yadaiah, N., Malladi, S. (2013) An optimized relation between Ti and Td in Modified Ziegler Nichols PID controller tuning. In: 2013 IEEE International Conference on Control Applications (CCA), pp. 1275–1280
9. Leva, A., Maggio, M.: Model-Based PI(D) Autotuning. In: Vilanova, R., Visioli, A. (eds.) PID Control in the Third Millennium: Lessons Learned and New Approaches, pp. 45–73. Springer, London, London (2012)
10. Pecharromán, R.R., Pagola, F.L.: Control design for PID controllers auto-tuning based on improved identification. IFAC Proc. Vol. **33**, 85–90 (2000)
11. Smirnov, N.I., Sabanin, V.R., Repin, A.I.: Sensitivity and robust tuning of PID controllers with real differentiation. Therm. Eng. **54**, 777–785 (2007)
12. Ozyetkin, M.M., Onat, C., Tan, N.: PID tuning method for integrating processes having time delay and inverse response. In: IFAC-PapersOnLine, 3rd IFAC Conference on Advances in Proportional-Integral-Derivative Control PID 2018, vol. 51, pp. 274–279 (2018)
13. Ayazyan, G.K., Novozhenin, A.Yu., Tausheva, E.V.: Parametric synthesis of PID regulators for a given degree of oscillation (in Russian). In: XII All-Russian Meeting on the Problems of Control (VSPU-2014). Russia, Institute of Control Sciences V.A. Trapeznikova RAS, pp. 147–159 (2014)
14. Ayazyan, G.K.: Calculation of automatic systems with typical control algorithms, p. 136. Publishing house UNI, Ufa (1989). (in Russian)
15. Ayazyan, G.K., Tausheva, E.V., Shaymuhametova, M.R.: Applying symbolic computing system maple for parametric synthesis of controllers (in Russian). In: VI International Conference.

Mathematics, Its Applications and Mathematical Education (MAME'17), pp. 59–64. Russia, Ulan-Ude (2017)
16. Volgin, V.V.: Concerning determination of optimum adjustment of PID-regulators. Avtomat. i Telemekh. **23**, 620–630 (1962)
17. Safronova, I.N., Volgin, V.V.: The method of multiple roots in optimizing a control-system with a PID-algorithm. Therm. Eng. **36**, 580–582 (1989)
18. Åström, K.J., Panagopoulos, H., Hägglund, T.: Design of PI controllers based on non-convex optimization. Automatica **34**, 585–601 (1998)
19. Bruce, J.W., Giblin, P.J.: Curves and singularities: a geometrical introduction to singularity theory, 2nd ed. ed. Cambridge University Press, Cambridge [England] ; New York, NY, USA, 321 p, (1992)
20. Lord, E.A.: The mathematical description of shape and form: E. A. Lord and C. B. Wilson, Ellis Horwood series in mathematics and its applications. (vol. 260). Halsted Press, Chichester, West Sussex, Ellis Horwood, New York (1984)

Application of Cut-Glue Approximation in Analytical Solution of the Problem of Nonlinear Control Design

A. R. Gaiduk, R. A. Neydorf and N. V. Kudinov

Abstract To create control systems for various objects, their mathematical models are used. They are obtained, often experimentally, by approximating arrays of numerical data. With significant non-linearity of the data, they are approximated at separate sites. However, such a fragmentary model of a nonlinear object as a whole is not analytical, which excludes the use of most methods for the synthesis of nonlinear controls. In such a situation, there is the prospect of applying the Cut-Glue approximation method, which allows us to obtain a common object model as a single analytical function. The chapter considers the theory and application of this method to the synthesis of nonlinear control. The mathematical model obtained by the Cut-Glue approximation method is reducing to a quasilinear form, which makes it possible to find a nonlinear control by an analytical method.

Keywords Experimental data · Fragmentary approximation · Analytical function · Multiplicative isolation function · Additive union · Mathematical model · Quasilinear form · Analytical synthesis · Control system

1 Background

As is known, to create a control system for various processes and objects in which these processes take place, the mathematical models (MM) are used. Quite often they are obtained by conducting special experiments, which are carried out by applying special test actions to the inputs of the object, varying in accordance with the design of the experiment [1–3]. The plan provides for the order of change and the values

A. R. Gaiduk (✉)
Southern Federal University, Taganrog, Russia
e-mail: gaiduk_2003@mail.ru

R. A. Neydorf · N. V. Kudinov
Don State Technical University, Rostov-on-Don, Russia
e-mail: ran_pro@mail.ru

N. V. Kudinov
e-mail: kudinov_nikita@mail.ru

A. G. Kravets et al. (eds.), *Cyber-Physical Systems: Industry 4.0 Challenges*, Studies in Systems, Decision and Control 260,
https://doi.org/10.1007/978-3-030-32648-7_10

of the input effects of the object, as well as fixing its responses to these impacts. As a result, the researcher receives a set of tables containing data on the dependence of the output variables of the object, both on external disturbances and on its other variables. The obtained point data on different intervals of the variables changes are fairly accurately approximated by suitable analytic (differentiable) functions. In some cases, these functions are selected based on the physically available information about the object under study.

However, the thus obtained fragmentary MM of the nonlinear object is not analytical, which does not allow the use of most of the known methods for the synthesis of the nonlinear controls. In order to overcome this complexity, it seems appropriate to apply the Cut-Glue approximation method, which allows us to obtain a common object's MM, which is an additive form of the analytic functions derived from fragmentary data [4–6].

In this chapter, after briefly covering the main theoretical principles of the CGA method and the algorithm for its application, the basic principles and an example of its application to the design of nonlinear controls are set out in sufficient detail [6, 7]. The analytical nature of the CGA MM allows, in particular, to design of the nonlinear controls based on the so-called quasilinear models [8, 9]. For this, it is not enough if MM has analytical properties, which is ensured when it is obtained using the Cut-Glue approximation method. The obtained CGA MM should be converted to the quasilinear form. This is because this form allows, as shown in [9, 10], to find the desired nonlinear control by an analytical method.

2 Research Problem

It is necessary to develop a theoretical substantiation and methodology for transforming MM obtained by the CGA method and therefore are the analytic functions, to a form known in the modern theory of automatic control as "quasilinear" [8–10]. This is because, due to the specificity of the CGA method, the final MM has a complex fractional-radical structure. It is necessary to show on a real-world example the possibility of applying the analytical method for the designing of a nonlinear control law using such quasilinear MM obtained for control objects using the CGA method based on experimentally obtained fragmentary models. At the same time, it is necessary to demonstrate the effectiveness of such non-linear control laws, despite their mathematical complexity inevitable when using CGA-models.

A meaningful, understandable and visual presentation of the task solution requires at least a brief presentation of both the essence and the results of applying the CGA method, and the concept of a quasilinear form of a nonlinear analytic function, methods for constructing this form, and also features of the method of the nonlinear controls design caused by the mathematical properties of this forms.

3 Construction of a Nonlinear MM of the Technical System Control Channel Using the Cut-Glue Approximation Method

For greater clarity, we consider the methodology and properties of applying this method to the task of developing an aerostat control system along the flight altitude stabilization channel at a constant course speed by affecting the angle of attack α [5–7]. When the aerodynamic interaction of the aerostat body with the oncoming flow occurs the effect of the so-called lifting—the appearance of a lift force, depending on the angle of attack. As a result of computer experimental studies of the 3D model of the aerostat, experimental data were obtained, identifying its static and dynamics.

The data needed to solve the problem are given in Table 1 as two fragments of this dependence. They are represented by the values of lifting L (in kN), and the aerostat angle of attack causing it α (in angular degrees). The left column shows the angular ranges selected during the experiment for the first and second fragments.

The data of the first fragment (5 pairs of variables) are given in the two upper lines, and for the second—in the two lower (8 pairs of variables). In the top row of table, in a detached cell highlighted with a gray background, the values obtained by extrapolating the angle $\alpha = 22.12°$ and the corresponding lifting value $L = 122.55$ kN are in bold type.

According to Table 1 in Fig. 1 plotted point dependencies, as for the fragments $L1(\alpha)$ and $L2(\alpha)$—Fig. 1a,b, and for the entire dependence $L(\alpha)$—Fig. 1c. They are shown contour squares. The graph clearly shows that in the investigated range of attack angles $0° \le \alpha \le 60°$, the static characteristic of the intended control channel consists of two branches that differ significantly in slope, each of which must have its own mathematical description, valid for the ranges α indicated in the first column of Table 1 and observed on the chart.

The kink in the characteristic $L(\alpha)$ is associated with the conditions for the interaction of the incident airflow with the body of the balloon and its exterior parts. When $\alpha_{cr} = 22.12°$ there is, probably, the so-called “stall”. As a result, the static characteristic of the aircraft has a point of discontinuity of the derivative, which makes it, in general, not only nonlinear but also non-analytical.

There are well-known approximation methods, which, according to the principles of solving the problem, can describe “kinks” with the desired accuracy. The simplest and most effective approach is the use of piecewise approximation [11–13], which makes it possible to approximate ED with any necessary accuracy fragmentary. However, the resulting MM will have discontinuities of derivatives along the boundaries of the fragments. This effect excludes analytical transformations. The approximation of fragments by splines [14–16] also does not allow the use of analytical transformations of MM.

In addition to the “fragment-oriented” approaches, there are methods aimed at building a unified analytical MM. Among these methods, the most well-known are: the regression analysis [17–20], polynomial expansions [21, 22], methods of radial basis functions [23, 24], and some others. It is important to note that such methods do

Table 1 Fragmented dependence of the lifting value $L(\alpha)$ on the aerostat angle of attack α

$0 \le U \le 22.12°$	U	0°	5°	10°	15°	20°		**22.12°**	
	$Y_1(U)$	0.894	25.31	51.63	80.45	111.6		**125.5**	
$22.12° \le U \le 60°$	U	25°	30°	35°	40°	45°	50°	55°	60°
	$Y_2(U)$	128.8	135.4	146.3	159.3	174.5	192.9	213.6	238.4

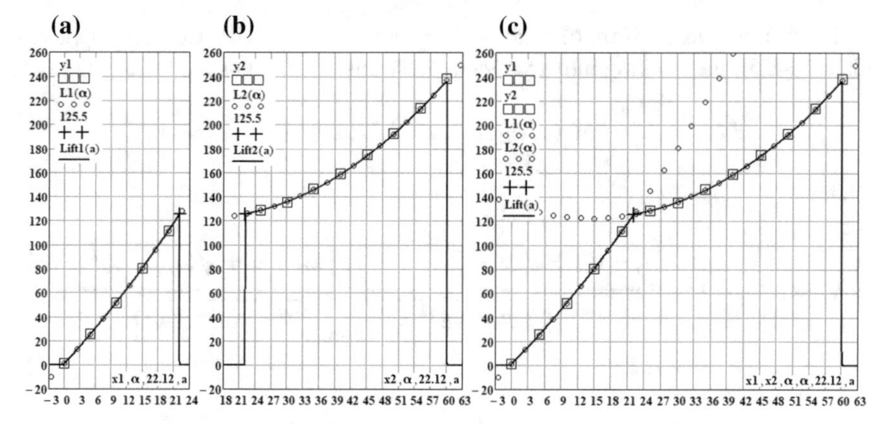

Fig. 1 Graphs of experimental and approximation dependencies $L(\alpha)$

not provide quantitative accuracy of data approximation in the presence of "kinks", breaks and multi-extremes of the describing ED.

The CGA method, as mentioned above, combines the advantages of both paradigms of the approach to approximating, and its multiplicative-additive paradigm eliminates their shortcomings. Therefore, the total ED arrays of the working range of change α is divided into two fragments, shown in the rows of Table 1 pointwise data, experimentally obtained for the ranges $0° \leq \alpha \leq 22.12°$ and $22.12° \leq \alpha \leq 60°$. They processed, in accordance with the CGA paradigm, using the methods of classical regression analysis [17–20]. As a result, polynomial regression equations obtained, which describe the simulated dependencies on these intervals with the following expressions with sufficient accuracy:

$$L^e(\alpha) = \begin{cases} \forall 0° \leq \alpha \leq 22.12° \;\; \rightarrow L_1^e(\alpha) = 0.894 + 4.614 \cdot \alpha + 0.046 \cdot \alpha^2; & (1) \\ \forall 22.12° \leq \alpha \leq 60° \;\; \rightarrow L_2^e(\alpha) = 133.6 - 1.584 \cdot \alpha + 0.055 \cdot \alpha^2. & (2) \end{cases}$$

The approximating curves (1) and (2) are shown in Fig. 1a, b contour circles. This is evident in the "legends" of graphs rendered in their upper left corners.

The experimentally obtained characteristic of the channel "angle of attack—lifting" with coordinate coordinates used by polynomials and (the corresponding parabolas are shown in Fig. 1 by solid lines) formally describe the dependence being studied, but only in the indicated ranges of the argument α. In general, this function is not analytic, since at a critical angle $\alpha_{cr} = 22.12°$, an abrupt change in the law of change of L(α) is observed, first of all, a discontinuity of the first kind in the derivative.

The Cut-Glue method of approximation, as shown in [7, 25, 26], can form an analytic function from such dependencies, giving the discontinuity of the derivative the character of a smoothed jump, moreover, with an "adjustable" degree of smoothing. This allows you to get a common channel MM "angle of attack—lifting" L(α) in the form of analytical dependence. To do this, we use the technique described in detail for CGA in [4].

If for getting in the form of analytical dependence we use the Cut-Glue approximation [4, 5], then, taking into account expressions (1), (2), has the following form

$$L(\alpha) = L_1^e(\alpha)M_1(\alpha, \alpha_{l1}, \alpha_{r1}, \varepsilon) + L_2^e(\alpha)M_2(\alpha, \alpha_{l2}, \alpha_{r2}, \varepsilon) \quad (3)$$

where

$$M_i(\alpha, \alpha_{li}, \alpha_{ri}, \varepsilon) = \frac{\left[\alpha - \alpha_{li} + \sqrt{(\alpha - \alpha_{li})^2 + \varepsilon^2}\right] \cdot \left[\alpha_{ri} - \alpha + \sqrt{(\alpha_{ri} - \alpha)^2 + \varepsilon^2}\right]}{\left[\sqrt{(\alpha - \alpha_{li})^2 + \varepsilon^2}\right] \cdot \left[\sqrt{(\alpha_{ri} - \alpha)^2 + \varepsilon^2}\right]},$$

$$i = 1, 2. \quad (4)$$

In expression (4) α_{il}, α_{ir} are the left and right borders of the i-th interval of change, respectively; ε is the tuning parameter of the function $\mathrm{M}_i(\alpha, \alpha_{li}, \alpha_{ri}, \varepsilon)$, which determines the marginal accuracy of the allocation of functions $L_1^e(\alpha)$ and $L_2^e(\alpha)$ in their areas of definition, given by intervals. Then closer ε to zero, the more clearly isolated analytical fragments $L_1^e(\alpha)\mathrm{M}_i(\alpha, \alpha_{li}, \alpha_{ri}, \varepsilon)$.

For fragmentary dependencies (1) and (2), the parameters specified in α (3) are taken to be the following values: $\alpha_{l1} = 0.1$; $\alpha_{r1} = 22.12$; $\alpha_{l2} = 22.12$; $\alpha_{r1} = 60.1$; $\varepsilon = 0.01$. Values α_{11} and α_{21} (outer edge) are shifted by 0.10 left and right, respectively. This is necessary to correctly reproduce the true boundary values of the function L(α) when $\alpha_{l1}^r = 0$ and $\alpha_{r2}^r = 60$ in which the function $\mathrm{M}_i(\alpha, \alpha_{li}, \alpha_{ri}, \varepsilon)$ is not singular. This circumstance needs to be explained separately.

An important property of the function $\mathrm{M}_i(\alpha, \alpha_{li}, \alpha_{ri}, \varepsilon)$ called by the authors as a multiplicatively isolating (or cutting out) function (MIF) is that within the interval $\alpha_{li} < \alpha\alpha_{ri}$ it is almost equal to one, and beyond its boundaries, it is practically zero. Its boundary values $\mathrm{M}_i(\alpha_{li}, \alpha_{li}, \alpha_{ri}, \varepsilon) = \mathrm{M}_i(\alpha_{ri}, \alpha_{li}, \alpha_{ri}, \varepsilon) = 0.5$ mean that it provides an almost "seamless" connection of neighboring fragments along their internal boundaries for the whole ED array. The shift in general terms of the additively formed function L(α) of the outer boundaries of the fragments to be merged by a small amount provides the necessary accuracy in the reproduction of boundary values since provides the location of the true boundary values in the inner regions of the fragments. This shift (in our case, it is assumed to be 0.1) depends on the value of the boundary parameter ε. It should be noted that in the general case the number of fragments could be much more than two, just as the dimension of the dependence can be more than one.

To illustrate the essence and result of the CGA method in Fig. 1, together with the experimental (y1 and y2) and approximating (L1(α) and L2(α)) curves, interval-isolated functions (IIF) are obtained, obtained from polynomials (1) and (2) by multiplying them by MIF. These IIFs (Lift1(α) and Lift2(α)) are shown by solid lines (see "legends" in the field of graphs). The plots of fragments 1a and 1b, as well as the combined function 1c, are spatially separated for greater clarity.

The graph of the resulting analytic function L(α) is constructed in Fig. 1c in the form of a solid line with practically no error coinciding with the half-branches of the interval isolated characteristics $L_1^e(\alpha)$ and $L_2^e(\alpha)$. You can make sure that this function is continuous and has a continuous first derivative (by the way, of a higher order too) in the whole range of α by analyzing it analytically. Graphically, this is confirmed by the fact that this function (in the legend of the graph—Lift (α)—lower, on the left) is built in the entire range of the x-axis on the one formed in the MathCAD, formula (3). This shows that it can be used in analytical transformations and is the single MM of the channel under study.

4 Quasilinear Form of Nonlinear Models

Thus, the Cut-Glue approximation of the experimental data of a fragmentary structure can be used to construct the analytical characteristics of any nonlinear objects channels. In connection with the essence of the problem of the present study formulated above, the first task of this article is to prove the possibility of converting CGA models to a quasilinear form, whose properties allow us to synthesize non-linear controls of nonlinear objects whose MMs contain only analytical nonlinearities.

A method for constructing a quasilinear representation of analytic nonlinear MMs was developed in [8–10] using the corollary of the integral independence theorem on the integration path [27, 28]. When constructing quasilinear models by the method outlined in these papers, the operations of differentiating nonlinearities and integrating derivatives with a shifted argument are used. However, due to the substantial non-linearity of the functions $M_i(\alpha, \alpha_{li}, \alpha_{ri}, \varepsilon)$ used in CGA and the bulk of the resulting expressions in general, this approach is not effective in this case. Therefore, another approach to constructing quasilinear models using the specifics of the CGA method, oriented, in the current version, to the use of polynomials in CGA is considered below.

5 Quasilinear Form of Polynomial CGA Lifting Models

In connection with the above orientation of the currently developed CGA method on the polynomial fragmentary approximation, this article solves the problem of constructing a quasilinear form exclusively for polynomial Cut-Glue models. Under certain conditions, this restriction makes it possible to exclude the differentiation of MM and the integration of these derivatives from the algorithm for constructing a quasilinear form, in contrast to the method described in [8–10]. Under the conditions of the cumbersomeness of nonlinear dependencies used in the CGA method [4–7], this solution greatly simplifies the transformations. Further, this approach is described in the example of the essentially nonlinear model (3) of the aircraft flight height control channel using the lifting phenomenon obtained in the previous section by the CGA method.

Turning to the transformation of function (3) to a quasilinear form, we will assume that $L(\alpha)$ describes the entire possible range of the variable α, while this range includes the value α_0 that corresponds to the steady-state of the nonlinear object under consideration.

Let x be the deviation of a variable α from the steady-state value α_0, i.e. $x = \alpha - \alpha_0$; wherein $\alpha = x + \alpha_0$. In expression (3), we add to each polynomial $L_i^e(\alpha)$ and subtract the value $L_i^e(\alpha_0)$, $i = 1, 2$. As a result, we get:

$$L(\alpha) = [L_1^e(\alpha) + L_1^e(\alpha_0) - L_1^e(\alpha_0)]M_1(\alpha, \alpha_{l1}, \alpha_{r1}, \varepsilon) \\ + [L_2^e(\alpha) + L_2^e(\alpha_0) - L_2^e(\alpha_0)]M_2(\alpha, \alpha_{l2}, \alpha_{r2}, \varepsilon)$$

Replacing in this equality α by $\alpha = x + \alpha_0$, we represent the previous equality as follows:

$$L(x + \alpha_0) = [L_1^e(x + \alpha_0) - L_1^e(\alpha_0)]M_1(x + \alpha_0) + L_1^e(\alpha_0)M_1(x + \alpha_0) \\ + [L_2^e(\alpha) - L_2^e(\alpha_0)]M_2(x + \alpha_0) + L_2^e(\alpha_0)]M_2(x + \alpha_0). \quad (5)$$

Here, for brevity, the replacement of the notation is introduced:

$$M_i(x+\alpha_0) \xrightarrow{\text{обозн.}} M_i(x+\alpha_0, \alpha_{li}, \alpha_{ri}, \varepsilon), \quad i = 1, 2. \quad (6)$$

Since $L_i^e(x+\alpha^0)$ is polynomials, the difference $L_i^e(x+\alpha_0) - L_i^e(\alpha_0)$ in expression (5) obviously have as common factor the variable x in the degree of not less than one. Then they can be represented as

$$\forall\, i = 1, 2; \;\; x + \alpha_0 \in [\alpha_{l1} \div \alpha_{r2}] \rightarrow R_i(x, \alpha_0)x = L_i^e(x + \alpha_0) - L_i^e(\alpha_0). \quad (7)$$

Using formulas (5), (7) allows us to represent function (3) in the form of the following expressions of pseudo-linear structure:

$$L(x + \alpha_0) = a(x)x + L_1^e(\alpha_0)M_1(x + \alpha_0) + L_2^e(\alpha_0)M_2(x + \alpha_0), \quad (8)$$

where

$$a(x) = R_1(x, \alpha_0)M_1(x + \alpha_0) + R_1(x, \alpha_0)M_2(x + \alpha_0). \quad (9)$$

For example, in the specific case of obtaining CGA mathematical model (3) for the polynomials (1) and (2), the differences $R_i(x, \alpha_0)$ take the following form:

$$R_1(x, \alpha_0) = 4.614 + 2 \cdot 0.046 \cdot \alpha_0 + 0.046x;$$

$$R_2(x, \alpha_0) = -19.214 + 2 \cdot 0.124 \cdot \alpha_0 + 0.124x$$

6 Quasilinear Model of Controlling of the Altitude Aerostat

For the unification of further transformations, we introduce for the variables involved in information transformations in the management of the universal, customary for the subject area designations:

- x_1 is the deviation of the variable from steady-state value α^0;
- x_2 is $d\alpha/dt = \omega = dx_1/dt$ the rate of the attack angle α or its deviations from α_0;
- x_3 is the deviation of the flight height h of the balloon from the set h_0;
- x_4 is $dh/dt = dx_3/dt$ the rate of change of height or its deviation from the set.

In addition, when building a control channel mathematical model, a number of additional designations must be entered to take into account the mechanical properties of the machine:

- A_m is torque acting on the body of the airship;
- I is the moment of inertia of the body of the airship relative to its horizontal transverse axis passing through the center of mass;
- $L(\alpha)$ is lifting force caused by the interaction of the hull with the incident flow depending on the angle of attack α;
- $G = m \cdot g$ is the aerostat weight; m is its mass, g is the acceleration of gravity.

Taking into account the introduced notation, the simplified equations of the dynamics of the aerostat considered in the problem of vertical movement will be written as follows:

$$\left.\begin{aligned} \dot{x}_1 &= x_2; \\ \dot{x}_2 &= A_m/I \\ \dot{x}_3 &= x_4; \\ \dot{x}_4 &= (L(\alpha) - G)/m. \end{aligned}\right\} \tag{10}$$

Realizing the state management paradigm, let us suppose that the rotating moment in the second Eq. (10) can be represented by a formula that assumes that is used to form the control by all four state variables:

$$A_m = k_\alpha u_\alpha(x); \quad k_\alpha = K_\alpha/I; \quad x = [x_1\ x_2\ x_3\ x_4]^T, \tag{11}$$

where k_α—reduced coefficient of the angle regulator; $u_\alpha(x)$—the law of control of the angle of attack; K_α—the transmission coefficient of the actuators, affecting on α (aerodynamic steering wheels).

This chapter discusses the use of mathematic model obtained by the CGA method for the synthesis of nonlinear control laws. Moreover, the adopted research paradigm is supposed to implementation of the airship motion control system, for example, by stabilizing a variable. Since the basis for the simulation was the dependence of the lifting on the angle of attack, you can set the task of stabilizing the flight altitude, which is controlled by a lift, since the weight of the aerostat is often a constant in flight. Thus, a further research problem is the transformation of the right-hand side of the fourth equation of system (10) to a quasilinear form.

To this end, we introduce the concept of a given angle α_0, the associated increment of the angle α, which we denote by x_1, and the constructed function (3) from this increment, i.e.:

$$\alpha = x + \alpha_0; \rightarrow x = \alpha - \alpha_0; \rightarrow C(x)x = L(\alpha_0) - L(\alpha). \tag{12}$$

Here, taking into account the notation (6), the functions are determined by the expressions:

$$C(x_1, \alpha_0) = R_1(x_1, \alpha_0)M_1(x_1 + \alpha_0) + R_2(x_1, \alpha_0)M_2(x_1 + \alpha_0), \tag{13}$$

$$L_0(x_1, \alpha_0) = L_1(\alpha_0) \cdot M_1(x_1 + \alpha_0) + L_2(\alpha_0) \cdot M_2(x_1 + \alpha_0), \tag{14}$$

Thus, the right side of the last equation of system (10) with allowance for (13), (14) can be represented by the following expression:

$$\begin{aligned}\frac{(L(\alpha) - G)}{m} &= \left\{\frac{C(x_1) \cdot x_1}{m} + \frac{L_0(x_1, \alpha_0)}{m} - g\right\} \\ &= k_m \cdot [C(x_1,\ \alpha_0) \cdot x_1 + u_h(x_1, \alpha_0, m)],\end{aligned} \tag{15}$$

where $k_m = 1/m$ is the coefficient of influence of the mass of the apparatus on the acceleration of vertical motion; $u_h(x_1, \alpha_1, m) = L_0(x_1, \alpha_0) - m * g$ is the law of nonlinear flight altitude control, ensuring its stabilization at the level corresponding to the given value α_0.

As can be seen from the above expressions (14) and (15), the control $u_h(x_1, \alpha_0, m)$ is uniquely determined by the sum $x_1 + \alpha_0$ and the parameters of the Cut-Glue approximation $L(\alpha)$ (3), therefore the task of the control system design of a nonlinear object (10) is to determine the control that directly affects the changes in the angle of attack.

To solve this problem, the Eq. (10) with regard to expressions (11) and (15) will be represented in a vector-matrix, quasilinear form:

$$\dot{x} = A(x)x + b(x)\, u_\alpha, \tag{16}$$

where

$$A(x) = \begin{bmatrix} 0 & 1\,0\,0 \\ 0 & 0\,0\,0 \\ 0 & 0\,0\,1 \\ C_1(x_1) & 0\,0\,0 \end{bmatrix}, \quad b(x) = \begin{bmatrix} 0 \\ k_\alpha \\ 0 \\ 0 \end{bmatrix} \tag{17}$$

7 Analytical Design of Nonlinear Control of Aerostat Altitude

The model (16), (17) describes the nonlinear plant, the parameters of which are known with some accuracy, i.e. this is an uncertain object. When designing control systems for such objects, adaptive learning control, and other methods are used [29–32]. However, in these cases, the control design using a quasilinear model is simpler.

For brevity of the description of this method, the notation is entered into (17). In accordance with the applied design method based on quasilinear models [8–10], control is also sought in the quasilinear form, i.e.

$$u_{\alpha}(x) = -k^T(x)x = -[k_1(x)\, k_2(x)\, k_3(x)\, k_4(x)]\, x. \tag{18}$$

From expressions (16) and (18) it follows that the equation of a closed system in the quasilinear form has the form

$$\dot{x} = D(x)x, \tag{19}$$

where

$$D(x) = A(x) - b(x)\, k^T(x). \tag{20}$$

The characteristic polynomial of the system matrix (20) is determined by the expression

$$D(s, x) = \det(sE - D(x)) = \det[sE - A(x) + b(x)k^T(x)], \tag{21}$$

where E is the identity matrix. Based on the properties of the determinants, equality (21) can be represented as follows

$$D(s, x) = \det(sE - A(x)) - k^T(x))\text{Adj}\,(sE - A(x))b(x)$$

or

$$D(s, x) = A(s, x)) + \sum_{i=1}^{n} k_i(x)\, V_i(x), \tag{22}$$

where

$$A(s, x) = \det(sE - A(x)), \tag{23}$$

$$V_1(s, x) = [1\;0\;0\;0]\;\text{Adj}\,(sE - A(x))b(x),$$
$$V_2(s, x) = [0\;1\;0\;0]\;\text{Adj}\,(sE - A(x))b(x),$$

$$V_3(s,x) = [0\,0\,1\,0]\ \mathrm{Adj}\,(sE - A(x))b(x),$$
$$V_4(s,x) = [0\,0\,0\,1]\ \mathrm{Adj}\,(sE - A(x))b(x) \tag{24}$$

Note that polynomials are the numerators of the "transfer functions" $W_{x_i u_\alpha}(s,x) = V_i(s,x)/A(s,x), i = \overline{1,4}$ of the object (16), (17) for each state variable x_i.

Thus, in accordance with expressions (22), (23), to determine the functional coefficients $k_i(x)$, from the expression for the desired control (18) it is necessary, first of all, to find the characteristic polynomial of the system matrix A(x) of the object. In this case, the polynomial A(s,x) $= s^4$, and the associated matrix

$$\mathrm{Adj}(sE - A(x)) = \begin{bmatrix} s^3 & s^2 & 0 & 0 \\ 0 & s^3 & 0 & 0 \\ C_1(x_1)s & C_1(x_1) & s^3 & s^2 \\ C_1(x_1)s^2 & C_1(x_1)s & 0 & s^3 \end{bmatrix}. \tag{25}$$

Therefore, substituting the matrix (25) and the vector $b(x)$ from (17) into formulas (24), we obtain:

$$V_1(s,x) = k_\alpha s^2,\ V_2(s,x) = k_\alpha s^3,\ V_3(s,x) = C_1(x_1)k_\alpha,\ V_3(s,x) = C_1(x_1)k_\alpha s \tag{26}$$

Let the desired characteristic polynomial of the system matrix $D(x) = A(x) - b(x)k^T(x)$ be equal to

$$D^*(s) = s^4 + \delta_3 s^3 + \delta_2 s^2 + \delta_1 s + \delta_0, \tag{27}$$

moreover, its roots $-\sigma_1$, $-\sigma_2, -\sigma_3$ и $-\sigma_4$are real, negative and different.

Find the difference between polynomials $D^*(s)$ (27) and $A(s,x) = s^4$:

$$H(s,x) = D^*(s) - A(s,x) = \delta_3 s^3 + \delta_2 s^2 + \delta_1 s + \delta_0. \tag{28}$$

Taking into account the difference (28) and expressions (26), equality (22) can be represented [9, 10] in vector-matrix form as a system of algebraic equations of the following form:

$$\begin{bmatrix} 0 & 0 & C_1(x_1)k_\alpha & 0 \\ 0 & 0 & 0 & C_1(x_1)k_\alpha \\ k_\alpha & 0 & 0 & 0 \\ 0 & k_\alpha & 0 & 0 \end{bmatrix} \begin{bmatrix} k_1(x) \\ k_2(x) \\ k_3(x) \\ k_4(x) \end{bmatrix} = \begin{bmatrix} \delta_0 \\ \delta_1 \\ \delta_2 \\ \delta_3 \end{bmatrix}. \tag{29}$$

Each column of the matrix of this system is composed of the polynomials' coefficients $V_i(s,x)$ (27), in ascending order s. The vector of unknowns includes the sought-for coefficients $k_i(x)$, and the column on the right-hand side of the system

(29) is composed of the coefficients of the difference $D(s) - A(s, \tilde{x})$ (28), also in ascending orders.

We emphasize that the design method of control systems based on quasilinear models can be used only in the case when a nonlinear object whose mathematical model is presented in a quasilinear form (16), (17) is completely controllable in a certain neighborhood of its equilibrium position, those if the condition is:

$$\det U(x) \neq 0, \quad x \in \Omega \tag{30}$$

where

$$U(x) = [b(x) \;\; A(x)b(x) \;\; \ldots \;\; A^{n-1}(x)b(x)]. \tag{31}$$

The condition (30) is due to the fact that the algebraic system (29) has a unique solution only when this condition is fulfilled [10].

Substituting the matrix $A(x)$ and the vector $b(x)$ (17) into equality (31) with $n = 4$ and calculating the determinant (30), we get $\det U(x) = k_\alpha^4 k_m^2 C^2(x_1, \alpha_0)$. In this case, the condition of controllability is obviously equivalent to the inequality $C(x_1, \alpha_0) \neq 0$, where $x_1 \in \Omega$. On the basis of polynomials (1)–(3), it is easy to establish that in the case under consideration the function $C(x_1, \alpha_0)$ (13) satisfies the obtained inequality with $0° \leq x_1 + \alpha_0 \leq 60°$. Therefore, a solution of the system (29) exists and leads to the following expressions: $k_1(x) = -\delta_2/k_\alpha$, $k_2(z) = \delta_3/k_\alpha$, $k_3(z) = \delta_0/C(x_1, \alpha_0)k_\alpha$, $k_4(z) = \delta_1/C(x_1, \alpha_0)k_\alpha$. Substituting these coefficients in (18), we obtain the desired nonlinear control of the angle of attack of the aerostat:

$$u_\alpha(x) = -k^T(x)x = -\left[\frac{\delta_2}{k_\alpha} \; \frac{\delta_3}{k_\alpha} \; \frac{\delta_0}{k_\alpha C(x_1, \alpha_0)} \; \frac{\delta_1}{k_\alpha C(x_1, \alpha_0)}\right]x. \tag{32}$$

It is not difficult to verify, substituting the matrix, vector (17) and vector from (32) into equality (20), that the characteristic polynomial (22) of the matrix of the synthesized system is equal to the polynomial (27).

The structural-functional scheme of the system of automatic stabilization of the flight altitude of aerostat that implements the control of the channel under investigation according to the law (32) with the control object MM obtained in the form of (10) and converted to (16), (17) is shown in Fig. 2. It demonstrates the informational transformations of both state variables and real physical variables. The scheme uses the notation for physical and mathematical variables, introduced in Sect. 6.

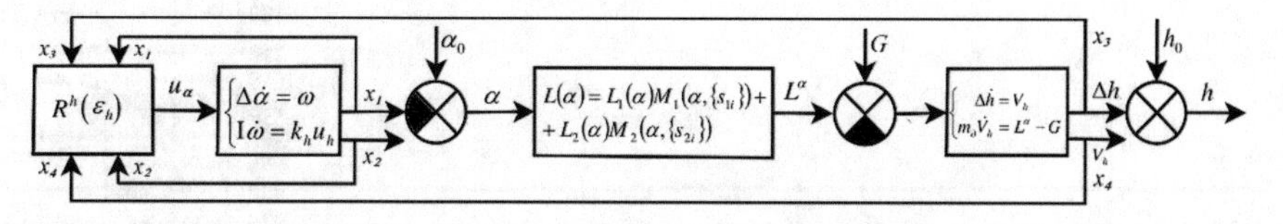

Fig. 2 Structural and functional scheme of the automatic stabilization of the altitude of the aerostat flight

8 Computer Simulation of the Designed System

The simulation model of the controlled system (16), (17), (32) is implemented in the high-level Matlab language using the open system of computer mathematics (SCM) GNU Octave. The simulation model consists of a functional set of subroutines, each of which solves a narrow problem:

– calculation of the right parts of the system of differential equations describing the dynamics of the control plant in deviations;
– calculation of the set of values of the multiplicative-distinguishing function (IMF) on various fragments of experimental data;
– calculation of the set of values of the function obtained by Cut-Glue approximation of the static characteristic of the aerostat;
– calculation of the quasilinear approximation value of the lifting characteristic.

The simplification of the structure of the program code of the simulation model is achieved by abstraction from the nature of the specific plant being modeled and the automatic controller.

The simulation results are shown in Fig. 3. In the steady-state the attack angle $\alpha°$, the lifting L, and the aerostat altitude h have the values: $\alpha = 22.12°$, $L = 125.55$ kN and $h = 1000$ m. This is clearly seen in Fig. 3a. Transients were simulated caused by a sharp decrease in the value of the attack angle by one angular degree. This is clearly seen in Fig. 3a. In accordance with Fig. 1c, while reducing the attack angle, the lifting value decreases sharply (in this case to 119 kN), which leads to the decrease in the altitude aerostat. These phenomena are reflected in the graphs in Fig. 3b, c at $t \approx 0$.

Further, under the action of the control (32) of the attack angle increases to a value 26.6°, and then decreases to a steady-state value $\alpha = 22.12°$ (Fig. 3a). In accordance with such change of the attack angle, the lifting increases to 127.7 kN, and then

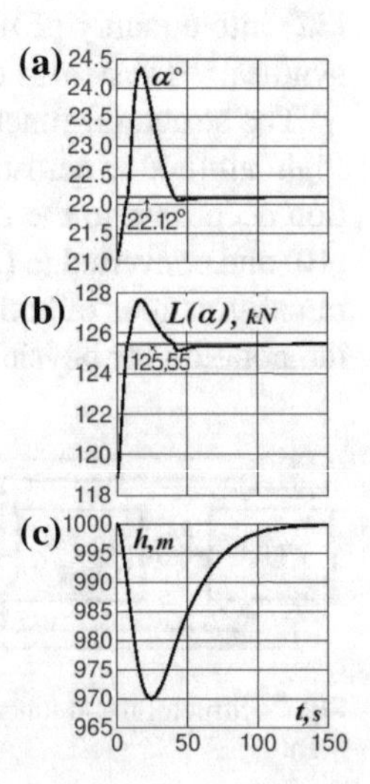

Fig. 3 Transients in the stabilization system of the altitude aerostat

approaches the steady value of 125.55 kN. Similarly, the aerostat altitude increases and approaches to the steady value—1000 m.

Thus, the results of simulation of the designed system testify, firstly, of its stability, secondly that the presence of a "dysfunctional" nonlinear characteristic when describing such nonlinearity using the Cut-Glue approximation method does not complicate control system of a nonlinear plant or its simulation.

9 Conclusion

It is shown that mathematical models of the essentially nonlinear plants, created by fragmentary data using the Cut-Glue method of approximation, are analytical functions.

These models can be used to not only the solve problems of the estimating and displaying their static properties, but also to serve as the basis for the building of the dynamic models of nonlinear control plants, as well as to ensure the solution of the design problems of nonlinear control systems by analytical methods.

In cases where mathematical models are built based on fragmentary data using the method of Cut-Glue approximation with the application of polynomials, the possibility arises of constructing the quasilinear forms of these models without the use of differentiation and integration operations. This opens up broad possibilities for applying the Cut-Glue approximation method for the analytical design of nonlinear control systems by the quasilinear and other approaches.

Acknowledgements The Russian Fund of Basic Research (grant No. 18-08-01178\19) supported this research.

References

1. Dyshlyuk, E.N., Kotlyarov, R.V., Pachkin, S.G.: Methods of structural and parametric identification of control objects on the example of the furnace emulator EP10. Equip. Technol. Food Prod. **47**(4), 159–165 (2017). https://doi.org/10.21179/2074-9414-2017-4-159-165
2. Wu, C.F.J., Hamada, M.S.: Experiments: Planning, Analysis, and Optimization, 743 p. Wiley (2009)
3. Voevoda, A.A., Bobobekov, K.M., Troshina, G.V.: The parameters identification of the automatic control system with the controller. J. Phys.: Conf. Ser. **1210**(art. 012021) (2019)
4. Neidorf, R.A., Neydorf, A.R.: Modeling of essentially non-linear technical systems by the method of multiplicatively additive fragmentary approximation of experimental data. Inf. Space **1**, 47–57 (2019)
5. Neydorf, R., Neydorf, A., Vučinić, D.: Cut-glue approximation method for strongly nonlinear and multidimensional object dependencies modeling. Ad. Struct. Mater. **72**, 155–173 (2018)
6. Neydorf, R., Neydorf, A.: Technology of cut-glue approximation method for modeling strongly nonlinear multivariable objects. Theoretical Bases and Prospects of Practical Application. SAE Technical Paper 2016–01-2035 (2016). https://doi.org/10.4271/2016-01-2035

7. Neydorf, R., Yarakhmedov, O., Polyakh, V., Chernogorov, I., Vucinic, D.: Cut-glue approximation based on particle swarm sub-optimization for strongly nonlinear parametric dependencies of mathematical models. In: Improved Performance of Materials. Design and Experimental Approaches, pp. 185–196. Springer (2018)
8. Gaiduk, A.R.: Algebraic synthesis of nonlinear stabilizing controls. In: Synthesis of Algorithms of Complex Systems, vol. 7, pp. 15–19. TRTI Publisher, Taganrog (1989)
9. Gaiduk, A.R., Stojković, N.M.: Analytical design of quasilinear control systems. Facta Univ. Ser.: Autom. Control Robot **13**(2), 73–84 (2014)
10. Gaiduk, A.R.: Theory and Methods of Analytical Synthesis of Automatic Control Systems (Polynomial Approach), p. 415. Phizmatlit Publisher, Moscow (2012)
11. Loran, P.-J.: Approximation and Optimization, p. 496. World Publisher, Moscow (1975)
12. Insung, I., Naylor, B.: Piecewise linear approximations of digitized space curves with applications. Computer Science Technical Reports, Report Number 90–1036, Paper 37 (1990)
13. Pinheiro, A.M.G., Ghanbari, M.: Piecewise approximation of contours through scale-space selection of dominant points. IEEE Trans. Image Process **19**(6) (2010)
14. Ahlberg, J.H., Nilson, E.N., Walsh, J.L.: The Theory of Splines and their Applications. Academic Press, New York (1967)
15. De Boor, C.: A Practical Guide to Splines. Springer (1978)
16. Micula, G., Micula, S.: Handbook of Splines. Kluwer Academic Publishers, Dordrecht, Boston, London (1999)
17. Rawlings, J.O., Pantula, S.G., Dickey D.A.: Applied Regression Analysis: A Research Tool, 2nd edn., 659 p (1998)
18. Bates, D.M., Watts, D.G.: Nonlinear Regression Analysis and Its Applications. Wiley, New York (1988)
19. Drapper, N.R., Smith, H.: Applied Regression Analysis, vol. 1. Wiley, New York (1981)
20. Drapper, N.R., Smith, H.: Applied Regression Analysis, vol. 2. Wiley, New York (1981)
21. Totik, V.: Orthogonal polynomials. Surv. Approx. Theory **1**, 70–125 (2005)
22. Khrushchev, S.: Orthogonal Polynomials and Continued Fractions From Euler's Point of View. Atilim University, Turkey: Cambridge University Press. www.cambridge.org/9780521854191 (2008)
23. Powell, M.J.D.: The theory of radial basis function approximation. Adv. Num. Anal. **II**. OUP, Oxford (1992)
24. Buhmann, M.D.: Radial Basis Functions: Theory and Implementations. Cambridge University Press. http://catdir.loc.gov/catdir/samples/cam033/2002034983 (2003)
25. Neydorf, R., Sigida, Y., Voloshin, V., Chen, Y.: Stability analysis of the MAAT feeder airship during ascent and descent with wind disturbances. SAE Technical Paper 2013–01-2111 (2013). https://doi.org/10.4271/2013-01-2111
26. Voloshin, V., Chen, Y., Neydorf, R., Boldyreva, A.: Aerodynamic characteristics study and possible improvements of MAAT feeder airships. SAE Technical Paper 2013-01-2112 (2013). https://doi.org/10.4271/2013-01-2112
27. Ku, Y.H., Puri N.N.: On Lyapunov Functions of Order Nonlinear Systems. Journal Franklin Institution (1963)
28. Ku, Y.H.: Lyapunov function of a fourth-Order system. IEEE Trans. Autom. Control, **9**, 276–278 (1064)
29. Chowdhary, G., Yucelen, T., Muhlegg, M., Johnson, E.: Concurrent learning adaptive control of linear systems with exponentially convergent bounds. Int. J. Adapt. Control Sig. Process. **27**(4), 280–301 (2013)
30. Zhu, Y., Hou, Z.: Controller dynamic linearisation-based model-free adaptive control framework for a class of non-linear system. IET Control Theory Appl. **9**(7), 1162–172 (2015)
31. Shao, Z., Zheng, C., Efimov, D., Perruquetti, W.: Identification, estimation and control for linear systems using measurements of higher order derivatives. J. Dyn. Syst. Measur. Control Am. Soc. Mech. Eng. **139**(12), 1–6 (2017)
32. Jayawardhana, R.N., Ghosh, B.K.: Kalman filter based iterative learning control for discrete time MIMO systems. In: Proceedings of 30th Chinese Control and Decision Conference, pp. 2257–2264 (2018)

Adaptive-Robust Control of Technological Processes with Delay on Control

Ivan V. Gogol, Olga A. Remizova, Vladislav V. Syrokvashin and Aleksandr L. Fokin

Abstract A method for the synthesis of adaptive systems is proposed, which ensures the robustness of the system with respect to the delay, technological objects with a delay in control in the presence of uncertainty in setting the latency and time-varying coefficients of the model of the linear inertial part changing in time, within a certain range of normal operation, arbitrary to the law. Adaptive identification type systems and direct adaptive control systems are considered. The basis for the synthesis of the main control loop of the control system is the method for determining the tuning parameters of the traditional laws of regulation (I, PI, PID). As the main contour, used a system, which is robust in relation to the change in the magnitude of the delay. The chapter presents an identification approach and direct adaptive control.

Keywords Linear SISO objects · Robust control systems · PID control laws · Predictor · Disturbing influence · Parametric uncertainty

1 Introduction

During the solution of problems of automation of technological processes, very often there is the problem of compensating for unknown limited disturbances in the presence of a delay in control. For technological processes are characterized by the situation when the value of delay is estimated very approximately (the error can reach up to 50% and more), and the value of delay is, as a rule, significant in relation to the rest of the dynamics of the technological object. Another problem is the presence of time-variable coefficients (parameters) of the mathematical model of the object [1–5].

Thus, the designing of the output regulator is reduced to a complex mathematical problem, which is solved in the literature under various assumptions. So there is an approach associated with the construction of a predictor of a perturbation with a

---

I. V. Gogol · O. A. Remizova (✉) · V. V. Syrokvashin · A. L. Fokin
Saint-Petersburg State Institute of Technology, Moskovsky Prospect 26, Saint-Petersburg 190013, Russia
e-mail: remizova-oa@technolog.edu.ru

A. G. Kravets et al. (eds.), *Cyber-Physical Systems: Industry 4.0 Challenges*, Studies in Systems, Decision and Control 260,
https://doi.org/10.1007/978-3-030-32648-7_11

known value of the delay [6–8]. Usually, the model coefficients are assumed here. In [9, 10], an approach based on the method of frequency identification was considered, which makes it possible to determine parameter estimates and delays over sufficiently long periods of time with subsequent control. An approach based on the extended error method was considered in [11]. In [12], an approach was considered based on the representation of the model as a partial differential equation with a known lag value. An algorithm without a predictor is proposed in [13], but under the assumption that the delay is known and the coefficients are constant, as in [11, 12].

In this chapter, was consider a linear object with a delay in control with one input and one output, the mathematical model of which is a link in delay connected in series with the inertial part of arbitrary complexity. The coefficients of the model of dynamics (parameters) of the inertial part can vary in time according to an arbitrary law, while remaining limited in a certain region, which corresponds to the normal mode of operation of the object. The amount of control delay belongs to the specified value range.

The main contour of the adaptive system is considered to be a robust system with respect to the uncertainty of the lag with traditional (I, PI, FID) control laws, which are most often (up to 95%) used in the automation of technological processes [14]. The task of the adaptation circuit is to adjust the parameters of the traditional laws regulating the main circuit when the inertial part parameters change over time. The chapter considers identification approach and direct adaptive control.

The structure of the control system does not contain a predictor of disturbances, since the magnitude of the delay is significant with respect to the rest of the dynamics and is given with uncertainty. The robust settings of the main contour provide roughness with respect to the delay, and the adaptation algorithm takes into account the time variation of the parameters of the inertial part.

2 Formulation of the Problem

Consider control objects whose dynamics are described by a linear equation

$$Q(p,t)y(t) = k_0(t)R(p,t)(u(t-\tau) + f_1(t)) + f_2(t) \tag{1}$$

where $Q(p,t) = q_0(t)p^n + q_1(t)p^{n-1} + \cdots + q_{n-1}(t)p + 1$, $R(p,t) = r_0(t)p^m + r_1(t)p^{m-1} + \cdots + r_{m-1}(t)p + 1$.

Assumptions:

1. Differential operator $R(p,t)$ must be stable.
2. Differential operator $RQ(p,t)$ must be stable, or it can be represented as $Q(p,t) = p^s\overline{Q}(p,t)$, where $\overline{Q}(p,t)$ stable differential operator, similar to $Q(p,t)$, s—given natural number.
3. Operator ratios $Q(p,t)$, $k_0(t)R(p,t)$ and disturbing influences $f_1(t)$, $f_2(t)$ at the input and output of object—limited functions with known ranges of change:

$\tau \in [\underline{\tau}, \bar{\tau}]$, $q_i(t) \in [\underline{q}_i, \bar{q}_i]$, $i \in [0, n-1]$, $r_i(t) \in [\underline{r}_i, \bar{r}_i]$, $i \in [0, m-1]$, $k_0(t) \in [\underline{k}_0, \bar{k}_0]$, $f_1(t) \in [\underline{f}_1, \bar{f}_1]$, $f_2(t) \in [\underline{f}_2, \bar{f}_2]$, which correspond to the normal mode of the technological object. For the synthesis of the main circuit uses a nominal model with constant coefficients of the form

$$Q^0(p)y(t) = k_0^0 R^0(p)u(t - \tau_0^0) \tag{2}$$

where $Q^0(p) = q_0^0 p^n + q_1^0 p^{n-1} + \cdots + q_{n-1}^0 p + 1$, $R^0(p) = r_0^0 p^m + r_1^0 p^{m-1} + \cdots + r_{m-1}^0 p + 1$,—differential operators with constant coefficients, $q_i^0 \in [\underline{q}_i, \bar{q}_i]$, $i \in [0, n-1]$, $r_i^0(t) \in [\underline{r}_i, \bar{r}_i]$, $i \in [0, m-1]$, $\tau_0^0 \in [\underline{\tau}, \bar{\tau}]$, $k_0^0 \in [\underline{k}_0, \bar{k}_0]$—nominal values of parameters.

In the case when the nominal object model has poles at the point zero $Q^0(p) = p^l \overline{Q}^0(p)$, where $\overline{Q}^0(p)$ stable differential operator, similar to $Q^0(p)$, approximation is used, of the form

$$\frac{1}{p} \approx \frac{1}{p + \varepsilon} = \frac{\gamma}{\gamma p + 1} \tag{3}$$

where $\varepsilon \ll 1$, $\gamma = \varepsilon^{-1} \gg 1$.

This replacement allows, in the future, in the synthesis of the law of regulation, to consider only stable differential operators $Q^0(p)$.

To simplify the regulator (reducing the number of adjustable parameters), required to reduce the nominal approximating model, this is carried out in two stages. At the first stage, the inertial part of the nominal model (2) is represented as a transfer function

$$k_0^0 \frac{R^0(p)}{Q^0(p)} \approx \frac{k_0^0 \exp(-\tau_1 p)}{G_l^0(p)} \tag{4}$$

where $G_l^0(p) = g_0^0 p^l + \cdots + g_{l-1}^0 p + 1$—reduced Hurwitz polynomial l order, $l \le \rho$, where ρ—relative degree of inertial part (2), usually $l = 1, 2, 3$, τ_1—additional (transient) delay.

The order of the polynomial $G_l^0(p)$ depends on the nature of the inertia dynamics on the left-hand side of (2) [15]. If this is aperiodic dynamics, then $l = 1$, if it is oscillatory, then $l = 2$, if an integrator is additionally present, then $l = 3$. In the latter case, formula (3) is used for approximation. The construction of the model (4) is easiest to implement on the basis of the transient characteristics of the object by modeling.

As a result, instead of (2), we obtain the model of the nominal dynamics:

$$W_0^0(p) \approx \frac{k_0^0 \exp(-\tau_0 p)}{G_l^0(p)} \tag{5}$$

where $\tau_0 = \tau_0^0 + \tau_1$—total lag time.

Similarly, for a real movement (1), a reduced model can be written within the framework of the same structure as

$$G_l(p, t)y_0(t) = k_0(t)u(t - \tau_0) \tag{6}$$

where $G_l(p, t) = g_0(t)p^l + \cdots + g_{l-1}(t)p + 1$—stable differential operator, $y_0(t)$—output of the reduced model.

The second stage of the model reduction is closely related to the use of the dynamic compensation method [15–18]. Here it is assumed further simplification (5), (6) due to the subsequent compensation of the dynamics of this model by the dynamics of the regulator. For small values of l in (4), this effectively simplifies the procedure for the synthesis of the regulator by allowing the use of the simplest (basic) model and provides quality indicators of the system. At the same time, the complexity of the model used for the synthesis and the complexity of the obtained regulator, despite the simplification, will correspond to the complexity of the model (5).

Therefore, further considered two basic models of the object. The first is a lag component

$$W_{B0}^0(p) = k_0^0 \exp(-\tau_0 p) \tag{7}$$

Two basic transfer functions (I and PI) of the regulator correspond to it. They are used in cases where in the dynamics of the inertial part (2) there is no dominant time constant. If not, then a different base model is used.

$$W_{B0}^0(p) = \frac{k_0^0 \exp(-\tau_0 p)}{T_0 p + 1} \tag{8}$$

where T_0—nominal value of the dominant time constant.

In this case (5) $G_l^0(p) = (T_0 p + 1)\overline{G}_{l-1}^0(p)$, $\overline{G}_{l-1}^0(p) = \bar{g}_1^0 p^{l-1} + \cdots + \bar{g}_{l-2}^0 p + 1$. This transfer function corresponds to the basic PID control law.

As the structure of the basic regulators, it is proposed to use the previously obtained [15–17] robust to the uncertainty values of the delay And, PI, PID control algorithms

$$W_{BP1}(p, t) = \frac{k_{P1}(t)}{k_0^0 \tau_0 p} \tag{9}$$

$$W_{BP2}(p, t) = \frac{k_{P2}(t)}{k_0^0 \tau_0} \cdot \frac{T_1(t)p + 1}{p} \tag{10}$$

$$W_{BP3}(p, t) = \frac{k_{P3}(t)}{\tau_0 k_0^0} \cdot \frac{T_1(t)p + 1}{p} \cdot \frac{T_2(t)p + 1}{T_E p + 1} \tag{11}$$

where k_{P1}, k_{P2}, k_{P3}, T_1, T_2—configurable settings, low time constant $T_E \ll \min(T_1, T_2)$ is introduced to ensure the condition of the physical realizability of the transfer function of the regulator.

Configurable parameters are changed during the operation of the adaptation algorithm. Based on the basic laws of regulation (9), (10) with the help of the method of dynamic compensation, real laws of regulation are obtained

$$W_P(p,t) = W_{BPi}(p,t)\frac{G_l(p,t)}{(T_E p + 1)^l} \tag{12}$$

where $i = 1, 2$.

But the variable parameters $g_0(t), \ldots, g_{l-1}(t)$ of the differential operator $G_l(p,t)$ from formula (6) are unknown, so instead of (12) the real control law is

$$W_P(p,t) = W_{BPi}(p,t)\frac{\theta_0 p^l + \cdots + \theta_{l-1} p + 1}{(T_E p + 1)^l} \tag{13}$$

where $i = 1, 2$, $\theta = \left[\theta_0 \ldots \theta_{l-1}\right]^T$—vector of configurable controller parameters. For the basic law of regulation (11) instead of (12) get

$$W_P(p,t) = W_{BP3}(p,t)\frac{\overline{G}_{l-1}(p,t)}{(T_E p + 1)^{l-1}} \tag{14}$$

where $\overline{G}_{l-1}(p,t) = \bar{g}_1(t)p^{l-1} + \cdots + \bar{g}_{l-2}(t)p + 1$, accordingly, the differential operator from (6) will be $G_l(p,t) = (T_0 p + 1)\overline{G}_{l-1}(p,t)$.

The real law of regulation in this case is

$$W_P(p,t) = W_{BP3}(p,t)\frac{\theta_1 p^{l-1} + \cdots + \theta_{l-2} p + 1}{(T_E p + 1)^{l-1}} \tag{15}$$

where $\theta = \left[\theta_1 \ldots \theta_{l-2}\right]^T$—vector of configurable controller parameters.

Thus, the vector of configurable parameters of the regulators (13), (15), taking into account the parameters of the control laws (9)–(11), will be $\hat{\theta} = [k_{P1}, k_{P2}, k_{P3}, T_1, T_2, \theta^{\mathrm{T}}]^T$. This is a fairly large number of parameters that can be reduced by replacing some of them with their nominal values.

For the formation of the desired movement (reference model), the nominal object model (5) and the transfer function of the regulators (13), (15) are used with the nominal values of adjustable parameters $g^0 = \left[g_0^0 \ldots g_{l-1}^0\right]^T$. The following formulas are used instead of (9)–(11)

$$W_{BP1}^0(p) = \frac{k_{P1}^0}{k_0^0 \tau_0 p} \tag{16}$$

where $k_{P1}^0 = 0.343q$, q—standard model adjustment parameter,

$$W_{BP2}^0(p) = \frac{k_{P2}^0}{k_0^0 \tau_0} \cdot \frac{T_1^0 p + 1}{p} \tag{17}$$

where $k_{P2}^0 = 0.343 \cdot 1.587q$, $T_1^0 = \tau_0/4$—settings obtained in [15, 16], can be also used the optimal settings [11],

$$W_{BP3}^0(p) = \frac{k_{P3}^0}{\tau_0 k_0^0} \cdot \frac{T_1^0 p + 1}{p} \cdot \frac{T_2^0 p + 1}{T_E p + 1} \tag{18}$$

where $T_E \ll \min\left(T_1^0, T_2^0\right), k_{P3}^0, T_1^0, T_2^0$—parameters that are configured on the basis of a compromise between robust and speed [17].

Thus, the desired motion can be described by the transfer function of a closed system

$$\Phi_m(p) = \frac{W_P^0(p) W_0^0(p)}{1 + W_P^0(p) W_0^0(p)} = \frac{k_0^0 W_P^0(p) \exp(-\tau_0 p)}{G_l^0(p) + k_0^0 W_P^0(p) \exp(-\tau_0 p)} \tag{19}$$

where $W_P^0(p) = W_{BPi}^0(p) \frac{G_l^0(p)}{(T_E p+1)^l}$, $i = 1, 2$, or $W_P^0(p) = W_{BP3}^0(p) \frac{\overline{G}_{l-1}^0(p)}{(T_E p+1)^{l-1}}$, $W_0^0(p)$ calculated by the formula (5).

Accordingly, for the desired nominal motion we get

$$\left\{G_l^0(p) + k_0^0 W_P^0(p) \exp(-\tau_0 p)\right\} y_m(t) = k_0^0 W_P^0(p) \exp(-\tau_0 p) r(t) \tag{20}$$

where $y_m(t)$—standard model output, $r(t)$—set at the system input.

The movement of a real closed reduced system (6), (13), (15) is described similarly

$$\{G_l(p, t) + k_0(t) W_P(p, t) \exp(-\tau_0 p)\} y_0(t) = k_0(t) W_P(p, t) \exp(-\tau_0 p) r(t) \tag{21}$$

where $y_0(t)$—output of the reduced model, $r(t)$—set at the system input.

Thus, a reduced model of the object (6), as well as reduced models of the standard and real motion of the closed system (20), (21) of the main contour, are obtained. The aim of the work is to develop algorithms of identification type and direct adaptive control based on reduced models. These tasks are discussed further. The use of reduced models is associated with the simplification of algorithms.

3 Synthesis of the Adaptation Algorithm (Identification Approach)

With the identification approach in real time, the estimation of the parameters $k_0(t), g_0(t), \ldots, g_{l-1}(t)$ of Eq. (6) is performed. In this case, the condition of quasistationarity is assumed to be satisfied. The maximum number of adjustable parameters when $l = 3$ is $4 + 4 = 8$. When using a gradient algorithm, this will lead to a significant adaptation time. In addition, additional filtering will be required here [19] to avoid measuring the derivatives of the output value $y_0(t)$.

The simplest solution is to configure only one parameter $k_0(t)$ at $g_0(t) = g_0^0, \ldots, g_{l-1}(t) = g_{l-1}^0$. An alternative approach is to apply the method of dynamic expansion of the regressor [20], which allows decomposing the $2l + 2$ multidimensional estimation problem to the $2l + 2$ one-dimensional problem, but this is a much more complicated algorithm that should be used if setting one parameter does not lead to a result.

The proposed synthesis algorithm is standard [21], the only difference is that it is applied to the reduced model. If there is only one adjustable parameter $k_0(t)$ model (6) will look like

$$y_0(t) = k_0(t)v(t) \tag{22}$$

where $v(t) = \frac{1}{g_0^0 p^l + \cdots + g_{l-1}^0 p + 1} u(t - \tau_0)$.

Consider the difference signal

$$e_1(t) = y_0(t) - y(t) \tag{23}$$

where $y(t)$—output of the original model (1), $y_0(t)$—output of the reduced model (6).

Then the gradient tuning algorithm will look like

$$\dot{k}_0 = -\gamma e_1 \tag{24}$$

where γ—positive constant.

It is known [19] that the parametric estimation error will exponentially tend to zero if the condition of nonvanishing excitation

$$\int_t^{t+T} v^2(\xi)d\xi \geq \alpha \tag{25}$$

where T, α—constants.

Further, the control problem is solved using standard regulation laws (9)–(11) at $T_1(t) = T_1^0, T_2(t) = T_2^0$ at T and at nominal values of parameters in (13), (15).

4 Synthesis of the Adaptation Algorithm (Direct Adaptation)

Here was used the well-known algorithm [21] (p. 143), obtained for an object without delay. In order to apply the known result, it is required for the delay component in (20), (21) to apply the Pade approximation. In practical implementation, a second order approximation was used

$$\exp(-\tau_0 p) \approx \frac{(\tau_0^2/12)p^2 - (\tau_0/2)p + 1}{(\tau_0^2/12)p^2 + (\tau_0/2)p + 1} \tag{26}$$

Further, for (20), (21), which have the same order, a transition to the equations of state is required. Let the equation of state of the desired motion of the closed system (20) be

$$\dot{x}_m = A_m x_m + b_m r \tag{27}$$

where x_m—desired motion state vector, A_m, b_m—own matrix and control vector of the corresponding dimensions.

The equation of state for (21) is represented as

$$\dot{x}_0 = A\left(g, \hat{\theta}\right)x_0 + b\left(g, \hat{\theta}\right)r \tag{28}$$

where $g = \left[k_0(t), g_0(t), \ldots, g_{l-1}(t)\right]^T, \hat{\theta} = [k_{P1}, k_{P2}, k_{P3}, T_1, T_2, \theta]^T, \theta = \left[\theta_0, \ldots, \theta_{l-1}\right]^T$—parameter vector that includes the undefined parameters of the reduced model (6) and the adjustable parameters of the regulators (9)–(11), (13), (15).

Matrices $A = A\left(g, \hat{\theta}\right), b = b\left(g, \hat{\theta}\right)$ must have the same structure and dimension with matrices A_m, b_m. When applying the method of dynamic compensation, the order of Eq. (20) may be less than in (21) due to the reduction of identical terms in (5) and $W_P^0(p)$. In this case, to preserve the dimension in $W_P^0(p)$ use the parameter values slightly offset from their nominal values at (5).

The model of a generalized custom object in accordance with the method [14] can be obtained on the basis of (21) in the form of the equation of state of the form

$$\dot{x}_0 = \left(A_0(g) + \Delta A\left(\hat{\theta}\right)\right)x_0 + \left(b_0(g) + \Delta b\left(\hat{\theta}\right)\right)r \tag{29}$$

where x_0—state vector, A_0, b_0—parts A, b, which depend on the vector of uncertain values of parameters g, and $\Delta A, \Delta b$ depend on the vector of adjustable parameters $\hat{\theta}$

For the synthesis of the adaptation algorithm, the velocity gradient method is used

$$\frac{d}{dt}\Delta A(t) = -\gamma P e(t) x_0(t)^T \quad \frac{d}{dt}\Delta b(t) = -\gamma P e(t) r(t)^T \tag{30}$$

where $e(t) = x_0(t) - x_m(t)$—error vector, γ—setting parameter, $P^T = P > 0$—positive definite $(l \times l)$ matrix, satisfying the Lyapunov equation $PA_m + A_m^T P = -Q$ for some matrix $Q = Q^T > 0$, which is selected when setting up.

It is known [21] that algorithm (29) provides stability, and also has the following identifying properties

$$A_0 + \Delta A(t) \to A_m, \qquad b_0 + \Delta b(t) \to b_m \text{ при } t \to \infty \tag{31}$$

In this case, it is assumed that the reference model is sufficiently fully activated by the input signal $r(t)$. To determine the vector of tunable parameters $\hat{\theta}(t)$ t each moment of time, we can consider the equality

$$A_0(g(t)) + \Delta A\left(\hat{\theta}(t)\right) = A_m, b_0(g(t)) + \Delta b\left(\hat{\theta}(t)\right) = b_m \tag{32}$$

But since the vector of variable parameters $g(t)$ is not known, it is necessary to use its nominal value g^0 to solve Eq. (32). This means that instead of model (21), an approximate model is considered

$$\left\{G_l^0(p) + k_0(t) W_P(p,t) \exp(-\tau_0 p)\right\} y^0(t) = k_0(t) W_P(p,t) \exp(-\tau_0 p) r(t) \tag{33}$$

where $y^0(t)$—output of approximate reduced model, $r(t)$—set at the system input.

Accordingly, instead of (29), (30) we get

$$\dot{x}^0 = \left(A_0\left(g^0\right) + \Delta A\left(\hat{\theta}\right)\right) x^0 + \left(b_0(g) + \Delta b\left(\hat{\theta}\right)\right) r \tag{34}$$

where x^0—approximate model state vector.

$$\frac{d}{dt}\Delta A(t) = -\gamma P \tilde{e}(t) x^0(t)^T \quad \frac{d}{dt}\Delta b(t) = -\gamma P \tilde{e}(t) r(t)^T \tag{35}$$

where $\tilde{e}(t) = x^0(t) - x_m(t)$—error vector.

Instead of (32) to determine the settings of the regulator

$$A_0\left(g^0\right) + \Delta A\left(\hat{\theta}(t)\right) = A_m, b_0\left(g^0\right) + \Delta b\left(\hat{\theta}(t)\right) = b_m \tag{36}$$

The system of Eq. (36) consists of a maximum of $2l^2$ equations. But, if, for example, to use the normal canonical form, then we get only $l + 1$ equation. The solution of this system finally forms the adaptation algorithm.

5 Example

Let the object model (1) have the form

$$(q_1(t)p+1)(q_2(t)p+1)^2(q_3(t)p+1)y(t) = k_0(t)(u(t-45)+f_1(t))+f_2(t) \tag{37}$$

where $k_0(t) = 1+\cos 0.0005t, q_1(t) = 10+2\cos 0.009t, q_2(t) = 20+4\cos 0.0002t$ $q_3(t) = 25+5\cos 0.0003t$, $f_1(t) = 0.5 \cdot 1(t)$, $f_2(t) = 0.1\cos 0.01t$, $k_0^0 = 1$, $q_1^0 = 10, q_2^0 = 20, q_3^0 = 25$.

The nominal model (5)

$$(40p+1)y(t) = u(t-70) \tag{38}$$

Reduced model of real movement (6) in the form:

$$(40p+1)y_0(t) = k_0(t)u(t-70) \tag{39}$$

The object basic model is considered in the form (7), and the basic transfer function of the regulator in the form (9), (16). The real transfer function of the controller:

$$W_P(p,t) = \frac{1}{70\hat{k}_0(t)} \frac{(40p+1)}{p} = \frac{u(p)}{\varepsilon(p)} \tag{40}$$

$\hat{k}_0(t)$—tunable parameter, accordingly in (9) $k_{P1}(t) = 1/\hat{k}_0(t)$ when $k_0^0 = 1$ in (7), (9).

In the identification approach, the transfer coefficient is adjusted according to the law (24) with $\gamma = 0.1$. The parameter $q = 0.5$ in (16) is selected based on the type of the desired transient process. In the absence of disturbances $f_1(t) = f_2(t) = 0$ the transient response is (Fig. 1).

Fluctuations caused by changes in the parameters $k_0(t)$, $q_i(t)$, they are placed in 5% zone. In the presence of disturbances at the input and output of the object, noted in (37), we obtain the step characteristic (Fig. 2).

it is obvious that the rise time has greatly increased. This is due to the constant perturbation at the entrance of the object. Also, there was an additional oscillatory component due to the harmonic noise at the output of the object.

With direct adaptation, the desired movement is formed on the basis of (20) with the approximation of the delay component (26). The equation of the desired reduced closed system has the form

$$\left(p^3 + \frac{6+q0.343}{70}p^2 + \left(1 - \frac{q0.343}{2}\right)\frac{12}{70^2}p + \frac{q0.343 \cdot 12}{70^3}\right)$$
$$\times y_m = \left(\frac{q0.343}{70}p^2 - \frac{q0.343 \cdot 6}{70^2}p + \frac{q0.343 \cdot 12}{70^3}\right)r \tag{41}$$

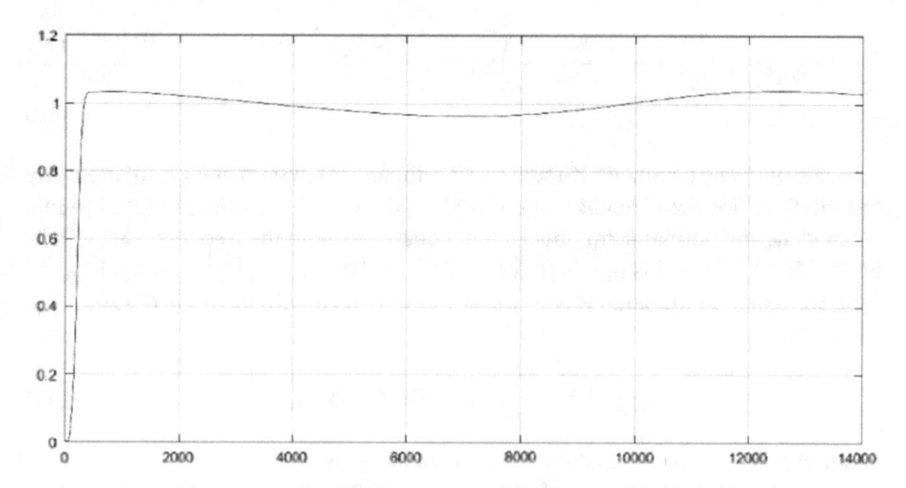

Fig. 1 Step response of a closed system

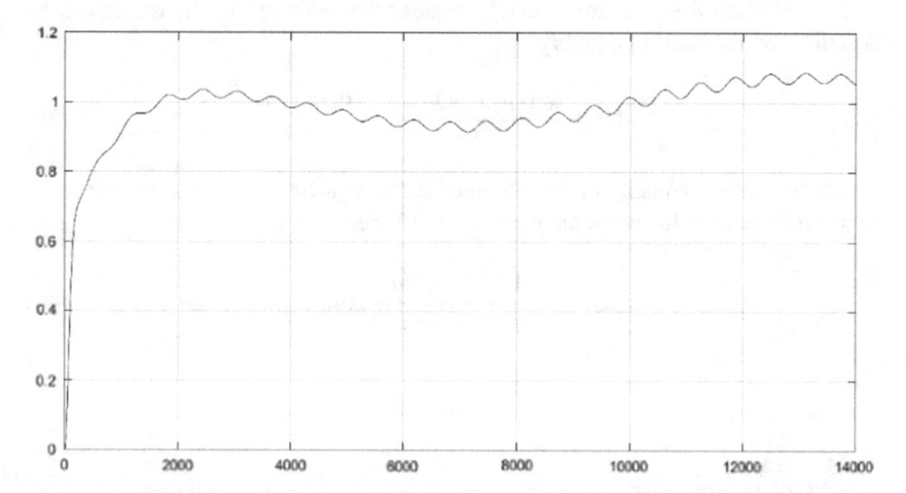

Fig. 2 Step response in the presence of disturbances

The approximate equation of the reduced motion (33) with $k_0^0 = 1$ is:

$$\left(p^3 + \frac{6 + k_0 q 0.343}{70} p^2 + \left(1 - \frac{k_0 q 0.343}{2}\right)\frac{12}{70^2} p + \frac{k_0 q 0.343 \cdot 12}{70^3}\right)$$
$$\times\, y = \left(\frac{k_0 q 0.343}{70} p^2 - \frac{k_0 q 0.343 \cdot 6}{70^2} p + \frac{k_0 q 0.343 \cdot 12}{70^3}\right) r \qquad (42)$$

Based on (42), an equation is formed for obtaining $A_0(g_0)$, $b_0(g_0)$ в (34), (36), in (34), (36), the corresponding equation can be considered as

$$\left(p^3 + \frac{6}{70}p^2 + \frac{12}{70^2}p + \frac{0.343 \cdot 12}{70^3}\right)y_m = \left(\frac{0.343}{70}p^2 - \frac{0.343 \cdot 6}{70^2}p + \frac{0.343 \cdot 12}{70^3}\right)r \tag{43}$$

Since only one parameter K requires an estimate, the system of Eq. (36) uses only one equation for the element (3.1). Based on the equation of the reduced motion, when using the normal form, this element coincides with the free term on the left-hand side of (42) and is $\alpha_{0P} = k_0 12 \cdot 0.343 \cdot q/70^3$. Accordingly, on the basis of (43), we obtain in the same way as $\alpha_0 = 12 \cdot 0.343/70^3$. On the basis of (36), (41), we obtain

$$\alpha_0 + \Delta\alpha(t) = \alpha_{0m} = 12 \cdot 0.343 \cdot q/70^3 \tag{44}$$

where $\Delta\alpha$ is calculated based on the gradient algorithm (35).

Figure 3 shows the process of convergence of the left part (44) to the value α_{0m} when $\gamma = 0.00001$ in (35)

To determine the estimate of $\hat{k}_0(t)$ consider the value of $\Delta\alpha(t)$, which can be described by the basis of equality

$$\Delta\alpha(t) = \frac{k_0(t)q0.343 \cdot 12}{70^3} - \frac{0.343 \cdot 12}{70^3} \tag{45}$$

The left side of equality (45) is obtained in the adjustment process, and the right, as the difference of the constant term of the left side (42), (43):

$$\hat{k}_0(t) = \frac{1}{q} + \frac{70^3/q}{12 \cdot 0.343}\Delta\alpha_0(t) \tag{46}$$

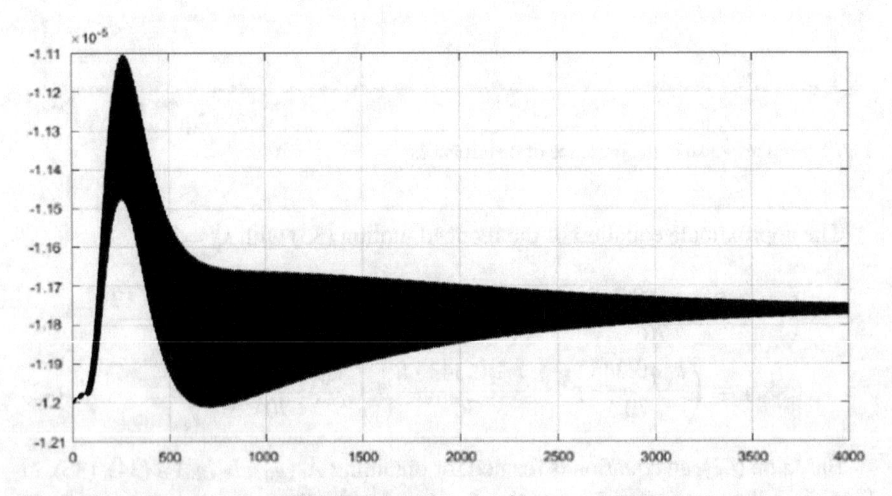

Fig. 3 Convergence process (44) in an adaptation algorithm

In the absence of disturbances $f_1(t) = f_2(t) = 0$, the step response takes the form (Fig. 4).

The quality of regulation is higher than when using the adaptation of the identification type (Fig. 1). The step response in the presence of disturbances is shown in Fig. 5.

The suppression of disturbances is worse here than when using the adaptation of the identification type (Fig. 2). The transient response does not fit into the 5% accuracy zone.

In general, the considered systems well suppress high-frequency and low-frequency disturbances. But there is a mid-frequency region where it is not possible

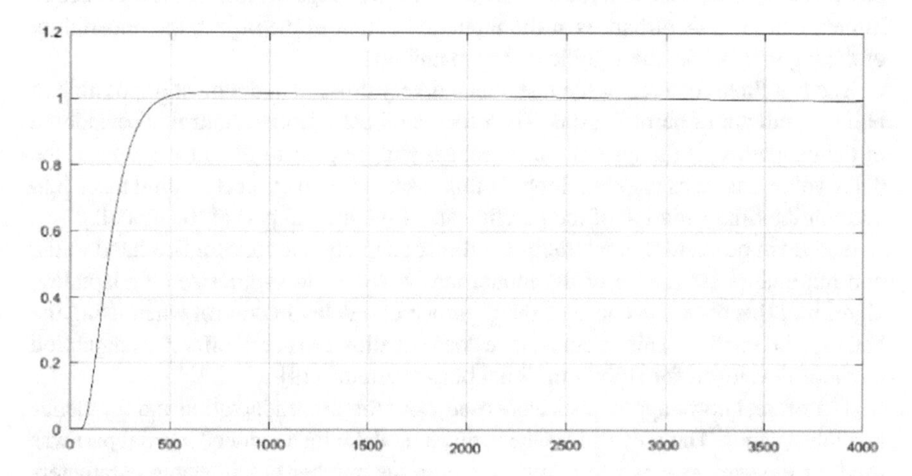

Fig. 4 Step response of a closed system

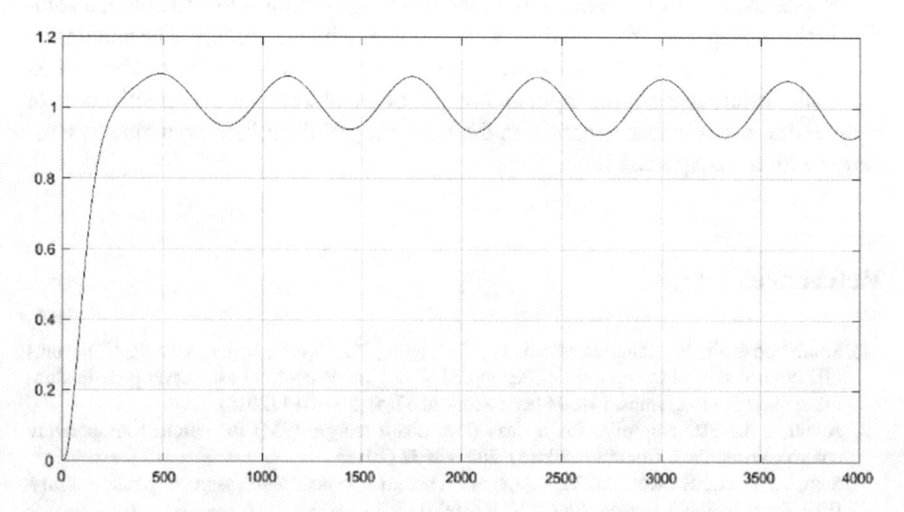

Fig. 5 Step response in the presence of disturbances

to ensure 5% accuracy due to the significant adaptation time for large values of delay. At the same time, the performance of the system is preserved with the uncertainty of setting the delay value to 50%.

6 Conclusion

The chapter considers the most general formulation of adaptive control of an object with a control delay, which is relevant for the problems of automation of technological processes. It is assumed that the model coefficients change over time, the presence of limited unknown disturbances at the input and output of the object, the uncertainty of the lag magnitude, the significant lag magnitude.

The last three features of the task make it very difficult and almost impossible to build a predictor of perturbations. Therefore, a robust-adaptive control is considered in the chapter, when a system that is robust with respect to the uncertainty of the delay value is used as the main loop. In this case, the adaptation algorithm takes into account the time variation of the coefficients of the inertial part of the model.

The main problem of constructing a direct adaptation algorithm is related to the non-minimal phase nature of the model that was used to synthesize an adaptation algorithm; this occurs because of the presence of a delay in control when using the Pade approximation. This circumstance does not allow the use of effective adaptation methods developed for strictly minimal phase systems [19].

Therefore, known approaches were used, both for direct adaptation and for identification systems. The only difference is that a model with a reduced inertial part was used for the synthesis, which allows reducing the number of adjustable parameters and simplifying the control algorithm.

Despite the traditional construction, the resulting systems allow to obtain a solution in the most general formulation of the problem with uncertainty. The simulation showed that the constructed adaptive systems allow to compensate for the limited harmonic disturbances at the input and output of the object, and the effectiveness of suppressing disturbances depends on the frequency of disturbances, which is associated with the adaptation time.

References

1. Mahmoodabadi, M., Taherkhorsandi, M., Talebipour, M., Castillo-Villar, K.: Adaptive robust PID control subject to supervisory decoupled sliding mode control based upon genetic algorithm optimization. Trans. Inst. Measur. Control **37**(4), 505–514 (2015)
2. Aldemir, A.: PID controller tuning based on phase margin (PM) for wireless temperature control. Wirel. Pers. Commun. **103**(3), 2621–2632 (2018)
3. Sato, T., Tajika, H., Konishi, Y.: Adaptive PID control system with assigned robust stability. IEEJ Trans. Electr. Electron. Eng. **13**(8), 1169–11181 (2018)

4. Suganya, S.: Design of PID controller for chemical process-heuristic algorithm approach. In: Proceedings of Third International Conference on Science Technology Engineering & Management (ICONSTEM) (2017)
5. Bensafia, Y.: Robust fractionalized PID controller design using the sub-optimal approximation of FOTF. In: Bensafia, Y. (ed.) Proceedings of 6th International Conference on Systems and Control (ICSC) (2017)
6. Pyrkin, A.A., et al.: Compensation of polyharmonic perturbations acting on the state and output of a linear object with a delay in the control channel. Avtomatika i telemehanika **12**, 43–64 (2015) (in Russian)
7. Paramonov, A.V., Gerasimov, D.N., Nikiforov, V.O.: Synthesis of an adaptive parameter tuning algorithm with improved convergence for a linear dynamic error model. Izvestija vysshih uchebnyh zavedenij. Priborostroenie **60**(9), 818–825 (2017) (in Russian)
8. Cykunov, A.M.: Robust Control with Disturbance Compensation, 300 p. M.:Fizmatlit (2012) (in Russian)
9. Vilanova, R., Arrieta, O.: Robust PI/PID controllers for load disturbance based on direct synthesis. ISA Trans. **81**, 177–196 (2018)
10. Nobuyama, E., Kami, Y.: Robust PID controller design for both delay-free and time-delay systems. IFAC Papersonline **51**(4), 930–935 (2018)
11. Parsheva, E.A., Tsykunov, A.M.: Adaptive decentralized control of multivariable objects. Autom. Remote Control **62**(2), 290–330 (2001)
12. Krstic, M.: Delay Compensation for Nonlinear, Adaptive, and PDE Systems, p. 466. Birkhauser, Springer (2009)
13. Furtat, I.B., Tsykunov, A.M.: Adaptive control of plants of unknown relative degree. Autom. Remote Control **71**(6), 1076–1084 (2010)
14. Denisenko V.V.: Raznovidnosti PID - reguljatorov [Varieties of PID—regulators]. Avtomatizacija v pro-myshlennosti **6**, 45–50 (2007)
15. Fokin, A.L.L.: Synthesis of robust process control systems with standard controllers. Izv. SPbGTI(TU) **27**, 101–106 (in Russian)
16. Remizova, O.A., Syrokvashin, V.V., Fokin, A.L.: Synthesis of robust control systems with standard controllers. Izv. vuzov. Priborostroenie. **58**(12), 12–18 (2015) (in Russian)
17. Gogol, I.V., Remizova, O.A., Syrokvashin, V.V., Fokin, A.L.: Synthesis of robust regulators to control technological processes in the class of traditional laws of regulation. Izv. SPbGTI(TU) **44**, 98–105 (2018) (in Russian)
18. Jakovis, L.M.: Simple ways to calculate typical regulators for complex industrial automation objects. Avtomatizacija v promyshlennosti **6**, 51–56 (2007) (in Russian)
19. Kim, D.P.: Multidimensional, Nonlinear, Optimal and Adaptive Systems, 400 p. M.:Fizmatlit (2007) (in Russian)
20. Aranovskiy, S., Bobtsov, A., Ortega, R., Pyrkin, A.: Improved transients in multiple frequencies estimation via dynamic regressor extension and mixing. In: 12th IFAC International Workshop on Adaptation and Learning in Control and Signal Processing, vol. 49, no. 13, pp. 99–104 (2016)
21. Andrievskij, B.R., Bobcov, A.A., Fradkov, A.L.: Methods of Analysis and Synthesis of Nonlinear Control Systems, 336 p. M. Izhevsk: Institut komp'juternyh issledovanij (2018) (in Russian)

Robust Control System Based on Traditional PID Control Laws

Ivan V. Gogol, Olga A. Remizova, Vladislav V. Syrokvashin and Aleksandr L. Fokin

Abstract The chapter considers a new approach for designing robust control systems based on traditional PID control laws for linear adaptive control systems for SISO objects (single input–single output), with a lag in control in the presence of parametric uncertainty and lag value. The technique of transition from real models of the object to reduced models depending on the influence of the dominant time constant is considered. An algorithm for the synthesis of tuning parameters of regulators based on a compromise between the requirements of speed and coarseness of a closed system is developed. To demonstrate the performance of this method, the main parameters of a traditional PI controller for a real control object with a dominant time constant and without, represented by a mathematical model in the form of a transfer function, and are calculated as examples.

Keywords Linear SISO objects · Robust control systems · PID control laws · Predictor · Disturbing influence · Parametric uncertainty

1 Introduction

Were considered linear SISO control objects with a delay in control in the presence of parametric uncertainty and uncertainty of the magnitude of the delay [1–5]. The above makes it impossible, to use a predictor in solving the problem of compensating disturbances within a two-loop system, other approaches are needed [6, 7]. On the other hand, in the practice of automating technological processes, the majority of control systems in the normal mode are built using the PID control laws, and designing a predictor for a disturbance is a difficult task; therefore, the synthesis of robust control in the class of traditional regulators is further considered [8–10].

I. V. Gogol · O. A. Remizova (✉) · V. V. Syrokvashin · A. L. Fokin
Saint-Petersburg State Institute of Technology, Moskovsky Prospect 26, Saint-Petersburg 190013, Russia
e-mail: remizova-oa@technolog.edu.ru

A. G. Kravets et al. (eds.), *Cyber-Physical Systems: Industry 4.0 Challenges*, Studies in Systems, Decision and Control 260,
https://doi.org/10.1007/978-3-030-32648-7_12

During the solution of this problem, there is a contradiction between the requirements for speed and the requirements for the robust of the system. Increased robustness leads to a loss of speed [11, 12]. Therefore, the actual task is the building a synthesis methodology to ensure a compromise between these indicators. For practical implementation, this technique should be fairly simple and, at the same time, universal in the sense that it allows calculating the controller for an arbitrary complexity model from the same positions [13, 14].

2 Formulation of the Problem

Consider the transfer function of the object

$$W_0(p) = k_0 \frac{\beta_m(p)}{\alpha_n(p)} \exp(-\tau p) \tag{1}$$

where $a_n(p)$, $\beta_m(p)$—Hurwitz n and m order polynomials, $a_n(0)$, $\beta_m(0) = 1$, k_0—transfer coefficient, $\underline{k_0} \leq k_0 \leq \overline{k_0}$, τ—delay $\underline{\tau} \leq \tau \leq \overline{\tau}$, polynomial coefficients $a_n(p)$, $\beta_m(p)$ may vary at specified intervals.

Along with the real transfer function [15, 16], the nominal transfer function of the object is considered, which coincides in structure with the transfer function (1) of the form

$$W_0^0(p) = k_0^0 \frac{\beta_m^0(p)}{\alpha_n^0(p)} \cdot \exp(-\tau_0 p) \tag{2}$$

where k_0^0, τ_0—nominal values of gain and delay, $\beta_m^0(p)$, $a_n^0(p)$—Hurwitz nominal numerator and denominator polynomials, $\left(m \leq n, \beta_m^0(0) = \alpha_n^0(0) = 1\right)$, the coefficients of the nominal model belong to the intervals of (1), $\underline{k_0} \leq k_0^0 \leq \overline{k_0}$, $\underline{\tau} \leq \tau_0 \leq \overline{\tau}$.

In technical applications [17], there is often a case where the transfer function of an object has poles at the point zero, then to ensure the Hurwitz polynomials of the denominator in (1), (2) when the controller is synthesized, considered approximation, which has the following form

$$\frac{1}{p} \approx \frac{1}{p + \varepsilon} = \frac{\gamma}{\gamma p + 1} \tag{3}$$

where $\varepsilon \ll 1$, $\gamma = \varepsilon^{-1} \gg 1$.

For the synthesis of the regulator, a reduced approximating model is used, which is formed in two stages. At the first stage, the traditional approach is used to solve automation problems, when the inertial part of the nominal model (2) is represented as

$$k_0^0 \frac{\beta_0(p)}{\alpha_0(p)} \approx \frac{k_0^0 \exp(-\tau_1 p)}{\chi_l(p)} \tag{4}$$

$\chi_1(p)$—Hurwitz low l—order polynomial, $l = 1, 2, 3$, τ_1—additional (transient) delay.

The exponent of the polynomial $\chi_1(p)$ depends on the nature of the inertia dynamics on the left side (4). If this is aperiodic dynamics, then $l = 1$, if it is oscillatory, then $l = 2$, if there is an additional integrator, then $l = 3$. In the latter case, the formula (3) is used for approximation. Building a model (4) is easiest to implement on the basis of the transient response by modeling.

As a result, instead of (2), we get a model of dynamics with an additional delay of the form

$$W_0^0(p) \approx \frac{k_0^0 \exp(-\tau_0 p)}{\chi_l(p)} \tag{5}$$

where $\tau_0 = \tau_0^0 + \tau_1$.

Next, it is required to form a model of the desired motion of a closed system, in which a compromise between speed and robust is ensured. This problem cannot be solved for all options for constructing a simplified model (5) within the framework of a simplified methodology. Therefore, there is a need for the second stage of the reduction of the object model. Two basic reduced models will be considered.

The first is a lag link

$$W_0^0(p) = k_0^0 \exp(-\tau_0 p) \tag{6}$$

It is used in cases when in the dynamics of the inertial part (2) there is no dominant time constant. If not, then the base model is used

$$W_0^0(p) = \frac{k_0^0 \exp(-\tau_0 p)}{T_0 p + 1} \tag{7}$$

where T_0—the nominal value of the dominant time constant.

For the transfer function (6), the basic PI law of regulation is considered. It has the following form

$$W_{P1}(p) = \frac{k_{P1}}{k_0^0 \tau_0} \cdot \frac{T_1 p + 1}{p} \tag{8}$$

where k_{P1}—tunable transfer coefficient, T_1—the tunable time constant.

For transfer function (7), the PID law is considered

$$W_{P2}(p) = \frac{k_{P2}}{\tau_0 k_0^0} \cdot \frac{T_1 p + 1}{p} \cdot \frac{T_2 p + 1}{T_E p + 1} \tag{9}$$

where k_{P2}, T_1, T_2—adjustable parameters, the small time constant $T_E \ll \min(T_1, T_2)$ is introduced to ensure the condition of physical reliability of the transfer function of the regulator.

Adjustable parameters of regulator models (8), (9) are selected based on a compromise between speed and robust requirements for a closed system (6), (8) or a system (7), (9). The resulting systems provide the desired motion for the base model. To realize the desired motion of a system with a nominal model (5) with an accuracy of small time constants, you can use the method of dynamic compensation. Then we get the real laws of regulation of the form

$$W_P(p) = W_{Pi}(p)\frac{\chi_l(p)}{(T_E p + 1)^l} \tag{10}$$

where $i = 1, 2$.

As is known, when applying the compensation method, the stable compensated dynamics of the approximated inertial part (5) is present in the dynamics of a closed system and manifests itself under non-zero initial conditions. But with real control, this does not occur, since the synthesized controller operates with a real (1) (2), and not with an approximated (5) object [18–20].

3 Research Results

Next, we consider the formation of the desired motion for the basic models of the object (6), (7). For the synthesis of robust regulators (8), (9), we apply the robust Nyquist criterion [21]. Let the following inequality be used as an additive measure of the uncertainty of the transfer function of an open-loop system

$$\left|W(j\omega) - W^0(j\omega)\right| \le \gamma \tag{11}$$

where $W(j\omega) = W_P(j\omega)W_0(j\omega)$, $W^0(j\omega) = W_P(j\omega)W_0^0(j\omega)$ frequency transfer functions of real and nominal open systems, $W_P(j\omega)$—frequency transfer function of the basic controller, $\gamma > 0$—positive number.

The robust Nyquist criterion requires that the Nyquist plot of an open-loop basic nominal system $W^0(j\omega)$ does not cover a circle of the radius γ with center at the point $(-1, j0)$. From Fig. 1 it can be seen that the coarseness of the system is related to the magnitude of the stability margin in amplitude h, but this quantity is not a measure of coarseness since it is possible that the Nyquist plot enters a circle of radius h.

The chapter proposes further to use the magnitude of stability in amplitude h only as a criterion for the synthesis of the system, since an increase in this value is a necessary condition for increasing the roughness, and a radius h, for evaluating the roughness of the system.

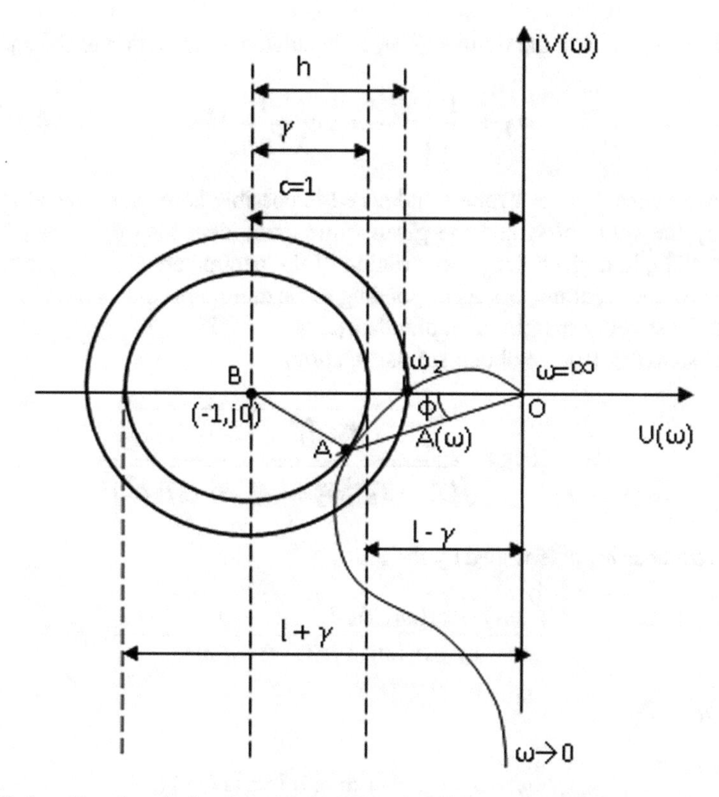

Fig. 1 Robust Nyquist criterion (construction of the forbidden region)

Figure 1 shows the procedure for constructing a forbidden region for the phase response, which guarantees a given value of the radius γ, which is similar to the well-known procedure for constructing a forbidden region, guaranteeing a given value of the oscillation index. In this case, the boundary of the forbidden area for Nyquist plot can be constructed on the basis of the ratio

$$\phi(\omega) = \arccos \frac{A^2(\omega) + 1 - \gamma^2}{2A(\omega)} \tag{12}$$

where $1 - \gamma \leq A(\omega) \leq 1 + \gamma$.

If the Nyquist plot relates to the forbidden area, then the roughness of the system is estimated by the value γ. For the synthesis, we set the value to h. Then for the system (6), (8) we obtain the equality constraint

$$h_1 = -20 \lg k \frac{\sqrt{1 + T_1^2 \omega_1^2}}{\omega_1} \tag{13}$$

where $k = k_{P1}/\tau_0$, and the frequency ω_1 is calculated as a solution to the equation

$$\tau_0 = \frac{1}{\omega_1}\left\{\pi - arctg\left(\frac{1}{T_1\omega_1}\right)\right\} \quad (14)$$

A compromise between robust and speed is possible here, since speed is determined by the value of k, and the given value of h_1 depends on the pair (k, T_1). Consequently, in a given range of variation of the parameters (k, T_1), one can find the values of these parameters corresponding to the minimum control time for a given value of the stability margin in amplitude h_1.

For system (7), (9), we obtain similar relations

$$h_2 = -20\lg k\frac{\sqrt{\left(1 - T_1T_2\omega_2^2\right)^2 + (T_1 + T_2)^2\omega_2^2}}{\sqrt{(T_0 + T_E)^2\omega_2^4 + \omega_2^2\left(1 - T_0T_E\omega_2^2\right)^2}} \quad (15)$$

where frequence ω_2 determined by the ratio

$$\omega_2 = \frac{\psi(\omega_1) - a_1(\omega_1, T_1, T_2)\cdot\omega_1 - a_2(\omega_1)\cdot\omega_1}{\tau_0 - a_1(\omega_1, T_1, T_2) - a_2(\omega_1)} \quad (16)$$

where $\omega_1 = \frac{\pi}{2\tau_0}$

$$a_1(\omega_1, T_1, T_2) = \frac{(T_1 + T_2)\left(1 + T_1T_2\omega_1^2\right)}{1 + \left(T_1^2 + T_2^2 + T_1^2T_2^2\omega_1^2\right)\omega_1^2},$$

$$a_2(\omega_1) = -\frac{(T_0 + T_E)\left(1 + T_0T_E\omega_1^2\right)}{1 + \left(T_0^2 + T_E^2 + T_0^2T_E^2\omega_1^2\right)\omega_1^2},$$

$$\psi(\omega_1) = arctg\frac{(T_1 + T_2)\omega_1}{1 - T_1T_2\omega_1^2} + arctg\frac{\left(1 - T_0T_E\omega_1^2\right)}{\omega_1(T_0 + T_E)} + \pi.$$

For the pair (k_{P2}, T_1), the value of T_2 is determined so that the condition (15) is satisfied. Further, for a triple (k_{P2}, T_1, T_2), transient response is constructed, which is used to determine speed and overshoot. In this case, the coarseness and speed depend on the selected pair (T_1, T_2), and the magnitude of the overshoot mainly depends on k_{P2}.

The coarseness of both systems γ is estimated based on the construction of the forbidden region (12).

4 Example

As an example, consider the nominal transfer function of an object.

$$W_0^0(p) = \frac{k_0^0 \exp(-10p)}{(60p+1)^2(6p+1)^2} \approx \frac{k_0^0 \exp(-26.5)}{(115p+1)(12p+1)} \quad (17)$$

It also shows the reduced transfer function (5). Since the dominant time constant is not taken into account, the real transfer function of the controller (10) (based on the PI law (8)) will have the form

$$W_{P1}(p) = \frac{0.61}{26.5 \cdot 1} \cdot \frac{10p+1}{p} \cdot \frac{115.0p+1}{0.1p+1} \cdot \frac{12p+1}{0.1p+1} \quad (18)$$

It gives a transient response with maximum speed $t_P = 90$ s at $h = 9$ dB (0.6452), at the same time robust $\gamma = 0.6285$.

Given the dominant time constant, the transfer function of the controller will be

$$W_{P2}(p) = \frac{0.49}{26.5 \cdot 1} \cdot \frac{150p+1}{p} \cdot \frac{14.6p+1}{0.1p+1} \cdot \frac{12p+1}{0.1p+1} \quad (19)$$

Now get $t_P = 75$ s at $h = 9$ dB (0.6452), at the same time robust $\gamma = 0.5429$. Speed increased by reducing robust. Transient response is shown in Fig. 2.

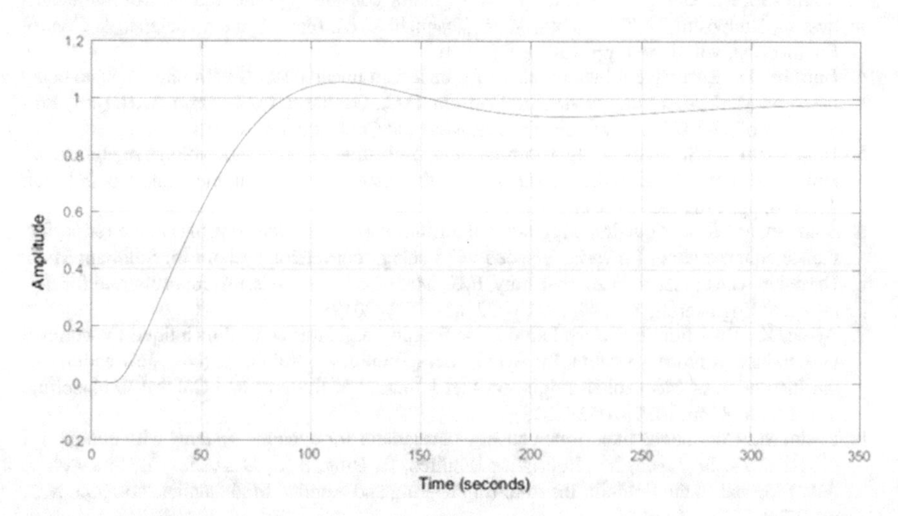

Fig. 2 Transient response nominal system

5 Conclusion

The selection of the regulator is carried out according to the specified indicators of robust (radius of the circle γ) and overshoot (5%), which in the considered class of regulators with 5% accuracy guarantees the maximum speed for a given coarseness γ. It is also desirable that the positive and negative amplitudes are approximately the same, as in Fig. 2. By changing the values of γ, it is possible to obtain the function of the dependence of the minimum regulation time on the value of γ.

A similar problem of regulator synthesis is implemented in the MATLAB package during the PID control of the control law, but the proposed method allows to obtain greater system robust to parametric uncertainty (for example, an object transfer coefficient) with a smaller value of control time.

References

1. Dean, S.: Regret bounds for robust adaptive control of the linear quadratic regulator. In: Dean, S., Mania, H., Manti, N., Recht, B., Tu, S. (eds.) Advances in Neural Information Processing Systems, vol. 31 (2018)
2. Escalante, F.: Robust Kalman filter and robust regulator for discrete-time markovian jump linear systems: control of series elastic actuators. In: Escalante, F.M., Jutinico, A.L., Jaimes, J.C., Siqueira, A.A.G., Terra, M.H. (eds.) IEEE Conference on Control Technology and Applications, pp. 976–981 (2018)
3. Rudposhti, M.: Design of robust optimal regulator considering state and control nonlinearities. In: Rudposhti, M.K., Nekoui, M.A., Teshnehlab, M. (eds.) Systems Science & Control Engineering, vol. 6, № 1, pp. 150–159 (2018)
4. Bortolin, D.: Recursive robust regulator for uncertain linear systems with random state delay based on Markovian jump model. In: Bortolin, D.C., Gagliardi, G.M., Terra, M.H. (eds) Proceedings of 57th IEEE Conference on Decision and Control (CDC) (2018)
5. Hekimoglu, B.: Sine-cosine algorithm-based optimization for automatic voltage regulator system. In: Hekimoglu, B. (ed.) Transactions of the Institute of Measurement and Control, vol. 41, № 6, pp. 1761–1771 (2019)
6. Soliman, M.: Linear-quadratic regulator algorithm-based cascaded control scheme for performance enhancement of a variable-speed wind energy conversion system. In: Soliman, M.A., Hasanien, H.M., Azazi, H.Z., El-Kholy, E.E., Mahmoud, S.A. (eds.) Arabian Journal for Science and Engineering, vol. 44, № 3, pp. 2281–2293 (2019)
7. Ayten, K.: Real-time implementation of self-tuning regulator control technique for coupled tank industrial process system. In: Ayten, K.K., Dumlu, A., Kaleli, A. (eds.) Proceedings of the Institution of Mechanical Engineers Part I-Journal of Systems and Control Engineering, vol. 232, № 8, pp. 1039–1052 (2018)
8. Basin, M.: Continuous fixed-time convergent regulator for dynamic systems with unbounded disturbances. In: Basin, M., Rodriguez-Ramirez, P., Ding, S.X., Daszenies, T., Shtessel, Y. (eds.) Journal of the Franklin Institute-Engineering and Applied Mathematics, vol. 355, № 5, pp. 2762–2778 (2018)
9. Pradhan, R.: Design of PID controller for automatic voltage regulator system using Ant Lion Optimizer. In: Pradhan, R., Majhi, S.K., Pati, B.B. (eds.) World Journal of Engineering, vol. 15, № 3, pp. 373–387 (2018)
10. Lian, K.: Robust fuzzy output regulator design for nonlinear systems without virtual desired variable calculation. In: Lian, K.Y., Liu, C.H., Chiu, C.S. (eds.) Control Engineering and Applied Informatics, vol. 19, № 1, pp. 27–36 (2017)

11. Ilyushin, Y.: Development of a technique for the synthesis of a pulsed regulator of a distributed control system. In: Ilyushin, Y.V., Novozhilov, I.M. (eds.) Proceedings of IEEE II International Conference on Control in Technical Systems (CTS), pp. 168–171 (2017)
12. Danik, Y.: The robustness of the stabilizing regulator for quasilinear discrete systems with state dependent coefficients. In: Danik, Y.E., Dmitriev, M.G. (eds.) Proceedings of International Siberian Conference on Control and Communications (SIBCON) (2016)
13. Singh, A.: An extended linear quadratic regulator and its application for control of power system dynamics. In: Singh, A.K., Pal, B.C. (eds.) Proceedings of IEEE First International Conference on Control, Measurement and Instrumentation (CMI), pp. 110–114 (2016)
14. Lian, K.: Simplified robust fuzzy output regulator design for discrete-time nonlinear systems. In: Lian, K.Y., Liu, C.H., Chiu, C.S. (eds.) Journal of Intelligent & Fuzzy Systems, vol. 31, № 3, pp. 1499–1511 (2016)
15. Victor, M.: Model-Reference Robust Tuning of PID Controllers. In: Victor, M., Vilanova, R. (eds.). Springer International Publishing (2016)
16. Papadopoulos, K.: PID Controller Tuning Using the Magnitude Optimum Criterion. Springer International Publishing, Heidelberg (2015)
17. Remizova, O., Syrokvashin, V.V., Fokin, A.L.: Sintez robastnyh sistem upravlenija s tipovymi reguljatorami (Synthesis of robust control systems with standard controllers). Izv. vuzov. Priborostroenie, vol. 58, №12, pp. 12–18 (2015) (in Russian)
18. O'Dwyer A.: A summary of PI and PID controller tuning rules for processes with time delay. Part 1: PI tuning rules. In: Preprints of Proceedings of PID '00: IFAC Workshop on Digital Control. Terrassa, Spain, pp. 175–180, April 2000
19. O'Dwyer, A.: Handbook of PI and PID controller tuning rules, 2nd edn. Imperial College Press, London (2006)
20. O'Dwyer, A.: Handbook of PI and PID controller tuning rules, 3rd edn. Imperial College Press, London (2009)
21. Gogol, I., Remizova, O., Syrokvashin, V., Fokin, A.: Sintez robastnyh reguljatorov dlja upravlenija tehnologicheskimi processami v klasse tradicionnyh zakonov regulirovanija (Synthesis of robust regulators to control technological processes in the class of traditional laws of regulation), Izv. SPbGTI(TU), № 44, pp. 98–105 (2018) (in Russian)

Features of Electromechanical Control of a Complex Power Plant with a Vortex-Type Wind-Conversion Device

V. A. Kostyukov, M. Yu. Medvedev, N. K. Poluyanovich and M. N. Dubygo

Abstract The possibilities of stabilizing the rotor speed of a vertical-axial wind power plant as, which can be included as an element in a complex power plant, which is a cyber-physical system (CPS), for additional and emergency power supply of surface robotic complexes are considered. For this purpose, it is proposed to use the method of stabilizing the angular velocity by controlling the position of the moving structural element of the installation in question. Received the corresponding law of regulation of the angular velocity of rotation of the rotor. The equations of the system for stabilizing the rotor rotation frequency with aperiodic wind disturbance are simulated. It is shown that the constructed regulator is able to effectively counter the influence of wind disturbances.

Keywords Vortex wind turbine · Cyber-physical system · Aerodynamic moment · Rotor · Variable geometry elements · Wind disturbances · Stabilization · Rotor speed

1 Introduction

Currently, there is a large group of offshore facilities, stationary and mobile, including robotic complexes, which need auxiliary autonomous energy sources. There are many international companies [1, 2].

V. A. Kostyukov (✉) · M. Yu. Medvedev · N. K. Poluyanovich · M. N. Dubygo
South Federal University, 105 Bolshaya Sadovaya, Rostov-on-Don 344006, Russia
e-mail: vakostukov@sfedu.ru

M. Yu. Medvedev
e-mail: medvmihal@sfedu.ru

N. K. Poluyanovich
e-mail: nik1-58@mail.ru

M. N. Dubygo
e-mail: w_m88@mail.ru

A. G. Kravets et al. (eds.), *Cyber-Physical Systems: Industry 4.0 Challenges*, Studies in Systems, Decision and Control 260,
https://doi.org/10.1007/978-3-030-32648-7_13

Many international companies are developing and introducing alternative energy sources on offshore surface platforms in order to reduce their overall fuel consumption.

One of the approaches, which allows economically perceptible (more than 10%) to reduce the consumption of conventional fuel, is the use of—an integrated power generating unit (CSES), which is a cyber-physical system. The CSEU consists of a wind power installation (wind turbine) and a solar-powered installation.

The key problems are: (1) the choice of the type of wind turbine, suitable for installation on a stationary/mobile platform as an element of the specified CSES; (2) the development of an electromechanical control system of the CSEU, which allows to optimize its operation by the criterion of the maximum power generated under severe constraints imposed by the requirements of safety and reliability of operation of all carrier systems [3–6].

This article discusses one of the aspects of the second of these two problems associated with the stabilization of the angular velocity of rotation of a wind turbine by adjusting the aerodynamic properties of this installation. Control in CPS is carried out with the help of variable elements of the geometry (IUE) of a wind turbine [7]. As an example, a promising CPS is considered: a vertical-axial wind turbine of a vortex type. This installation, its principle of operation and aerodynamic properties are discussed in the articles [7–9] and the patent [10]. The efficiency of using such a wind turbine as part of a CSEU for a small displacement boat was shown in [11].

2 Statement of the Problem on the Development of Wind WPP Control System

Consider the task of controlling the rotor speed of a vertical-axis vortex WPP (see Fig. 1a) containing the static, doubly connected part of the structure, the stator, axisymmetric with the rotor.

In the works [7–9, 10], the aerodynamic advantages of such an installation are shown, both in terms of generated power, and in terms of minimizing the noise level, as compared with analogs [12–17]. The design of this installation was obtained as

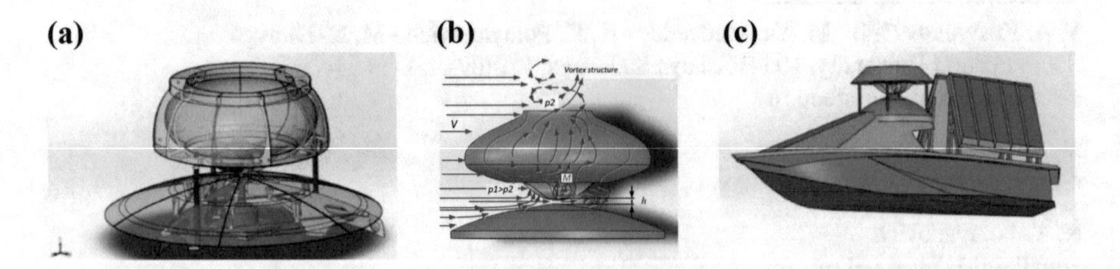

Fig. 1 Using the developed WPP for power supply of mobile objects: **a** 3d model of the vortex type used by WPP; **b** to the explanation of the control variable; **c** a simplified model of a small boat with WPP installed on it and solar panels

a result of aerodynamic optimization, which was carried out using the Ansys Fluent and Fine Hexa software packages.

The position of the lower movable part of the stator, characterized by the value of h, can vary vertically relative to the rotor (see Fig. 1b). Therefore, this part of the stator is a variable element of the WPP geometry [11]. The deviation of the position of this element leads to a change in the aerodynamic properties of the installation.

A possible application of the WPP under consideration as part of a complex power plant for a small-displacement boat is shown in Fig. 1c.

Let us consider the main technical characteristics of the developed IPP with the vortex type WPP proposed above for a boat of length $L = 10\,\text{m}$.

For air velocity relative to WPP V = 3 m/s, the average power of IPP is $P_{IPP} = 2\,\text{kW}$ [11], which is 14% of the specified power of the propulsion system $P_{PR} = 10\,\text{kW}$. Moreover, WPP accounts for the calculated power $P_{WPP} = 0.1\,\text{kW}$, and on solar panels (SP)—$P_{SP} = 1.3\,\text{kW}$, calculated on the basis that an average power of 190 W is removed from 1 m^2 SP, which corresponds to the efficiency of the SP $\eta_{SP} \approx 19\,\%$. At V = 10 m/s, the power values of WPP, IPP are respectively equal: $P_{WPP} = 3.7\,\text{kW}$, $P_{IPP} = 5\,\text{kW}$, which, taking into account P_{SP}, is already 56% of the required propulsion power.

WPP dimensions are 2.4 × 1.6 × 2.4 m, SP—2.2 × 2.5 × 2.9 m. The vertical size of the wind turbine can be reduced by using the existing structural elements of the boat as the lower part of the stator (Fig. 1c).

In contrast to the article [11], in the present work, the task is to stabilize the angular velocity of the rotor rotation not only taking into account the inertia of the actuation of the VEG actuator but also in accordance with a certain control error equation.

3 Synthesis of Wind Turbine Regulator

Let the equation of the controlled object is given—WPP, which is the equation of the rotational motion of its rotor with an angular velocity ω:

$$J\frac{d\omega}{dt} = M(V, \omega, h) + M_c(\omega), \tag{1}$$

where V—wind speed, the changes of which in this problem represent external disturbance; h—control value determining the position of the VEG; J—a reduced moment of inertia of the rotor; $M(V, \omega, h)$ и $M_c(\omega)$—useful aerodynamic moment on the rotor and moment of resistance determined by the following empirical dependencies [3]:

$$M(V, \omega, h) = V\Big[a_1 + a_2\Big(V - \tilde{V}\Big)\omega\Big] f_u(h), \tag{2}$$

$$M_c(\omega) = -b\omega, \tag{3}$$

$$f_u(h) = \begin{cases} a_3 h^{-1} + a_4, if h \in [h_1, \ h_2]; \\ h_1, \ \ if h < h_1; \\ h_2, \ \ if h > h_2, \end{cases} \quad (4)$$

where $a_1, \ a_2, \ a_3, \ a_4, \ b, \ \tilde{V}$—constant coefficients, $h_1, \ h_2$—control value boundary values h, defined by design features WPP.

The equation of the actuator is made as follows [18]:

$$q\frac{dh(t)}{dt} + rh(t) = f^*(t), \quad (5)$$

where q and r—some constant coefficients, the function of the right side is $f^*(t) = q\dot{h}^*(t) + rh^*(t)$. The solution of Eq. (5) in the general case is [19]:

$$h(t) = (1/q)e^{-(r/q)t}\left[\int\limits_0^t e^{(r/q)t} f^*(t)dt + qh_0\right]$$
$$= h^*(t) + \left[h_0 - h^*(0)\right]\exp[-(r/q)t], \quad (6)$$

where $h_0 = h(0)$—given an initial value of h.

In the synthesis of the desired controller, we will proceed from the following error equation [18]:

$$\frac{d^2\varepsilon}{dt^2} + A\frac{d\varepsilon}{dt} + B\varepsilon = 0, \quad (7)$$

where $\varepsilon = \omega_g - \omega$—regulation error; ω_g—target value of angular velocity ω; $A, \ B$—some constant coefficients that determine the nature of the transition process on the angular frequency, in particular, the decay time of this process and the degree of its overshoot.

Based on the equation of state (1) and the equation of error (7), the expression:

$$\frac{1}{J}\left(M'_V\dot{V} + \left(M'_\omega + M'_{c,\omega}\right)\dot{\omega} + M'_h\dot{h}\right) + \frac{A}{J}(M + M_c) + B\omega = f(t), \quad (8)$$

where $f(t) = \ddot{\omega}_g + A\dot{\omega}_g + B\omega_g$, $M'_V, \ M'_\omega, \ M'_h$—corresponding partial derivatives of the moment $M(V, \omega, h)$, $M'_{c,\omega} = dM_c/d\omega$.

Combining (1)–(8), we can come to the following expression for the control variable:

$$h(t) = \begin{cases} \tilde{h}(t) \equiv h^*(t) + [h_0 - h^*(0)]\exp[-rt/q], if\tilde{h}(t) \in [h_1; \ h_2], \\ h_1, if\tilde{h}(t) < h_1; \\ h_2, if\tilde{h}(t) > h_2. \end{cases} \quad (9)$$

Here the function h(t) is found as a result of solving a differential equation:

$$\frac{dh^*}{dt} = \frac{1}{M'_h}\{BJ\varepsilon - F(V, \dot{V}, \omega, h^*)\} \quad (10)$$

in conjunction with the WPP Eq. (1). The function $F(V, \dot{V}, \omega, h^*)$ and the partial derivative $M'_{h\ \mathrm{B}}$ (10) are determined by the expressions:

$$F(V, \dot{V}, \omega, h^*) = \left[a_1 + a_2\omega(2V - \tilde{V})\right]f_u(h^*)\dot{V} + \frac{1}{J}\left[a_2V(V - \tilde{V})f_u(h^*) + AJ - b\right](M + M_c), \quad (11)$$

$$M'_h = V\left[a_1 + a_2\left(V - \tilde{V}\right)\omega\right]\left(a_4 - a_3h^{-2}\right). \quad (12)$$

The full control law, taking into account both the regulator and the inertia of the actuation of the IEG actuator, is given by (1), (9), (10) with (2)–(4), (11) and (12).

The block diagram of the stabilization system of the rotor angular velocity is shown in Fig. 2. The driver determines the target law of variation of the angular velocity ω_g, taking into account the current and voltage on the load and the operator's target settings for the required power supply for this load. Further, on the basis of ω_g and the output w of the angular velocity meter w of the rotor (encoder), an error signal $\varepsilon = \omega_g - \omega' \approx \omega_g - \omega$ is generated, which is fed to the controller. The latter, in accordance with the expressions equalities (1)–(4), (10)–(12), forms the value $h^*(t)$, which arrives at the actuator, which changes the position of the VEG, compensating for changes in the wind speed.

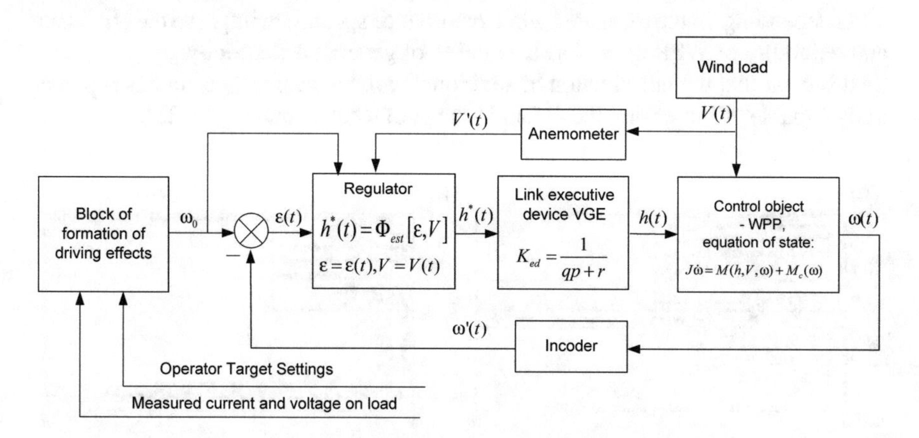

Fig. 2 Block diagram of the system of stabilization of the rotor speed of CPS

4 Modeling

In the study of the developed system, the following dependence of the perturbing wind effect was adopted:

$$V(t) = V^*(t) = V_0 + A_V \exp(\alpha_V t) \sin(\Omega_V t), \tag{13}$$

where A_V—amplitude oscillations of wind, α_V, Ω_V—constant parameters, $\alpha_V < 0$.

The simulation was carried out at three values of the initial amplitude of velocity oscillations: $A_V = 1; 2; 2.5; 3.36$ m/s. Limit values of the control value range: $h_1 = 0.01$ m, $h_2 = 0.65$ m. The target value of the rotor speed $\omega_g = 2\pi$ rad/s, допустимое transition completion time $T_{\max} = 6$ s; initial conditions $\omega_0 = 3\pi$ rad/s, $h_0 = 0.015$ m. Parameters in expression (13) of external perturbation: $\alpha_V = -0.1\,\text{s}^{-1}$, $\Omega_V = 2\pi$ rad/s; time constant of the actuator—$q/r = 0.2$ s.

The values of the constant parameters included in the expression of moments (aerodynamic and resistance) are as follows: $a_1 = 0.0716$, $a_2 = 1.704 \times 10^{-4}$, $a_3 = 0.202$, $a_4 = 0.692$, $b = 0.0019$.

In Fig. 3a shows the time dependences of the angular velocity $\omega(t)$, and Fig. 3b of the control value $h(t)$.

From these graphs, it can be seen that the duration of the transition process up to the amplitude of wind oscillations $A_V = 2.5$ m/s is permissible: $t_{\Pi\Pi} \approx 5\,\text{s} < T_{\max}$. Moreover, the degree of overshoot is not more than 20%. However, at $A_V = 3.36$ m/s the same time $t_{\Pi\Pi} \approx 10\,\text{c} > T_{\max}$ is already unacceptable.

Therefore, the proposed WPP rotor speed regulator, which changes the position of VEG depending on the measured wind disturbance speed, can improve the efficiency and reliability of WPP, as well as the quality of generated electricity.

It is clear that the introduction of additional variable geometry elements is potentially capable of increasing the dynamic range of torque control [20–22].

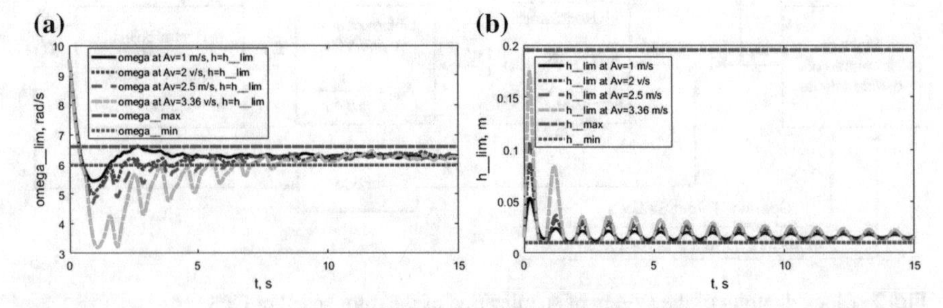

Fig. 3 Simulation results: **a** the dependence of the angular velocity $\omega(t)$; **b** dependence of the control value $h(t)$

Consequently, the mechanical regulation of the WPP rotor speed by controlling the variable elements of the geometry in combination with conventional methods of electrical stabilization of this frequency can improve the efficiency and reliability of WPP, as well as the quality of the generated electricity [23]. As a result, the efficiency of complex power plants, which use the considered WPP with variable geometry elements, is increased.

5 Conclusions

In this work, a cyber-physical system regulator is synthesized, which stabilizes the rotor speed of a vortex-type wind turbine with the use of variable elements of the wind turbine geometry.

This regulator is able to maintain the value of the angular velocity of the rotor with wind disturbances with an amplitude of the aperiodic component of no more than 2.5 m/s with a transient duration of no more than 6 s. This is achieved with a significant dynamic range of variation of the control value h.

The application of the proposed method of controlling the rotor speed of the WPP will significantly improve the adaptability of the control system of its output characteristics; significantly expand the dynamic range of torque control on the rotor of a wind power installation, as well as increase the robustness of these characteristics to external wind and internal structural parametric perturbations.

The use of a vertically axial vortex WPP with the aerodynamic control of the rotor speed discussed in this article as part of a complex power plant can significantly improve the efficiency of the latter.

The use of vertical-axial vortex wind turbines as part of cyber-physical systems with the considered aerodynamic control of the rotor speed can significantly improve the efficiency of these complexes.

Acknowledgements This work was supported by the RFBR grant No. 1-KV-1- 2018-18-08-00473 “Development and research of methods for optimizing and managing energy conversion processes in power plants of a complex type, including those that convert a continuous medium flow”.

References

1. Innovative Technologies & Solutions for Sustainable Shipping. Eco Marine Power. URL: ecomarinepower.com/en/about-us (date of access: 26.10.2018)
2. Ocius Technology Limited (Australia). URL: ocius.com.au (date of access: 26.10.2018)
3. Baniotopoulos, C., Borri, C., Stathopoulos, T.: Environmental Wind Engineering and Design of Wind Energy Structures (CISM International Centre for Mechanical Sciences). Springer, Heidelberg, 358 p (2011). ISBN: 3709109523
4. Pinto, T.: Electricity Markets with Growing Levels of Renewable Generation: Structure, Operation, Agent-Based Modeling and New Projects, 640p. Springer, Heidelberg (2018)

5. Gasch, R.: Wind Power Plants Fundamentals, Design, Construction and Operation, 2nd edn. Springer, Heidelberg, 567 p (2012). ISBN: 3642229379, 978-3-642-22937-4
6. Hau, E.; Wind Turbines—2013 Fundamentals, Technologies, Application, Economics, 3rd translated edn. Springer, Heidelberg, 879 p (2013). ISBN: 978-3-642-27151-9
7. Kostjukov, V., Maevskiy, A., Poluyanovich, N., Dubyago, M.: Adaptive mechatronic management system of wind-driven power-plant with variable geometry. In: 18th International Conference of Young Specialists on Micro/Nanotechnologies and Electron Devices (EDM), Erlagol, pp. 460–464 (2017)
8. Kostyukov, V., Medvedev, M., Mayevsky, A., Poluyanovich, N., Savchenko, V.: Investigation of a promising wind power plant with a rotor in a bell type of arrangement. Bull. Don State Tech. Univ. **1**(88), 85–91 (2017)
9. Kostyukov, V., Medvedev, M., Mayevsky, A., Poluyanovich, N., Savchenko, V.: Optimization of the geometry of the flare of a wind turbine of the "rotor in a bell" type. Bull. Don State Tech. Univ. **4**(91), 61–68 (2017)
10. Kostyukov, V.A., Medvedev, M.Yu., Mayevsky, A.M., Poluyanovich, N.K., Savchenko, V.V.: Patent for Utility Model "Device for Converting Kinetic Energy of Wind into Mechanical Energy Using the Lower Guide Structure". Dated 08/11/2016, No. 175397
11. Medvedev, M., Kostyukov, V., Mayevsky. A., Pavlenko, D.: Development of a Complex Power Plant for Surface Robotic Platforms. News of SFU. Technical Science. № 01, pp. 194–208 (2018)
12. Ying, P., Chen, Y., Xu, Y., Tian, Y.: Computational and experimental investigations of an omni-flow wind turbine. Appl. Energy **146**, 74–83 (2015)
13. Wróżyński, R., Sojka, M., Pyszny, K.: The application of GIS and 3D graphic software to visual impact assessment of wind turbines. Renew. Energy Part A **96**, 625–635 (2016)
14. Wanga, L., Liub, X., Koliosa, A.: Renewable and Sustainable Energy Reviews, vol. 64, pp. 195–210
15. Mikhnenkov, L.: Wind power plant of planetary type. Sci. Bull. MSTU, No. 125, pp. 22–24 (2008)
16. Liu, W.: Design and kinetic analysis of wind turbine blade-hub-tower coupled system. Renew. Energy **94**, 547–557 (2016)
17. Gorelov, D.: Energy Characteristics of the Rotor Daria (Review). Publishing House of the Siberian Branch of the Russian Academy of Sciences, pp. 325–333 (2010)
18. Pshihopov, VKh: Mathematical Models of Manipulation Robots: Textbook, p. 117. TTI SFU Publishing House, Taganrog (2008)
19. Matveyev, N.: Methods of Integration of Ordinary Differential Equations. M: Publishing house "High School", 555p (1967)
20. Pinson, P., Medsen, H.: The Integration of Renewable Energy Sources in the Electricity Markets. Springer, Heidelberg, 429 pp (2015)
21. Jiang, J., Tang, Ch., Ramakumar, R.: Control and Operation of Grid-Connected Wind Farms. Springer International Publishing, Heidelberg, 139p (2016). ISBN: 978-3-319-39135-9
22. Shabalina, O., Vorobkalov, P., Kataev, A., Kravets, A.: Educational computer games development: methodology, techniques, implementation. In: Proceedings of the 2013 International Conference on Advanced ICT and Education. Advances in Intelligent Systems Research, vol. 33, pp. 419–423 (2013)
23. Ushakov, V.: Electric Power Industry: Current State, Problems, Prospects. Springer, Heidelberg, 234p. (2014)

On Systemological Approach to Intelligent Decision-Making Support in Industrial Cyber-Physical Systems

Alexey V. Kizim and Alla G. Kravets

Abstract The presented study solves the actual scientific and technical problem of developing a new approach related to the creation of models, methods, and algorithms, as well as an application development platform for intelligent decision support in man-aging the process of maintenance, repair and upgrading to increase the efficiency of industrial equipment at all stages life cycle. Systematization of tasks, methods and means of maintenance and repair of Industrial Cyber-Physical Systems in the light of the equipment life cycle was build. Input effects and response systems, components, internal communications was identified. Systemological model of the process of maintenance and repair was formalized. To improve the efficiency of maintenance and repair, a method of continuous improvement of the process of maintenance and repair of the maintenance program has been developed. As a result of approbation, an increase in the efficiency and quality of equipment maintenance, a reduction in costs up to 15%, and an increase in the overall efficiency of the organization of maintenance and repair processes up to 20% were obtained, which is confirmed by the implementation certificates.

Keywords Maintenance and repair · Industrial cyber-physical systems · Intelligent decision-making support · Systemological approach · Continuous improvement · Equipment ontology · Multi-agent system

1 Introduction

The performance of any enterprise directly depends on the performance of its equipment. In the manufacturing sector, an important part of the process of organizing production is to ensure the operation of equipment for various purposes. Ensuring the operation of equipment means not only the performance of production operations

A. V. Kizim (✉) · A. G. Kravets
Volgograd State Technical University, Lenin av., 28, Volgograd 400005, Russian Federation
e-mail: kizim@mail.ru

A. G. Kravets
e-mail: agk@gde.ru

A. G. Kravets et al. (eds.), *Cyber-Physical Systems: Industry 4.0 Challenges*, Studies in Systems, Decision and Control 260,
https://doi.org/10.1007/978-3-030-32648-7_14

but also the maintenance and repair of equipment (MRE). Uninterrupted operation of the equipment is provided by the mechanical services of the enterprise or specialized contract service organizations. The efficiency of the equipment is ensured by the quality, timely and safe maintenance and repair of equipment with the rational use of resources [12].

Of particular importance to the topic of research is the fact that the Government of the Russian Federation approved the national program "Digital Economy of the Russian Federation" in 2018. The goal of the program is system development and introduction of digital technologies in all areas of life. The global concept of modern production "Industry 4.0" implies not only the widespread use of information technologies in production but also the creation of new generations of equipment combined into one digital ecosystem—Industrial Cyber-Physical Systems (ICPS). The basic components of management in the manufacturing industry are the modernization and digital transformation of production operations, ensuring the operation of equipment, decision-making processes [4].

At the same time, it is necessary to take into account if there is more than 50% of fully depreciated machinery and equipment in production, such production is recognized as degraded and is declared bankrupt. Unfortunately, a significant part of industrial production in Russia is in such a state [25]. On the other hand, there are examples of modernization and organization of new industries. At the same time, there are the tasks of organizing the effective operation of equipment from scratch, without the availability of statistical data and sometimes without the full support of the equipment manufacturer after the warranty period [2]. It is necessary to use effective systems for organizing the rational use of equipment. One of the ways to improve the competitiveness of an enterprise is to increase the efficiency of maintenance and repair of equipment or the so-called principle of a lean approach to production [19].

Under new conditions, many industrial enterprises are forced to independently solve the problem of ensuring the continuous operation of their equipment [18]. The efficiency of the equipment directly affects the production and financial performance of the enterprise and its economic condition. As part of the cost of organizing the production of equipment, a repair can be up to 30–40% [27]. Therefore, the solution to the problem of ensuring the organization of effective equipment operation is quite significant [20].

Thus, there was a problem situation, the essence of which lies in the fact that:

- methodological foundations of support for maintenance and repair have significant imperfections in the context of the current state and new challenges,
- there are no scientific and methodological approaches to ensuring the maximum efficiency of equipment throughout its entire life cycle,
- research and development on the organization of the maintenance and repair of industrial equipment are of a private nature;
- there is no uniform software-information and methodical platform for ensuring the effective operation of industrial equipment.

The presented study solves the actual scientific and technical problem of developing a new approach related to the creation of models, methods, and algorithms, as

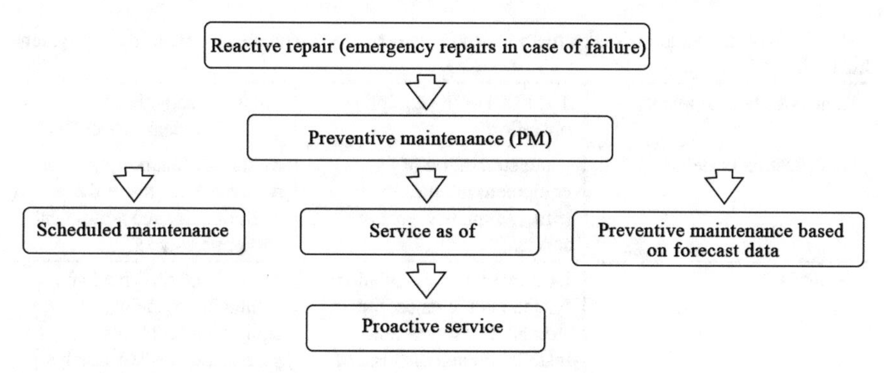

Fig. 1 Scheme of the evolution of MRE organization approaches

well as an application development platform for intelligent decision support in managing the process of maintenance, repair and upgrading to increase the efficiency of industrial equipment at all stages life cycle. The solution to this problem has scientific and practical value for building effective decision support systems for managing the state of technical objects in various subject areas.

The analysis of methods and approaches for the organization of maintenance and repair of equipment, the basic organization of maintenance and repair (planned, reactive and proactive based on the forecast and monitoring [3]). Briefly, the evolution of approaches can be described by the following scheme in Fig. 1. Improving the efficiency of the organization of MRE performance is possible through the use of modern advanced MRE methodologies, such as proactive maintenance [13].

2 Systematization of Tasks, Methods and Means of MRE

The analysis of the problem of organizing the effective functioning of ICPS was carried out; an analysis of the modern methods and management systems of the organization of technical maintenance and repairs of industrial equipment, the main problems in the area of scientific support for the organization of maintenance and repair work on ICPS was carried out [1, 14]. The problems, goals, tasks of the MRE organization are investigated. The basic concepts and definitions of the subject area are given, the concept of an MRE as an object is defined. The classes of ICPS are allocated: metal-cutting, woodworking, forge-pressing, foundry electric furnace, crushing and grinding, screening, electrical machines, compressor-refrigeration equipment, and pumps, ventilation, air conditioning, general pipelines, rolling stock of vehicles.

As a result of the analysis (Table 1), the requirements for the equipment operation support system are determined: to ensure reliable [6] operation of the equipment with the limitations imposed by the production program.

Possible events occurring with the equipment (installation, burn-in, normal operation), failure flows and reliability indicators are described.

Table 1 Systematization of tasks, methods and means of MRE with an account of the equipment life cycle

Stages of life equipment	The tasks of the equipment operation	Solution methods and methodological support tools
Identification of needs	Characterization of equipment maintenance (time, frequency, cost of service)	Methods of marketing research, analysis of the operation of such equipment and/or prototypes
Designing	Development of regulations for the maintenance, repair, storage, transportation, installation and disposal of equipment	Methods of analytical and simulation modeling of equipment reliability parameters, FMEA (analysis of causes of failures). Methods for determining the rules of maintenance and repair, storage, transportation, installation and disposal of equipment
Production	Monitoring compliance with TK	Monitoring compliance with manufacturing technology, equipment parameters
Testing and refinement	Determination of the reliability of the product Adjustment of the maintenance and repair regulations	Testing Statistical processing of results, the determination of the laws of failure Adaptation of MRE regulations
Storage and transportation/Purchase	Compliance with storage and transport requirements Selection of the necessary equipment for the purchase	Selection/adaptation of the structure of the repair service in the creation of production/ordinary purchase of equipment/modernization. PPR in the selection of equipment on the characteristics of the manufacturer (for effective maintenance and repair)
Commissioning	Control of compliance with the requirements of commissioning tests (installation, commissioning)	Planning for commissioning Commissioning tests

(continued)

Table 1 (continued)

Exploitation	(Tasks are described separately [24])	(Solution methods are described below)
Recycling	Investigation of the properties/characteristics of recycled equipment to adjust the maintenance and repair regulations Determination of the applicability of components and parts for further operation in other equipment	Fault detection Diagnostics of parameters of workable nodes for reuse Selection of analogs in the parameters

3 MRE Process Systemological Model

Let's describe the MRE system of ICPS, formalize the processes, and develop a systemological model for organizing the process of maintenance and repair of the ICPS. After that it's possible to build private models, formulate a set of tasks, and provide the concept of software and information support for the MRE of ICPS. Based on the study of the types of maintenance and repair, a characteristic of the maintenance and repair system is given: the composition of the maintenance and repair process, the maintenance and repair facilities, maintenance and repair facilities, subjects, including executors of maintenance and repair, and persons acting as its customers, the composition of the maintenance and repair documentation. The analysis of the needs for software and information support and the availability of specialized tools—MRE support tools (software, methodical, sources, standards).

The provisions of the application of a systematic approach to the organization software of MRE are formulated. The system analysis [11] of the maintenance and repair software system was made. Input effects and response systems, components, internal communications was identified. Systemological model of the process of MRE was formalized. Graphical interpretation of the model is shown in Fig. 2.

The systemological model of the MRE process in general:

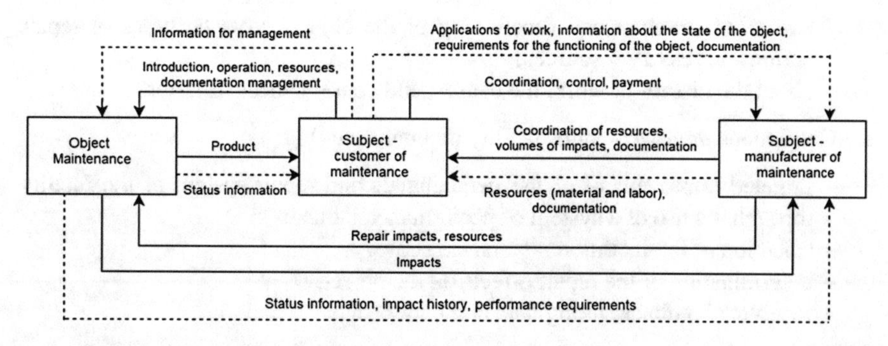

Fig. 2 Graphic interpretation of the system MRE process model

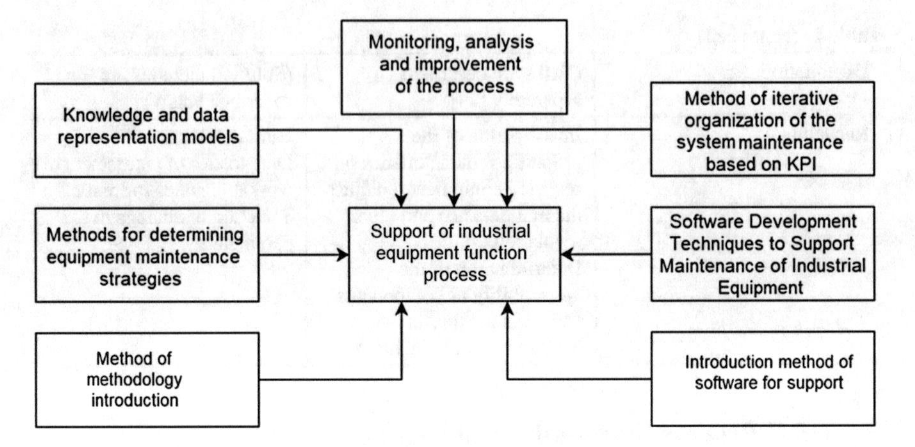

Fig. 3 Diagram of the concept of the decision making support to ensure the performance of ICPS

$$Pr = \{E,\ O,\ I,\ S,\ N\},$$

where E is the structural elements of the process of maintenance and repair (stages, works, operations, repair effects),

O is the maintenance and repair objects (industrial equipment),
I are the subjects of maintenance and repair,
S are the maintenance and repair, incl. means, machines, and tools,
N are methodical materials [forms of documents, standards, instructions, standards, regulations, guidelines, software and methodical software (methods and algorithms, software)].

The constructed formal models made it possible to isolate the composition of the structural elements of the MRE process. Based on this, the concept of intelligent support for the MRE of ICPS was proposed (Fig. 3) [14]:

1. Formalization of data and knowledge about the specific process of operation and maintenance and repair of a specific object:
   - about the structure and functioning of the object, subjects, types of repair actions, necessary resources;
   - about the history of work, the causes, and consequences of failures.
2. Continuous improvement of facility maintenance (Fig. 4):
   - targeted improvement of the maintenance and repair system of the facility through the use of a system of performance indicators;
   - selection of the facility maintenance strategy;
   - determination of the repair effects on the object;
   - receiving feedback, fixing data and knowledge.
3. Support for building a new equipment maintenance system.

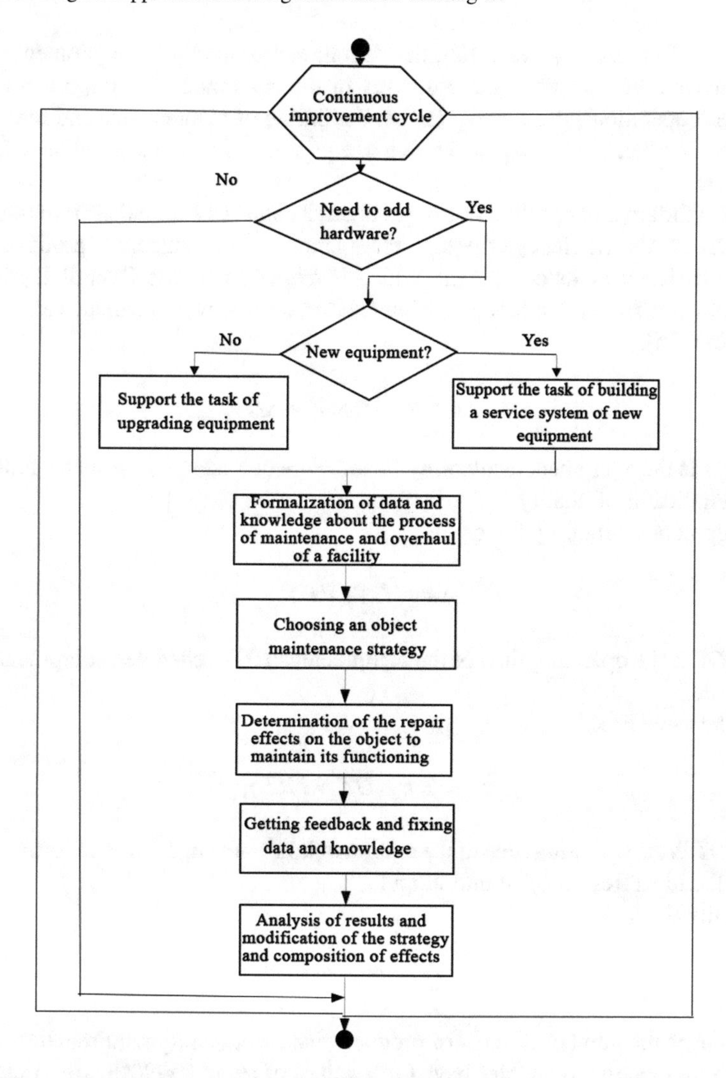

Fig. 4 The generalized algorithm of the continuous improvement

4. Support for equipment upgrades.

4 MRE Differentiation

For more efficient MRE management, it is necessary to apply different MRE organization approaches to different groups of equipment. For the segmentation of the equipment stock into groups, in accordance with the approach applied to its MRE, a number of criteria for evaluating the operational properties of equipment can be distinguished. For example, the criticality of the equipment (the importance of its continued operation), the cost of repairs, the timing of maintenance and repair, etc. At the same time, use the expert ranking of equipment by groups based on a number of criteria.

One of the areas of optimization of productivity and costs for industrial enterprises is to increase the efficiency of equipment operation. The company's profit depends on the efficiency of its equipment, which is described by the Overall Equipment Efficiency (OEE), the product of equipment availability by performance and quality indicators [15].

$$OEE = Av \times P \times Q,$$

where Av is the equipment availability factor; P—equipment performance indicator; Q is an indicator of quality.

Equipment availability factor

$$Av = OT/PPT,$$

where OT is the operating time of the equipment; PPT—scheduled equipment operation time.

At the same time,

$$P = TP/(OT * IRT),$$

where OT is the operating time of the equipment; TP—the number of units produced; IRT is the ideal frequency of unit output.

Quality Score

$$Q = GP/TP,$$

where GP is the number of units of products that meet quality requirements.

Leading modern companies have OEE values of about 85–90%. To estimate the return on investment, ROI is used—the coefficient of return on investment, the rate of return [3, 5].

5 MRE Decision-Making Support

One of the main objectives of this study is to support decision-making on the choice of strategies for maintenance and repair of equipment. The need for this arises in situations of setting up a new equipment position, key changes in the state of equipment, and its modernization.

5.1 The Decision-Making Problem Statement

From a formal point of view, the decision-making task can be characterized by the following tuple [17]:

$$\langle A, E, S; T\rangle,$$

where A is the set of alternatives (decision options), E is the environment of the decision-making task, S is the system of preferences of the person making the decision (DM). It is required to perform some action T over the set of alternatives A, for example, to find the most preferred alternative, to linearly order the set of permissible alternatives. Moreover, the decision-making environment is

$$E = \langle C_R, C_o, P\rangle,$$

where C_R is the cost of repairs, C_o is the cost of maintenance, L_P is the production loss. Then the objective function can be written in the form:

$$\begin{cases} f(C_R, C_O) \rightarrow \min \\ f(L_P) \rightarrow \min \\ OEE \rightarrow \min \\ ROI \rightarrow \min \end{cases}$$

When using a systemological approach to the task of increasing the efficiency of maintenance and repair, it is necessary to organize a continuous process for evaluating the effectiveness of the maintenance and repair process and adaptive introduction of changes to the maintenance and repair process in order to obtain the optimal value for the performance of the repair service and the repair and repair equipment. The task of monitoring and controlling [8–10] a MRE system is described with the help of a control system for targeted improvement of equipment operation. For the purposeful improvement of the enterprise's MRE system, it is recommended that the management of the MRE efficiency with the Balanced Key Performance Indicators (KPI) system is recommended.

5.2 KPI—Based Model

To improve the efficiency of maintenance and repair, a method of continuous improvement of the process of maintenance and repair of the maintenance program has been developed, which is characterized by the implementation of balanced scorecards and key performance indicators (Fig. 5). The main stages of the method:

1. Formation of a system of indicators MRE:

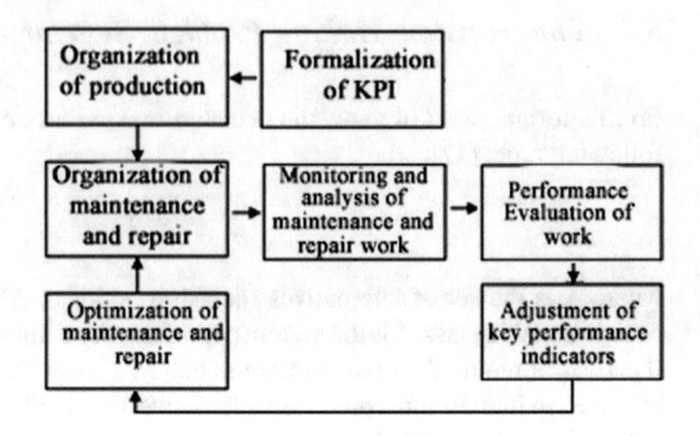

Fig. 5 KPI-based MRE system organization diagram

- Formalization of objectives, development of a strategy for applying the KPI MRE system and making decisions on the application of KPIs for the list of subjects and equipment. In the case of the presence of a complex of KPIs of a super-system (enterprise), it is necessary to arrange docking with them.
- Development of a KPI balanced scorecard system (the composition of indicators, calculation methods, allowable intervals, target levels of indicator values and methods for their interpretation, assessment periods) and identification of a subset of indicators that should be focused on first.
- Decomposition of the KPI system to units of organizational structure and equipment.

2. Application and modernization of KPI:

   - The direct operation of equipment and maintenance of MRE.
   - Periodic collection of source data and calculation of KPI values.
   - Continuous monitoring of KPI deviations.
   - Targeted KPI improvement. Modernization of the composition, methods of calculation, intervals of acceptable values and target levels of indicator values.
   - Improving the system of maintenance and repair of equipment.

The basic characteristics of KPIs are defined:

$$KPI_i = \{O, K, E, B, G, V, I, T, C\},$$

where

$O = \{MRE\ object,\ MRE\ subject,\ and\ MRE\ process\}$,
K is an indicator type (qualitative, quantitative),
E is a unit of measurement (hours, pieces, points, currencies, interest, etc.),
B is a trend (maximization *, minimization, constancy),
G—boundary values,
V—a method of calculation,
I—a way of interpreting values,
T—periods of calculation,

C—links with goals.

More than 100 industrial equipment KPIs have been identified according with the KPI-based model.

5.3 A Formal Statement of the MRE Strategy Selection Problem

Let at discrete instants of time there is a vector random process $Z(t) = \overline{X(t)} + \overline{N(t)}$ with known statistical characteristics of the components $\overline{X(t)}$ and $\overline{N(t)}$, where $\overline{N(t)}$ is the state vector (characterized by, for example, a set of output controlled parameters) of the equipment; $\overline{N(t)}$ is the additive vector of measurement errors (Gaussian random process). Denote by $\overline{R(t)}$ a vector nonrandom function, having the same dimension as $Z(t)$, and being discretely observed realizations (trajectories) of the vector $\overline{Z(t)}$. On the observed trajectories of the random process $\overline{Z(t)}$, we define some loss functional $\Phi\{R(t)\}$, which are defined as process control losses $\overline{Z(t)}$ to keep it in a given domain $D(t)$. Physically, the functional $\Phi\{R(t)\}$, means the loss of time or cost to maintain a complex system during its operation, and the area $D(t)$ means the operational tolerances for all monitored system parameters (as a rule, $D(t) = D$). Denote by $\delta t_i(Zt_i)$, i = 0, 1, 2, ..., the solution for controlling the process $Z(t_i)$ at the time ti. Let us call the solution $\delta m\ (Z_T) = \{\delta_1\ (Z_1), \delta_2\ (Z_2), \ldots, \delta_k\ (Z_k = ZT)\}$ the strategy of service management in the period (0, *T*), where T is the period of operation complex [7] system. Physically, the strategy $\delta_T(Z_T)$ means all the impacts that a complex system undergoes during operation by the operating personnel (adjustment, replacement of system blocks, etc.). The task is to choose the MRE strategy for the software $\delta_T(Z_T) = \delta{*}T(Z_T)$, which minimizes the expectation of the loss functional during the operation period (*O*, *T*), i.e. providing $min(\delta{*}T\ M[\Phi\{R(t)\}])$.

5.4 Cognitive Modeling

At the initial stages of the intellectual support of maintenance and repair of the software using cognitive modeling [26] (Fig. 6), it is possible to lose various scenarios for changing the system of organization of maintenance and repair in the enterprise when individual factors change.

Cognitive modeling is used for: primary acquisition of the model and when individual factors change; studies of various scenarios for changing the maintenance and repair system; assess the situation and analyze the influence of existing factors. Various scenarios for changing the MRE organization system are explored.

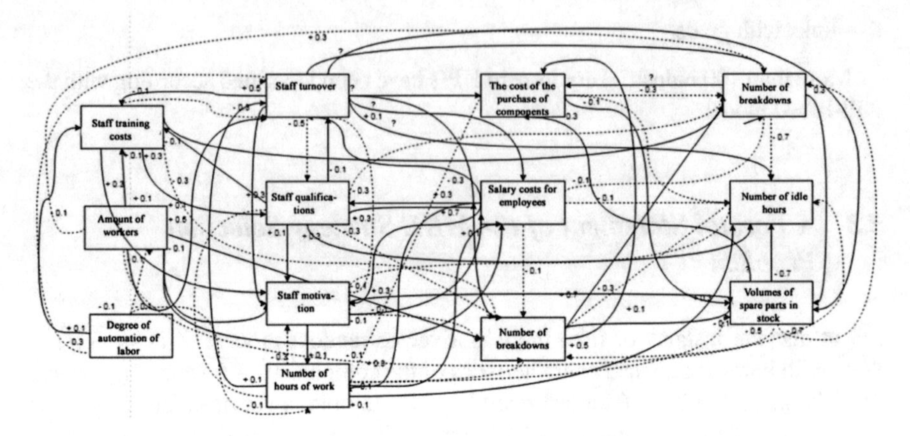

Fig. 6 The cognitive model of maintenance and repair of the software (fragment)

5.5 *ICPS MRE Meta-Ontology*

To implement the intellectual core of decision-making support, ontological models of data and knowledge about the process of maintenance and repair of ICPS are proposed at various levels of description and stages of the life cycle. A meta-ontology of ICPS MRE has been compiled (Fig. 7) [14].

Meta-ontology contains the following ontologies [16]: equipment classification, enterprise organizational structure, equipment production structure, typing (classifying types) of maintenance and repair, maintenance and repair process structure, maintenance and repair stage tasks, maintenance and repair information on the ICPS,

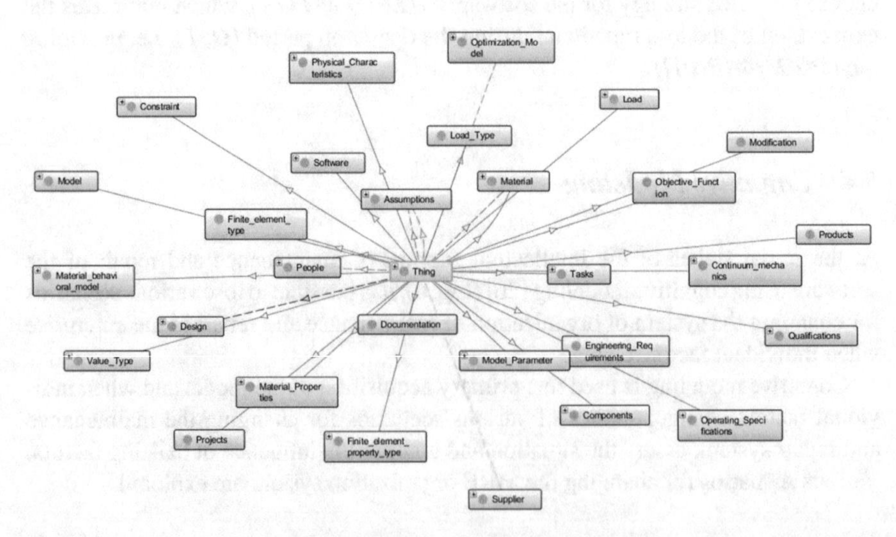

Fig. 7 Structure and classes of meta-ontology of ICPS MRE (fragment)

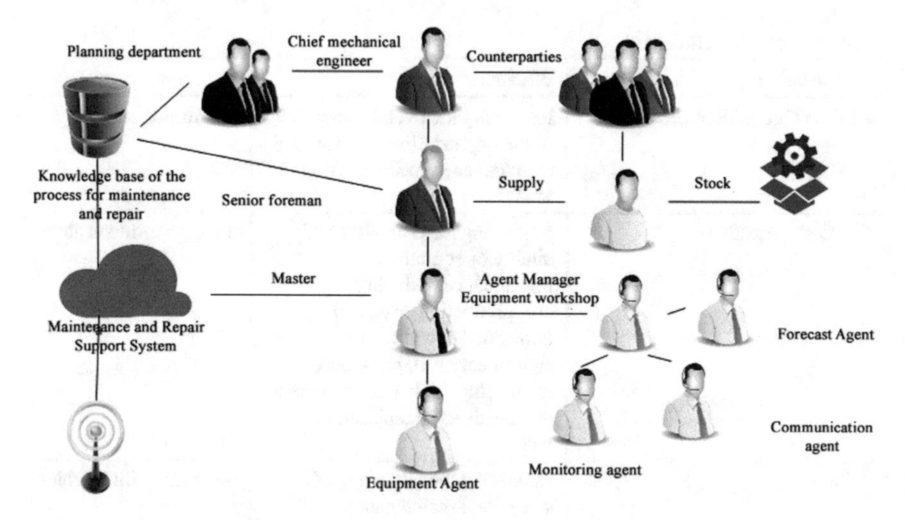

Fig. 8 The architecture of the multi-agent MRE support system

the staffing structure of the organization, information on contractors, maintenance and repair documentation, technological resources and interaction between them, performance indicators for maintenance and repair of the ICPS, reference books and classifiers.

5.6 MRE Multi-agent System

To solve the problems of organizing interaction, simplifying the process of managing objects of a similar structure, agents are used that are combined into a multi-agent system (MAS, Fig. 8) [11]. Agents have characteristics that make them indispensable in MRE tasks [22]. The ability to correctly respond to dynamically changing conditions makes multi-agent systems flexible [23] for their use in the maintenance of the software. Issues of corporate network security are considered in [21].

6 Software Implementation and Results

The following software tools have been developed on the platform: system for supporting the preparation of work plans, system for automating planning based on preventive work, accounting system for completed work, system for supporting the development and implementation/maintenance.

Table 2 Achieved efficiency

Organization	Efficiency	Impact per year
LLC «Volgogradservice»	Increasing the overall efficiency of the organization of repair and maintenance processes by 15–20%	3.4 million rubles
OJSC «Barricady»	Improving the efficiency and quality of equipment maintenance, reducing non-production costs and unplanned downtime of equipment by 10–15% and identifying the list of equipment that needs to be replaced or upgraded	up to 15 million rubles
CCB «Titan»	Improving the efficiency of equipment maintenance, optimizing its use, obtaining information on the need to upgrade equipment, reducing maintenance and repair costs by 8–12%, minimizing equipment downtime	up to 20 million rubles
Research and production company LLC "Magni"	Increasing the overall efficiency of the organization of repair and maintenance processes by 17–19%	up to 1.9 million rubles

Table 2 provides information on the effectiveness obtained in accordance with the acts of implementation. The results obtained make it possible to estimate the increase in the efficiency of the MRE organization p to 19–20%.

Approbation of the approach to automation at the service repair enterprise, industrial enterprises, for training purposes is described. The elements of the developed methodology and software systems were applied in the form of methods, techniques, technologies and software and information tools of the maintenance and repair organization and its software and information support in organizations.

7 Conclusions

The main result of the work is the scientific and methodological foundations of intelligent decision-making support, which increase the efficiency of the process of maintenance and repair of ICPS, which leads to the conclusion that this goal has been achieved. The results of the study are as follows.

A systematization of the stages, tasks, methods, and means of maintenance and repair of MRE has been carried out.

A set of models and methods of software and information support for making management decisions to ensure the performance of industrial equipment has been developed, implementing an integrated approach to the organization of its maintenance, repair and upgrading system.

A systemological model of the MRE process was created.

The concept of software and information support for proactive maintenance and repair of industrial cyber-physical systems that implements a continuous improvement approach for the maintenance and repair process is proposed, for which the following has been developed:

- method of continuous improvement of the process of maintenance and repair of software;
- method of decision making support of choosing a maintenance and repair strategy with the algorithms for determining repair effects.

A meta-ontological model of MRE of industrial equipment has been developed, which includes the classification of industrial equipment, formalized knowledge of the MRE process and the methods of supporting MRE at each stage of the life cycle.

A model of an intelligent multi-agent system has been developed for the support and maintenance of industrial equipment.

As a result of approbation, an increase in the efficiency and quality of equipment maintenance, a reduction in costs up to 15%, and an increase in the overall efficiency of the organization of repair and maintenance processes up to 20% were obtained, which is confirmed by the implementation certificates.

Acknowledgements This research was supported by the Russian Fund of Basic Research (grant No. 19-07-01200).

References

1. Aliseichik, A.P., Pavlovsky, V.E.: The model and dynamic estimates for the controllability and comfortability of a multiwheel mobile robot motion. Autom. Remote Control **76**(4), 675–688 (2015)
2. Arlazarov, V.L., Slavin, O.A., Shustov, A.V.: Evaluation of value of information and technical complex of complex system. Proc. Inst. Syst. Anal. Russ. Acad. Sci. **29**, 152–182 (2007)
3. Azarov, V.N., Kaperko, A.F.: General topics of metrology and measurement technology-analysis of the state, development trends, and new developments of transducers and information converters for measurement, monitoring. Meas. Tech. **41**(1), 2–9 (1998)
4. Bogomolov, A. V., et al.: Information-logical modeling of information collection and processing at the evaluation of the functional reliability of the aviation ergate control system operator. In: 2018 Third International Conference on Human Factors in Complex Technical Systems and Environments (ERGO) s and Environments (ERGO), IEEE, pp. 106–110 (2018)
5. Bosenko, V.N., Kravets, A.G., Kamaev, V.A.: Development of an automated system to improve the efficiency of the oil pipeline construction management. World Appl. Sci. J. **24**(24), 24–30 (2013)

6. Bukin, A.G., Lychagov, A.S., Sadekov, R.N., Slavin, O.A.: A computer vision system for navigation of ground vehicles: hardware and software. Gyroscopy Navig. **7**(1), 66–71 (2016)
7. Chernyshev, S.L., Lyapunov, S.V., Wolkov, A.V.: Modern problems of aircraft aerodynamics. Adv. Aerodyn. **1**(1), 7p (2019)
8. Chistyakova, T.B., et al.: Decision support system for optimal production planning polymeric materials using genetic algorithms. In: 2016 XIX IEEE International Conference on Soft Computing and Measurements (SCM), IEEE, pp. 257–259 (2016)
9. Galyaev, A.A., Miller, B.M., Rubinovich, E.Y.: Optimal Impulsive Control of Dynamical System in an Impact Phase. Analysis and Simulation of Contact Problems, pp. 385–386. Springer, Heidelberg (2006)
10. Karpenko, A.P., Leshchev, I.A.: Nature-Inspired Algorithms for Global Optimization in Group Robotics Problems. Smart Electromechanical Systems, pp. 91–106. Springer, Cham (2019)
11. Kizim, A.V., et al.: Developing a model of multi-agent system of a process of a tech inspection and equipment repair. In: Joint Conference on Knowledge-Based Software Engineering, pp. 457–465. Springer, Cham (2014)
12. Kizim, A.V., et al.: Development of the intelligent platform of technical systems modernization at different stages of the life cycle. Proc. Comput. Sci. **121**, 913–919 (2017)
13. Kizim, A.V., et al.: Predictive modeling as a basis for monitoring, diagnosis, forecasting and upgrading of a technical system. In: 2017 IEEE 11th International Conference on Application of Information and Communication Technologies (AICT), IEEE, pp. 1–5 (2017)
14. Kizim, A.V., et al.: Cretion and use of ontology of subject domain 'electrical engineering'. In: 2015 9th International Conference on Application of Information and Communication Technologies (AICT), IEEE, pp. 25–29 (2015)
15. Kizim, A.V.: The developing of the maintenance and repair body of knowledge to increasing equipment maintenance and repair organization efficiency. Inf. Resour. Manage. J. (IRMJ) **29**(4), 49–64 (2016)
16. Kizim, A., Matokhina, A., Nesterov, B.: Development of ontological knowledge representation model of industrial equipment. In: Proceedings of the Creativity in Intelligent Technologies and Data Science, CIT&DS 2015, Volgograd, Russia, 15–17 Sept 2015, vol. 535, p. 354. Springer, Heidelberg (2015)
17. Kozlov, V.N.: The system analysis, optimization and decision-making: study guide. Prospect, Moscow (2010)
18. Kravets, A., Kozunova, S.: The risk management model of design department's PDM information system. Commun. Comput. Inf. Sci. **754**, 490–500 (2017)
19. Kravets, A., Shumeiko, N., Lempert, B., Salnikova, N., Shcherbakova, N.: "Smart Queue" approach for new technical solutions discovery in patent applications. Commun. Comput. Inf. Sci. **754**, 37–47 (2017)
20. Kravets, A.G., Belov, A.G., Sadovnikova, N.P.: Models and methods of professional competence level research. Recent Patents Comput. Sci. **9**(2), 150–159 (2016)
21. Kravets, A.G., Bui, N.D., Al-Ashval, M.: Mobile security solution for enterprise network. Commun. Comput. Inf. Sci. **466** CCIS, 371–382 (2014)
22. Kravets, A.G., Fomenkov, S.A., Kravets, A.D.: Component-based approach to multi-agent system generation. Commun. Comput. Inf. Sci. **466** CCIS, 483–490 (2014)
23. Kravets, A.G., Kravets, A.D., Korotkov, A.A.: Intelligent multi-agent systems generation. World Appl. Sci. J. **24**(24), 98–104 (2013)
24. Matokhina, A.V., Kizim, A.V., Nikitin, N.A.: Technical system modernization during the operation stage. In: Conference on Creativity in Intelligent Technologies and Data Science, pp. 350–360. Springer, Cham (2017)
25. Moshev, E.R., Meshalkin, V.P.: Computer-based logistics support system for the maintenance of chemical plant equipment. Theoret. Found. Chem. Eng. **48**(6), 855–863 (2014)

26. Parygin, D., Sadovnikova, N., Kravets, A., Gnedkova, E.: Cognitive and ontological modeling for decision support in the tasks of the urban transportation system development management. IISA 2015—6th International Conference on Information, Intelligence, Systems and Applications, Art. no. 7388073 (2016)
27. Shcherbakov, M., Groumpos, P.P., Kravets, A.: A method and IR4I index indicating the readiness of business processes for data science solutions. Commun. Comput. Inf. Sci. **754**, 21–34 (2017)

Industrial Cybersecurity

Cybersecurity Risks Analyses at Remote Monitoring of Object's State

T. I. Buldakova

Abstract The problem of data protection at remote monitoring systems of the complex object state is considered. Such systems are robust real-time systems with high requirements for performance and reliability. Currently, they belong to the class of cyber-physical systems. The features of remote monitoring systems are noted. Solutions developed during the monitoring process are based on the received data and the analysis results. Therefore, in the remote monitoring systems it is necessary to ensure the noise immunity of information processes. It is shown that the degree of data security is assessed by the level of cybersecurity risk. Therefore, when remotely monitoring the state of a complex object, it is necessary to continuously assess the level of cybersecurity and take protective measures when this level exceeds a predetermined threshold value. The main stages of information risk management during remote monitoring are given. Requirements to the choice of the most effective methods for realization of this process are formulated. A methodology for managing cybersecurity risks in remote monitoring systems, which are assessed of the object's state, and the results of its testing, using the example of a remote monitoring system for a human state, are proposed.

Keywords Monitoring systems · Information security · Cybersecurity risks · Risk management

1 Introduction

The development of information and communication technologies has led to the emergence of the remote monitoring system of the object's state (RMSOS). Such systems allow automated control of technological processes, as well as remotely

T. I. Buldakova (✉)
Bauman Moscow State Technical University, 2-nd Baumanskaya st., 5, Moscow 105005, Russia
e-mail: buldakova@bmstu.ru

A. G. Kravets et al. (eds.), *Cyber-Physical Systems: Industry 4.0 Challenges*, Studies in Systems, Decision and Control 260,
https://doi.org/10.1007/978-3-030-32648-7_15

assess the state of complex objects (technical devices and complexes, human operators, etc.). They are used in many application areas that include critical national infrastructures, such as heat and power, transport, health care, and defense.

RMSOS are real-time robust systems with high requirements on performance and reliability. Currently, such systems are classified as cyber-physical systems [1, 2].

The implementation of RMSOS allows quickly to obtain timely information about the working ability and security of complex objects, to control technological processes, to predict industrial safety risks, time to prepare and make the necessary decisions [3, 4].

Since the decisions made during the monitoring process are based on the obtained data and the results of their analysis, it is necessary to ensure noise immunity of information processes in the RMSOS. Even minor, at first glance, violations of any of the information processes can lead to serious consequences—loss of confidentiality, integrity and/or availability of information in RMSOS, serious damage or compromise of the organization. As a result, ineffective or even wrong decisions can be made.

The foregoing necessitates the continuous monitoring of the level of cybersecurity risks (the potential for data distortion in remote monitoring) and the development of protective countermeasures to reduce this level when a given threshold value is exceeded [5–7]. This chapter is devoted to the actual problem of protecting information. This problem is associated with managing cybersecurity risks in RMSOS, the level of which characterizes the degree of data security.

2 Features of Remote Monitoring Systems

At present, remote monitoring systems of the object's state are widely used in many areas [8–11]. They allow automated control of various processes in power system, telemedicine, oil and gas industry, geology, agriculture, etc., as well as remotely assess the state of complex objects (technical devices and equipment, human operator, biological populations, etc.). Such systems, like many remote monitoring systems, implement the processes of remote collection, transmission, processing, storage of data on the characteristics of the observation object and making decisions about its state [12].

Depending on the features of the object of observation and the monitoring system itself, the data transmission channel can be implemented in different ways: via a dedicated DSL channel, a wired channel, a radio channel, a GSM channel, the Internet, and Ethernet. However, regardless of the data transmission channel used, information flows in RMSOS can be distorted due to unauthorized impacts on information, including targeted ones [13].

Wrong decisions can be made on the basis of corrupted data. Therefore, the organization of remote monitoring should take into account the increased requirements for noise immunity of the information processes occurring in the system to ensure reliable operation and development of a justified solution [14–17].

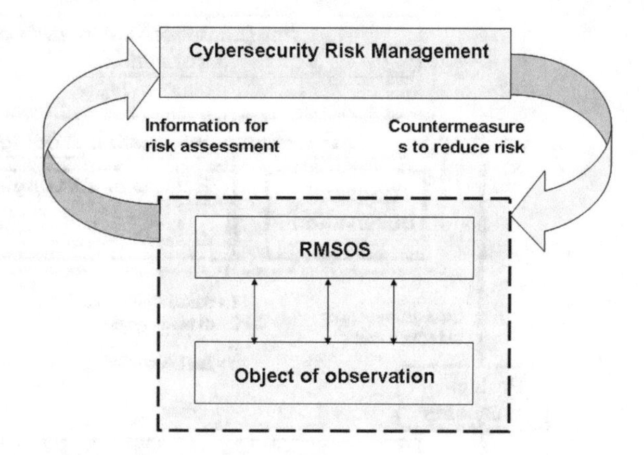

Fig. 1 Diagram of the cybersecurity risk management process

Thus, remote monitoring systems, like many cyber-physical systems, working with important information of limited access, are subject to both random and deliberate influences from the external environment, which creates risks of distortion, disclosure or destruction of information that can lead to serious consequences.

At the same time, the risk R is understood as a complex value, defined as a function (or functional) of a number of factors: $R = f(X_1, X_2, X_3)$, where X_1—information security threats; X_2—potential damage; X_3—automated system vulnerabilities. Effective cybersecurity risk management (based on an assessment of risk level) will improve the reliability of RMSOS operation (Fig. 1).

Consequently, the process of managing cybersecurity risks in RMSOS requires continuous monitoring of their level and the adoption of protective countermeasures to reduce risks if the resulting estimates exceed the threshold value. This is the problem of managing cybersecurity risks, the solution of which is required to ensure the security of information flows.

However, this risk depends not only on the operating conditions of RMSOS, but also on possible factors of information distortion that may be unknown. Therefore, development of a methodology for managing cybersecurity risks is required.

3 Cybersecurity Risk Management Stages in RMSOS

The scheme of interaction between RMSOS and the external environment is presented in Fig. 2. The effects of the external environment (unintended and focused) have an impact on both the object of observation and the monitoring system itself.

As a result, this can lead to a distortion of information flows associated with the remote collection of information about the characteristics of the object of observation, its transmission over communication channels, processing and storage of the received data. Thus, the data on the basis of which the decision on the state of the object is

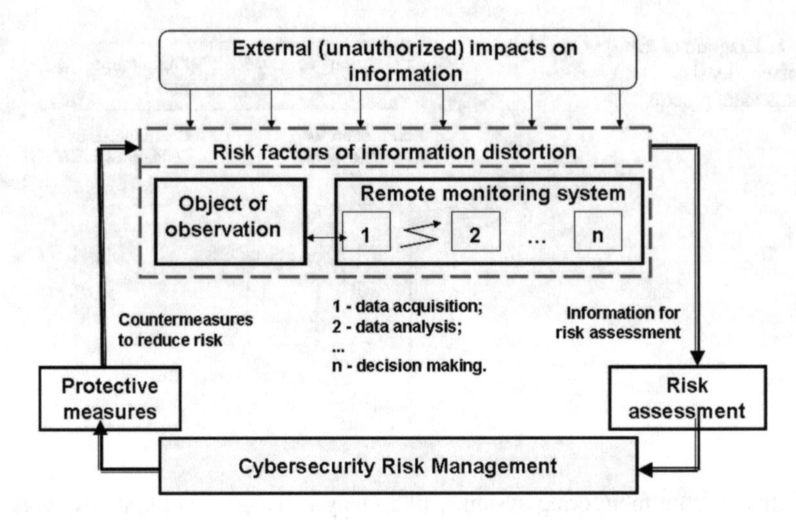

Fig. 2 The interaction of SMSOS with the external environment

made may be distorted. Therefore, it is necessary to timely detect and analyze the risks of information distortion.

The challenges of managing cybersecurity risks are directly related to difficulties and errors in assessing risk factors. Therefore, it is necessary to determine as fully as possible all the factors affecting the risk level.

Since the risk factors for information distortion may be unknown, their detection and identification are required. One of the approaches to the identification of risk factors is proposed in [5].

The analysis of existing methods and means of information risk management made it possible to formulate the main problems and identify the stages of information risk management [5, 7, 18–20]. At the same time, the main stages include: identification of risk factors; risk factor assessment; cybersecurity risk assessment; assessment of the ratio of protective measures and possible damage; implementation of the risk management process (Fig. 3). These stages were taken into account at development of risk management methodology.

Studies have shown that at each stage of the cybersecurity risk management process, their own methods and tools can be applied. Therefore, priority should be given not to the effectiveness of the methods as a whole, but to their effectiveness at a particular stage of the process, to the possibilities of their combinations, and also to the means of transition from one method to another, ensuring correct interpretation of results.

Thus, in the management of cybersecurity risks it is necessary to apply methods that satisfy the following performance indicators:

- the greatest consistency and adequacy of risk factor assessments;
- maximum adaptability to quality data;

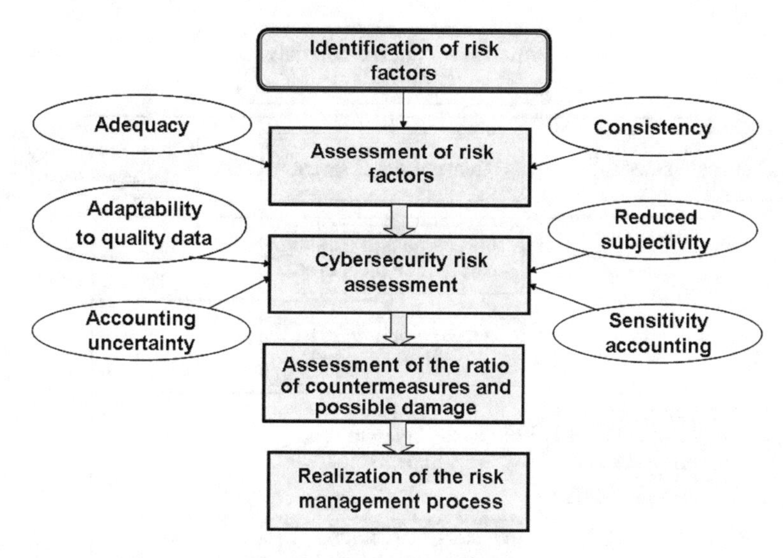

Fig. 3 Cybersecurity risk management stages

- minimal subjectivity and uncertainty of risk assessment;
- consideration of unequal risk sensitivity to various factors.

In fact, the cybersecurity risk management process involves the solution of two tasks: assessing existing risks and deciding how to reduce them when a certain threshold level is exceeded (Fig. 4).

The developed methodology that implements this process is based on the use of various methods at each stage of cybersecurity risk management.

4 Choosing Methods to Manage Cybersecurity Risks

The cybersecurity risk analysis, despite its specific nuances in various fields of activity, is an ordered algorithm consisting of the same stages, each of which can be applied its own methods. The main difficulties encountered at cybersecurity risk analysis in monitoring systems include: the difficulty of creating an automated system model for assessing its vulnerability; rapid loss of relevance of assessment results; the difficulty of aggregating data from various sources, including qualitative information; the need to involve several risk analysts to improve the adequacy of assessments.

Therefore, in the general case the task is to choose from a set of methods Y for assessing the cybersecurity risk of such a method y*, that would provide the maximum likelihood of adequate assessment, taking into account adaptability to qualitative data on risk factors X_1, X_2 и X_3:

$$y^* \in Y \;\Leftrightarrow\; p_1^* = \max p_1(X, p_2(y)), \tag{1}$$

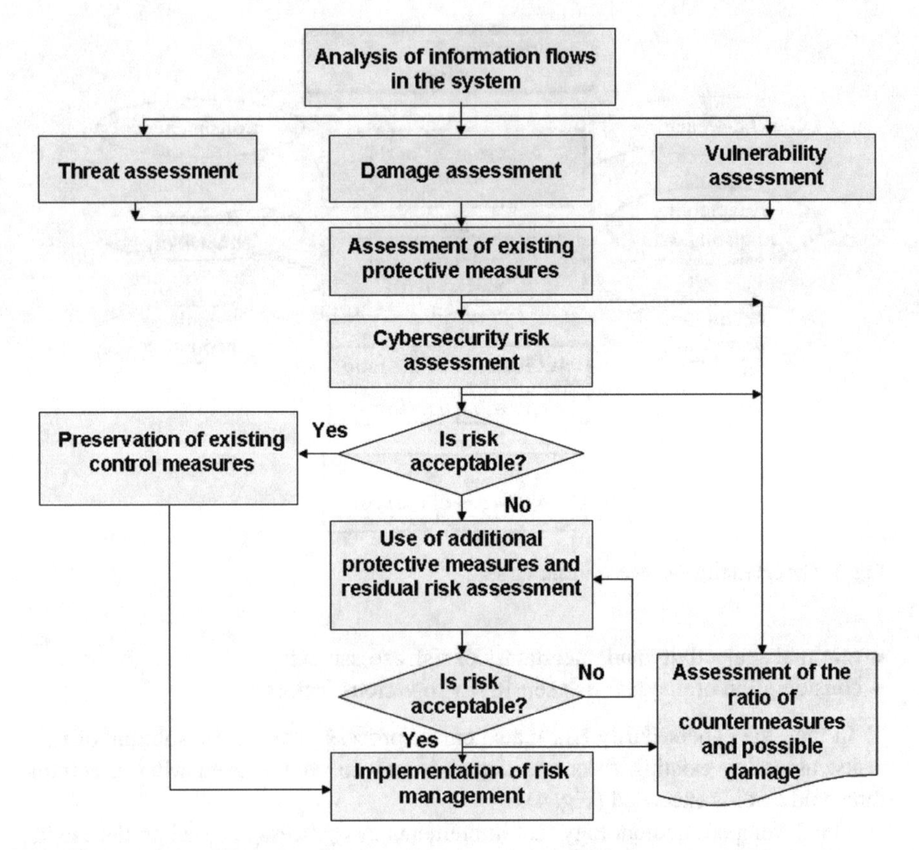

Fig. 4 Information risk analysis process

where X—set of risk factors; p_1—probability of adequate risk assessment; p_2—indicator of the adaptability of the method to qualitative data.

However, the solution of the problem is associated with answers to a number of questions:

- how to get an estimate of indicator p_1, for which you need to know the vector X;
- how to form a vector X taking into account those risk factors that may appear in real conditions of the system;
- how to ensure sufficient value of indicator p_2.

Therefore, it is proposed for each stage of cybersecurity risk management to choose its most appropriate method. The risk assessment method used must meet the requirement (1).

Identification of risk factors in the monitoring system is inextricably linked with the construction of its model and the study of information flows circulating in it. A comparative study of the structural, functional and object-oriented analysis methodologies, allowing to create such models, is showed that the IDEF0 methodology most fully takes into account and displays all the system elements necessary for analyzing data flows—all input and output information, applied mechanisms and control actions, control over the implementation of the process and feedback. Therefore, it is advisable to use the IDEF0 methodology to identify risk factors and compile their list.

From the results of [6, 18, 21, 22], it follows, when analyzing information risks, the simulation methods have the most efficiency indicators, among which are neuro-fuzzy networks (NFN) that can identify and adequately assess information security risk due to the neural network component (indicator p1). They are also adaptable to quality data (indicator p2) through the use of fuzzy logic.

As a result, a neuro-fuzzy network with the following configuration has been proposed for cybersecurity risk assessment. The structure consists of five layers: fuzzification, aggregation, activation, accumulation and defuzzification. The type of fuzzy inference system is the Takagi-Sugeno-Kang model, focused on accuracy rather than interpretability, unlike the Mamdani model. Three variables are fed to the input: threat, damage and vulnerability, the values of which are obtained using an expert survey. The output calculates the cybersecurity risk level.

Since the developed neuro-fuzzy network is based on the fuzzy Takagi-Sugeno-Kang model, the output variable has no membership function. Instead, a set of values between 0 and 1 inclusive is used. Based on this set, a 9-level scale was formed, which allows interpreting the resulting value of the risk level and drawing conclusions about its acceptability or the need to introduce additional countermeasures to reduce it. If the risk reduction option is chosen, then it is necessary to proceed to the stage of assessing the ratio of countermeasures and possible damage, which is connected with the solution of the problem of developing a method for the optimal choice of efficient and cost-effective countermeasures.

Thus, the developed cybersecurity risk management methodology is based on the joint use and interaction of the IDEF0-model, expert survey, neural network, game theory methods and allows you to most effectively implement the compilation of a list of risk factors, their assessment, risk-level calculation and choice of countermeasures [5]. The proposed technique was tested in telemedicine systems in assessing the state of a human operator. Its use has reduced the risk level by 10–15%.

5 Results of Testing the Proposed Methodology

Testing of the proposed methodology was performed for the remote monitoring system of the human state (RMSHS).

In this case, the sources of objective information about the functional state of a person are mobile measuring systems made using the "sensor-on-a-chip" and "laboratory-on-a-chip" technologies [23–25]. In RMSHS, the registered bio-signals are transmitted via communication channels to the medical monitoring and data processing centers, where an in-depth assessment of the human state is carried out.

The inclusion of a mobile measuring system in a single information space allows continuous monitoring of a person's state, regardless of its location. However, this poses a problem with ensuring the integrity, confidentiality and availability of data. Therefore, cybersecurity risk analysis is required.

Using the IDEF0-model, a list of cybersecurity risk factors for RMSHS was compiled, which were assessed by a group of experts. As a result of processing expert assessments of these risk factors, three input values were obtained for the NFN: (1) threat—8.3 (very high); (2) potential damage—6.3 (high); (3) vulnerability—4.3 (medium). After submitting these values to the NFN input, a risk level was obtained equal to 0.627 (above average), which showed that the monitoring system could continue to work, but the corrective action plan should be applied as soon as possible. In other words, the obtained level of cybersecurity risk reflected the weak security of information flows, and appropriate decisions should be made on data protection in RMSHS.

As a result of the re-assessment of risk factors, taking into account the implementation of the selected countermeasures, three input values were obtained for assessing the residual risk using the NFN: (1) threat—1.68 (very low); (2) potential damage—1.7 (very low); (3) vulnerability—0.68 (very low). After submitting new values to the NFN input, a residual risk level of 0.165 (very low) was obtained at the output, which means a high degree of data protection. This demonstrates the effectiveness of the proposed cybersecurity risk management methodology.

6 Conclusion

In this chapter the problem of data protection at remote monitoring is considered. It is shown, that the creation of the data protection technology for remote monitoring of the object's state remains an urgent problem and requires the development of new mathematical methods, models and algorithms to ensure the integrity, availability and confidentiality of information.

The conducted studies allowed us to identify the features of cybersecurity risk management in RMSOS, which allows to increase the level of data security, and to propose a methodology for the implementation of this process.

Cybersecurity risk management methodology should combine the most effective methods for each stage of this process. Its application significantly expands the possibilities of integrating various methods, using their strengths and ensuring a flexible and smooth transition from one method to another, as well as providing quality input information and reliability (confidence level) of data sources.

References

1. Lee, J., Bagheri, B., Kao, H.A.: A cyber-physical systems architecture for industry 4.0-based manufacturing systems. Manufact. Lett. **3**, 18–23 (2015)
2. Romero, D., Bernus, P., Noran, O., Stahre, J., Fast-Berglund, Å.: The operator 4.0: human cyber-physical systems & adaptive automation towards human-automation symbiosis work systems. In: IFIP Advances in Information and Communication Technology (2016). https://doi.org/10.1007/978-3-319-51133-7_80
3. Loktev, D.A., Loktev, A.A.: Diagnostics of external defects of railway infrastructure by analysis of its images. In: Global Smart Industry Conference (GloSIC). Chelyabinsk, Russia (2018). https://doi.org/10.1109/glosic.2018.8570083
4. Suyatinov, S.I.: Criteria and method for assessing the functional state of a human operator in a complex organizational and technical system. In: 2018 Global Smart Industry Conference (GloSIC), Chelyabinsk, Russia, pp. 1–6 (2018). https://doi.org/10.1109/glosic.2018.8570088
5. Buldakova, T.I., Mikov, D.A.: Comprehensive approach to information security risk management. In: CEUR Workshop Proceedings, vol. 2081, paper 05, pp. 21–26 (2017)
6. Lee, M.-Ch.: Information security risk analysis methods and research trends: AHP and fuzzy comprehensive method. Int. J. Comput. Sci. Inf. Technol. (IJCSIT) **6**(1), 29–45 (2014)
7. Appari, A., Johnson, M.E.: Information security and privacy in healthcare: current state of research. Int. J. Internet Enterp. Manage. **6**(4), 279–314 (2010)
8. Idhate, S., Bilapatre, A., Rathod, A., Kalbande, H.: Dam monitoring system using wireless sensor networks. Int. Res. J. Eng. Technol. **4**(4), 1767–1769 (2017)
9. Ibrahim, A., Muhammad, R., Alshitawi, M., Alharbi, A., Almarshoud, A.: Intelligent green house application based remote monitoring for precision agricultural strategies: a survey. J. Appl. Sci. **15**(7), 947–952 (2015)
10. Bashi, N., Karunanithi, M., Fatehi, F., Ding, H., Walters, D.: Remote monitoring of patients with heart failure: an overview of systematic reviews. J. Med. Internet Res. **19**(1), e18 (2017)
11. Neusypin, K., Selezneva, M., Proletarsky, A.: Nonlinear information processing algorithm for navigation complex with increased degree of parametric identifiability. In: Dolinina, O., et al. (eds.) Recent Research in Control Engineering and Decision Making. ICIT-2019. Studies in Systems, Decision and Control, vol. 199, pp. 37–49 (2019). https://doi.org/10.1007/978-3-030-12072-6_4
12. Xu, M., Sun, M., Wang, G., Huang, S.: Intelligent remote wireless streetlight monitoring system based on GPRS. In: Xiao, T., Zhang, L., Fei, M. (eds.) Communications in Computer and Information Science, vol. 324, pp. 228–237. Springer, Berlin, Heidelberg (2012)
13. Buldakova, T., Krivosheeva, D.: Data protection during remote monitoring of person's state. In: Dolinina, O., et al. (eds.) Recent Research in Control Engineering and Decision Making. ICIT-2019. Studies in Systems, Decision and Control, vol. 199, pp. 3–14. Springer, Cham (2019). https://doi.org/10.1007/978-3-030-12072-6_1
14. Kure, H.I., Islam Sh., Razzaque, M.A.: An integrated cyber security risk management approach for a cyber-physical system. Appl. Sci. **8**(6), 898 (2018). https://doi.org/10.3390/app8060898
15. Buldakova, T.I., Dzhalolov, ASh: Analysis of data processes and choices of data-processing and security technologies in situation centers. Sci. Tech. Inf. Process. **39**(2), 127–132 (2012). https://doi.org/10.3103/S0147688212020116
16. Buldakova, T.I., Suyatinov, S.I.: Reconstruction method for data protection in telemedicine systems. In: Progress in Biomedical Optics and Imaging—Proceedings of SPIE, vol. 9448, Paper 94481U (2014). https://doi.org/10.1117/12.2180644
17. Banerjee, A., Gupta, S.K.S., Venkatasubramanian, K.K.: PEES: physiology-based end-to-end security for mHealth. In: Proceedings of the 4th Conference on Wireless Health, Article No. 2 (2013)
18. Barabanov, A., Markov, A., Tsirlov, V.: Procedure for substantiated development of measures to design secure software for automated process control systems. In: 2016 International Siberian Conference on Control and Communications, SIBCON 2016—Proceedings, pp. 1–4 (2016). https://doi.org/10.1109/sibcon.2016.7491660

19. Yuan, C., Li, J., Zhang, R., Liu, J.: Grey and fuzzy evaluation of information system distress recovery capability. In: 2nd International Conference on Advances in Computer Science and Engineering, CSE, pp. 298–302 (2013)
20. Chang, P.T., Hung, K.C.: Applying the fuzzy weighted average approach to evaluation network security systems. Compu. Math. Appl. **49**, 1797–1814 (2005)
21. Liu, F., Dai, K., Wang, Z.Y.: Research on the technology of quantitative security evaluation based on fuzzy number arithmetic operation. Fuzzy Syst. Math. **18**(4), 51–54 (2004)
22. Jin, Y.: Fuzzy modeling of high-dimensional systems: complexity reduction and interpretability improvement. IEEE Trans. Fuzzy Syst. **8**(2), 212–221 (2000)
23. Buldakova, T.I., Suyatinov, S.I.: Registration and identification of pulse signal for medical diagnostics. In: Proceedings of SPIE—The International Society for Optical Engineering, vol. 4707, paper 48, pp. 343–350 (2002)
24. Mundt, C.W., Montgomery, K.N., Udoh, U.E., Barker, V.N.: A multiparameter wearable physiologic monitoring system for space and terrestrial applications. IEEE Trans. Inf. Technol. Biomed. **9**(3), 382–391 (2005)
25. Paradiso, R., Loriga, G., Taccini, N.: A wearable health care system based on knitted integrated sensors. IEEE Trans. Inf. Technol. Biomed. **9**(3), 337–344 (2005)

Industrial Cyber-Physical Systems: Risks Assessment and Attacks Modeling

Alla G. Kravets, Natalia Salnikova, Kirill Dmitrenko and Mikhail Lempert

Abstract The chapter is devoted to the attacks modeling for Cyber-Physical systems of industrial enterprises with regard to risk assessment. In this chapter the analysis of corporate systems of industrial is held; systems' attacks and risk assessment techniques were studied; software for attacks modeling are compared. Information models of corporate networks and attacks are developed; describes the design and basic functions of the module for assessing the risks of attacks in the corporate system. Corporate networks of more than 70% of industrial enterprises are potentially vulnerable to hacker attacks. Today, according to research by Positive Technologies analysts, hackers can cross the perimeter and get into the corporate network of 73% of the companies in the industrial segment. In 82% of companies, penetration from the corporate network to the technological one is possible. One of the main opportunities for obtaining unauthorized access to the enterprise network turned out to be administrative control channels. Solving the problem of ensuring the information security of Cyber-Physical systems is an urgent task today.

Keywords Informational resources · Cyber-physical systems · Industrial enterprise · Industrial cybersecurity · Information security threats · Attack modeling · Unauthorized access · Protection systems

A. G. Kravets (✉)
Volgograd State Technical University, 28 Lenin av., Volgograd 400005, Russia
e-mail: agk@gde.ru

N. Salnikova
Volgograd Institute of Management—Branch of the Russian Presidential Academy of National Economy and Public Administration, 8 Gagarin St., Volgograd 400131, Russia
e-mail: ns3112@mail.ru

K. Dmitrenko
I.M. Sechenov First Moscow State Medical University, 8b2 Trubetskaya St., Moscow 119991, Russia
e-mail: kirill.dmitrenko.msc@gmail.com

M. Lempert
Technion—Israel Institute of Technology, Haifa 3200003, Israel
e-mail: lempertmi@gmail.com

A. G. Kravets et al. (eds.), *Cyber-Physical Systems: Industry 4.0 Challenges*, Studies in Systems, Decision and Control 260,
https://doi.org/10.1007/978-3-030-32648-7_16

1 Introduction

The leading role in the modern economy is played by industries, especially industrial enterprises, associated with the production of high-tech products (auto, aircraft, and shipbuilding, production of nuclear, space, military equipment, etc.). The high-tech industry faces many new challenges, the most important of which is to provide flexible and stable maintenance services through the development of reliable automated Industrial Cyber-Physical Systems [1, 2]. It is necessary to take into account that the competitiveness of industrial enterprises, the size of their income, their position in the market substantially depend on the correct functioning of their information infrastructure, the integrity of the main information resources, and the protection of confidential information from unauthorized access, against cybersecurity industrial systems [3, 4]. On this basis, requirements for protection systems are increasing, which should ensure not only passive blocking of unauthorized access to internal resources of an enterprise's network from external networks, but also detect successful attacks, analyze the causes of threats to information security and, to the extent possible, eliminate them in automatic mode [5, 6]. Today, as a basic element of information security in the modeling and design of information systems protection systems, as a rule, the threat of attack is considered, while designing a protection system is reduced to solving the problem of protecting against actual threats [7, 8]. This approach is also declared by the relevant regulatory documents in the field of information security [9, 10]. However, in practice, expert assessment, which does not imply any modeling, any quantitative assessment of the level of relevance, is used to identify actual threats of attacks, which in each case casts doubt on the results of such design [11, 12]. At its core, the implementation of a successful attack on an information system in the theory of information security with significant reservations can be interpreted as a failure in the theory of reliability [13, 14].

The potential threat of an attack on an information system should be discussed in the case when an attack is technically feasible on the corresponding system, the real threat of an attack in the case when conditions are created in the system for the offender to realize it, and the actual threat of attack, in that case, the real threat can be realized by the intruder, when the intruder is interested in the implementation of this attack of the corresponding complexity on the information system and can carry it out [15, 16].

Modeling the threat of attack, in order to obtain a quantitative assessment of its relevance for an information system (naturally, for a specific, rather than some abstract system), involves solving two interrelated tasks with regard to a potential threat of attack: calculating quantitative characteristics that determine the creation of conditions for the appearance of a real threat of attack, and quantitative characteristics that determine the ability of the intruder to use a real threat of attack [17, 18]. Thus, the failure of information security characteristics depends not only on the technical properties of the system, which determine the creation of conditions for the realization of an attack, but also on the characteristics of the intruder, determining his interest and the ability to realize the corresponding attack—use the conditions created by the system—the real threat of attack.

2 Research Goals and Objectives

Due to the development of Internet technologies, more and more processes are being automated [19]. For their integrated and optimal management of enterprises, information systems are created, the main components of which are corporate networks [20, 21]. They use information in business processes to facilitate management decisions and business. Dependence on the information in the business environment is extremely high, where many trading operations are carried out electronically via the Internet [22]. Such information dependence has led to a significant increase in the impact of the security level of information systems on the success, and sometimes simply the possibility of doing business. Therefore, the security of information systems is one of the most important issues that attract much attention from analysts, engineers and other specialists in the field of information security [23, 24].

Currently, there are two main approaches to ensuring the information security of corporate networks—providing a basic level of information security, as well as an approach based on attack modeling. At the basic level of ensuring information security, tasks are set related to verifying the compliance of components of a corporate network with all the mandatory requirements of standards and regulatory documents and checking the security of components of a corporate network against major threats.

One of the main tasks in the field of attack modeling is to assess the possible damage to a specific component or the entire system without any risk to the information system.

In this regard, the purpose of this work is to develop a module for modeling attacks for the Industrial Cyber-Physical Systems (ICPS), taking into account the risk assessment. To achieve this goal, the following tasks were solved:

- to study the composition of corporate systems industrial enterprises;
- study the attacking effects;
- study network attacks and risk assessment techniques;
- develop information models of Industrial Cyber-Physical Systems and attacks;
- design and develop a module for assessing the risks of attacks in the Industrial Cyber-Physical Systems.

3 Corporate Systems of Industrial Enterprises

The industrial enterprise corporate system (IECS) is a stable formal socio-economic system that receives resources from the environment and uses them for the purposes of manufacturing products that have a certainly added value for the producer and a value for the consumer. To achieve its goals, such a system should include an efficiently operating control subsystem, the effectiveness of which can be significantly improved through automation based on computer and telecommunication technologies [25].

IECS do not necessarily associate with economic and, in particular, with commercial activities. These can be any organizational systems created for manufacturing, commercial or non-commercial purposes [26].

Modern IECSs differ from traditional ones by the presence of an automated management subsystem, i.e. almost full support for information technology in the organization and management, as well as comprehensive automation of all levels of organizational management, including production, and integrated into Cyber-Physical Systems. Cyber-Physical Systems are systems consisting of various natural objects, artificial subsystems, and controllers that allow such a formation to be represented as a whole. The architectural model of such systems includes: business architecture (strategy, leadership, organization, strategic management process); cyber-physical architecture; information and data architecture; application architecture (application components); industrial architecture; the architecture of control technics and management of information technology environment.

Practically all significant business connections in such organizations are realized in digital, computer form, and the main working processes are carried out in the infrastructure of electronic computer networks covering the entire organization, or connecting several organizations, forming the so-called industrial or advanced corporate networks. The key corporate resources are intellectual property, basic skills and knowledge of employees, material, financial, human and information resources that are managed in an automated (electronic) way.

Speaking about the differences of modern MECSs from traditional organizations, it should be noted that they are a much faster and more flexible adaptation to the conditions of a rapidly changing and complicating market environment, they more easily survive in difficult times. Thanks to electronic capabilities and rationalization of work, they have great potential for achieving unprecedented levels of profitability and competitiveness and have an exceptional willingness to work in a global economy.

The IECS is an industrial software-hardware communication system owned and/or controlled by a single organization in accordance with the rules of that organization.

Reasonable use of the potential of the corporate network in the enterprise provides the following benefits:

- a clear increase in labor productivity due to the competent organization of parallel computing processes, which is impossible to achieve in the presence of powerful but autonomous computing devices;
- resistance to failures and failures of individual elements of the system is transferred much less painful in the case of the combination of these elements into a single network;

- duplication of data on various types of network media, the ability to switch requests and processes to healthy internal network segments, dynamic and static reconfiguration capabilities—this is what the use of the corporate network gives in case of failures;
- multiple acceleration and facilitation of intra-network communication. Performing simultaneously a large number of different tasks aimed at a common result is greatly simplified when uninterrupted communication is established between various structures and departments of the organization;
- control of commercial and technical security, protection of important corporate data is easier to implement, having access to all software and hardware elements and peripheral devices at the same time.

4 Analysis of Modeling Attacks Tools

4.1 ADTool

Developer: Cybati Works [27]. License: not required. Type: building trees attacks. Visualization: tree diagrams.

The Attack Tree tool (ADTool) allows users to model and analyze attack protection scenarios represented by attack protection trees and attack protection conditions. It supports the anti-attack tree methodology developed in the framework of the ATREES, TREsPASS and ADT2P projects. Key features of the tool: creation and editing of defense trees against attacks and successive attack trees; a quantitative bottom-up analysis of attack scenarios; a modular display of attack protection trees, which allows you to simulate large real-life scenarios; ranking of possible attacks for certain attribute domains; printing, exporting to various formats and saving tree-like models of protection against attacks; customizable layouts. Disadvantages: can only be used as an automation and visualization tool; there are no attack banks.

4.2 Trike Tool

Developer: Octotrike [28]. License: not required. Type: questionnaire table of the formation of a threat model. Visualization: tabular.

The Trike is an open source threat modeling tool. The project began in 2006 as a new attempt to improve the effectiveness of existing threat modeling methodologies and is actively used and developed. There are two implementations of Trike: as a table; in the form of a module, the development of which has never been completed. Disadvantages: since 2012 there is no new information about Trike; implementation files cannot be obtained from the official site.

4.3 Threat Modeling Tool

Developer: Microsoft [29]. License: not required. Type: process modeling. Visualization: data flow diagrams.

SDL Threat Modeling Tool is the first threat modeling tool that is not intended for security experts. This product simplifies threat modeling for all developers by providing guidance on how to create and analyze threat models.

As part of the SDL design phase, the threat modeling allows software developers to identify and mitigate potential security problems in a timely manner when they are relatively easy and economical to solve. Thus, it helps to reduce the total cost of development.

The SDL Threat Modeling Tool connects to any problem tracking system, making the threat modeling process part of the standard development process.

The SDL Threat Modeler is different from other tools and approaches in two key areas: it is intended for developers and software oriented; it is focused on the analysis of the program architecture. The disadvantage is a narrow focus on the interaction of processes in the system.

4.4 Threat Modeler

Developer: MyAppSecurity [30]. License: proprietary. Type: process modeling. Visualization: communication diagrams.

This cloud solution has its own updated library of threats, its own system for generating reports, its own bank of protective measures and many other functions that differ depending on the tariff plan. This product uses link diagrams for building models. Disadvantages: a narrow focus on the interaction of processes in the system; no integration of a vulnerability bank.

4.5 GRC-Platform R-Vision Security Tool

Developer: R-Vision [31]. License: proprietary. Type: modeling and analysis of system processes. Visualization: process diagrams and reporting diagrams.

The R-Vision Security GRC Platform (SGRC) is a software platform for automating the risk management process and conducting an internal audit. The disadvantage is a narrow focus on the interaction between the elements of the system.

Table 1 Comparative characteristics of software for modeling attacks

Tools	Special features			
	Visualization	Process modeling	Risk assessment	Using vulnerability banks or attacks
Trike (Octotrike)	Tabular	Missing	Missing	Missing
ADTool (Cybati Works)	Attack trees	Present	Missing	Missing
SDL thread modeling tool (Microsoft)	Process diagrams	Present	Missing	Missing
Thread modeler (MyApp Security)	Link diagrams	Missing	Missing	Present
SGRC platform (R-Vision)	Process diagrams	Present	Present	Missing

4.6 Attack Modeling Software Comparison

Table 1 shows the comparative characteristics of attack modeling software.

5 Network Attacks and Risk Assessment Techniques

Risk assessment began to be used at nuclear power plants in Europe and America in the early 1960s, and later developed and used aerospace engineering, chemical industry, environmental protection, health, sports, national economic development, and many other areas. In information security, risk assessment techniques have emerged in order to predict the possible damage associated with the realization of threats, and accordingly, assess the required amount of investment in building information protection systems. When implementing an approach based on assessing and managing information security risks, a wider range of tasks is set: assessment of information security risk factors; assessment of the relevance of threats; reducing the level of information security risk to an acceptable level.

The urgency of applying an approach based on assessing and managing information security risks is due to the implementation of effective mechanisms for determining actual threats and choosing adequate measures to protect the information in the corporate network from the standpoint of expected damage.

The information security risk is the potential risk of damage to an organization as a result of the realization of a certain threat using vulnerabilities. Risk assessment methods are divided into three classes: qualitative, quantitative and analytical.

Qualitative methods take into account the many parameters of elements and attacking influences for risk assessment. Quantitative methods rely on the assessments of several experts regarding the likelihood of threats to be realized. Analytical methods rely on statistical data. There are two ways to assess risk: two-factor and three-factor.

A network attack is an information destructive effect on a distributed computing system carried out by software over communication channels. The most common types of attacks are shown in Table 2.

To implement the attacks needed vulnerabilities in the computing system. The vulnerability is a property of the system that can be used to violate the information security of the system.

The purpose of the attacks is the assets of the enterprise. Assets—everything that has value for the organization and is at its disposal: all kinds of resources, products, and services of the enterprise.

6 Attacks Modeling for the ICPS Based on Risk Assessment

Functional requirements for the program: creating a network model in the form of a graph; saving the network model to a file; loading the network model from a file; configuring the parameters of each network element; assessing the threats and risks of the created network model; outputting the simulation result in Fig. 1.

Main steps of attacks modeling tool are:

- creating a network model in the form of a graph. The network graph is displayed on a special graphic panel in the central part of the form. The nodes of the graph are the network elements, and the arcs are the connections between them. Adding nodes occurs by clicking on the special buttons of the elements. To add links, select two nodes and click on the link building button;
- save the model to file. The model data is stored in a text file in the CSV format and in UTF-8 encoding;
- load model from file. Model data is loaded from a text file of the CSV format and UTF-8 encoding into the program and displayed as a graph on a special panel;
- setting parameters for each element of the network. When a network element is selected, its parameters are displayed on a special table where they can be changed. When creating an element, each parameter is set to its default value.

The result of the simulation is displayed in a special text box. The results should indicate the calculation time, the overall system risk, the parameters of the element with the highest risk (identifier, name, operating system, risk, security rating, significance and number of similar elements). The functional structure of the tool must comply with the scheme presented in Fig. 2.

The tool is used according to the diagram shown in Fig. 3.

The tool has four main functions: the creation of a network model in the form of a graph; assessment of the model threats and risks; save the model to file; load model from file.

Table 2 Types of network attacks

The cause of the anomaly (source)	Type of manifestation	Effects
Attacks at the application level	It exploits known vulnerabilities and weaknesses in software that usually exist on standard Telnet, POP3, IMAP, SMTP, HTTP, FTP, and other servers	Attackers can gain access to a computer that allows launching applications for a specific account, which is usually the privileged system account
Autorooters	The capture process involves installing a rootkit on the computer and using the captured system to automate the invasion process	Allows an attacker to scan hundreds of thousands of systems in a short period of time
Backdoors	These are entrances to systems that can be created during an invasion or by using a specially designed Trojan horse	An attacker can use his entry into the computer or network over and over. The intruder will use the computer to gain access to other systems or launch a denial of service (DoS) attack if it no longer needs to use this computer
Denial of Service (DoS) and Distributed Denial of Service (DDoS) attacks	They create conditions under which system users cannot get data to certain resources or services	There are disruptions in the normal functioning of the system, which is usually complemented by a lack of resources required for the operation of a network, operating system or applications
TCP SYN flood	Data relating to all pending connections are in server memory, which can overflow if you intentionally create too many partially open connections	Violations of the normal functioning of the system
Ping of death attacks	Causes systems to respond in unpredictable ways when receiving too large IP packets	Crash, hang and reboot the system
IP spoofing attacks	IP address spoofing	An intruder inside or outside the network is impersonating a computer you can trust

(continued)

Table 2 (continued)

The cause of the anomaly (source)	Type of manifestation	Effects
Man-in-the-middle attacks	Such attacks are often carried out using sniffing (interception) of network packets, routing protocols, and transport protocols	Information theft, hacking the current communication session to gain access to private network resources, analyzing traffic to obtain information about the network and its users, DoS attacks, distorting the transmitted data and including new information in the network session
Network intelligence	Intelligence can take the form of queries to a DNS server, scanning a range of IP addresses (ping sweeps), and scanning ports	Attackers can learn the characteristics of applications running on hosts. This information will be useful when a hacker tries to break the system
Packet sniffers	Packet sniffers are used legally in networks for traffic analysis and troubleshooting. However, since some network applications send plaintext data, packet sniffing can even provide critical information, such as usernames and passwords	An attacker can gain access to a system user account, which a hacker can use to create a new account, and thus have access to the network and its resources at any time
Port redirection attacks	A hacked host is used to pass traffic through the firewall, which otherwise would be lost	Attackers install software to redirect traffic from an external host directly to an internal host
Virus and Trojan attacks	Use dangerous programs that are attached to other programs in order to perform a destructive function on the user workstation	An example of a virus is a program that attaches to command.com, deletes some files and infects all other versions of the command file.com that can detect. The Trojan is able to disguise itself as another program. An example, an application that runs a simple game on a user workstation. While the user is busy with the game, the Trojan horse sends a copy of itself to all addresses from its address book

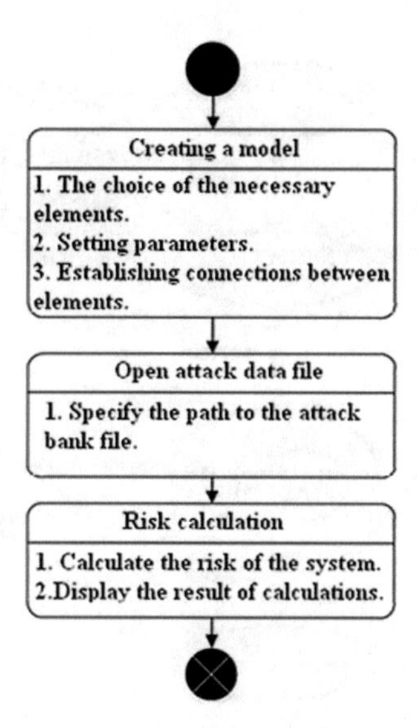

Fig. 1 The algorithm of attacks modeling in ICPS

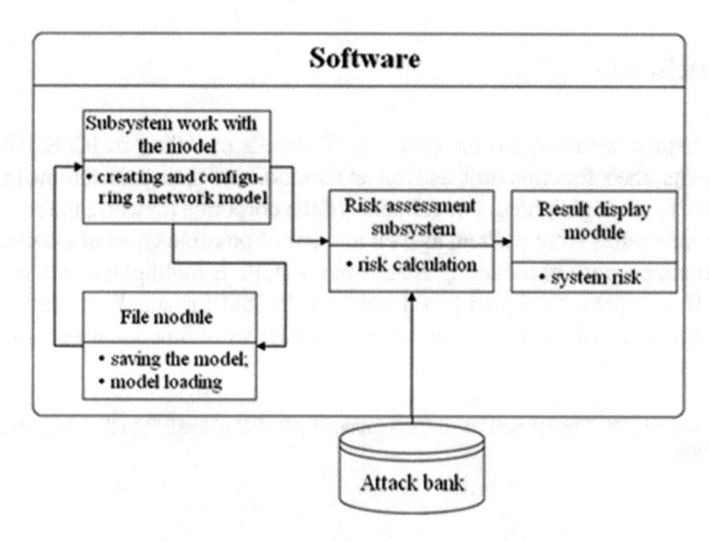

Fig. 2 The functional structure of the software

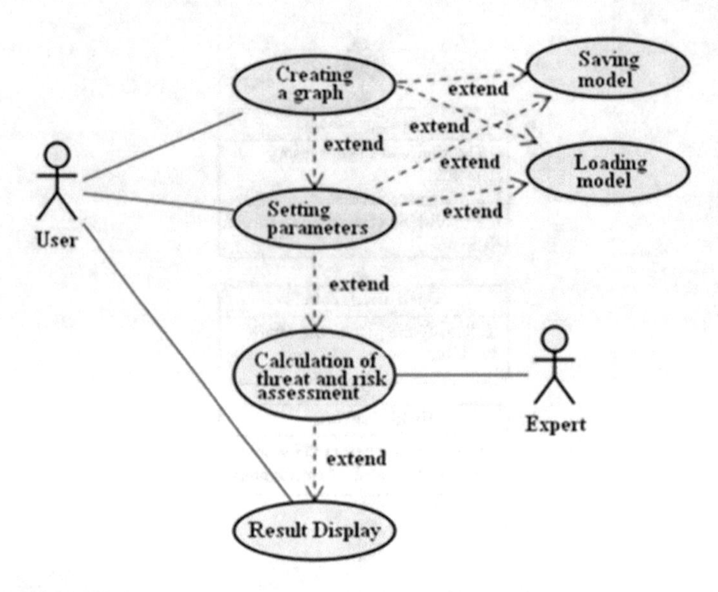

Fig. 3 USE-CASE chart

7 Conclusion

In this chapter, we analyzed the problem of attacks modeling in ICPS. The main components, their functionality, as well as variations of their connections (network architecture) are highlighted. For elements of the corporate network, the reasons for network anomalies were written, as well as a list of possible types of attacks.

The main purpose of modeling attacks in the ICPS is highlighted and the way to achieve it is chosen. Designed and developed the tool for attacks modeling in the ICPS, according to the written requirements. A number of functional tests and a fault tolerance test were carried out.

Acknowledgements The reported study was funded by RFBR according to the research project # 19-07-01200.

References

1. Vasilev, D., Kravets, E., Naumov, Y., Bulgakova, E., Bulgakov, V.: Analysis of the data used at oppugnancy of crimes in the oil and gas industry. Stud. Syst. Decis. Control **181**, 249–258 (2019)
2. Salnikova, N.A., Lempert, B.A., Lempert, M.B.: Integration of methods to quantify the quality of medical care in the automated processing systems of medical and economic information. In: Communications in Computer and Information Science (CIT&DS 2015), vol. 535, pp. 307–319. Volgograd, Russia Federation. Springer, Heidelberg (2015)

3. Prokofieva, E., Mazur, S., Chervonnykh, E., Zhuravlev, R.: Internet as a crime zone: Criminalistic and criminological aspects. Stud. Syst. Decis. Control **181**, 105–112 (2019)
4. Kravets, A.G., Skorobogatchenko, D.A., Salnikova, N.A., Orudjev, N.Y., Poplavskaya, O.V.: The traffic safety management system in urban conditions based on the C4.5 algorithm, In: Moscow Workshop on Electronic and Networking Technologies (MWENT 2018), Art. No. 8337254, pp. 1–7. Proceedings (2018)
5. Gurlev, I., Yemelyanova, E., Kilmashkina, T.: Development of communication as a tool for ensuring national security in data-driven world (Russian far north case-study). Stud. Syst. Decis. Control **181**, 237–248 (2019)
6. Kravets, A.G., Bui, N.D., Al-Ashval, M.: Mobile security solution for enterprise network. In: Communications in Computer and Information Science (CCIS 2014), vol. 466, pp. 371–382 (2014)
7. Kravets, A., Kozunova, S.: The risk management model of design department's PDM information system. In: Communications in Computer and Information Science (CIT&DS 2017), vol. 754, pp. 490–500. Volgograd, Russia Federation. Springer, Heidelberg (2017)
8. Orudjev, N.Y., Poplavskaya, O.V., Lempert, L.B., Salnikova, N.A.: Problems of medical confidentiality while using electronic documents in psychiatric practice. In: Atlantis Press, Proceedings of the 2016 Conference on Information Technologies in Science, Management, Social Sphere and Medicine (ITSMSSM 2016), vol. 51, pp. 120–125 (2016)
9. Goroshko, I., Toropov, B., Gurlev, I., Vasiliev, F.: Data analysis of the socio-economic factors' influence on the state of crime. Stud. Syst. Decis. Control **181**, 71–84 (2019)
10. Urban, V., Kniazhev, V., Maydykov, A., Yemelyanova, E.: Implementation of the law enforcement function of the state in the field of countering crimes committed using the internet. Stud. Syst. Decis. Control **181**, 113–120 (2019)
11. Quyên, L.X., Kravets, A.G.: Development of a protocol to ensure the safety of user data in social networks, based on the Backes method. In: Communications in Computer and Information Science (CCIS 2014), vol. 466, pp. 393–399 (2014)
12. Latov, Y., Grishchenko, L., Gaponenko, V., Vasiliev, F.: Mechanisms of countering the dissemination of extremist materials on the internet. Stud. Syst. Decis. Control **181**, 145–161 (2019)
13. Kravets, A, Shumeiko, N., Shcherbakova, N., Lempert, B., Salnikova, N.: "Smart queue" approach for new technical solutions discovery in patent applications, communications. In: Computer and Information Science (CIT&DS 2017), vol. 754, pp. 37–47. Volgograd, Russia Federation. Springer, Heidelberg (2017)
14. Lebedev, A., Shmonin, A., Vasiliev, F., Korobko, V.: Principles of mathematical models constructing based on the text or qualitative data of social systems. Stud. Syst. Decis. Control **181**, 29–38 (2019)
15. Kondratiev, V., Shchepkin, A., Irikov, V.: Mechanisms for ensuring road safety: The Russian Federation case-study. Stud. Syst. Decis. Control **181**, 183–203 (2019)
16. Shcherbakov, M.V., Kachalov, D.L., Kamaev, V.A., Shcherbakova, N.L., Tyukov, A.P., Strekalov, S.D.: A design of web application for complex event processing based on hadoop and java servlets. Int. J. Soft Comput. **10**(3), 218–219 (2015)
17. Kravets, A.G., Kanavina, M.A., Salnikova, N.A.: Development of an integrated method of placement of solar and wind energy objects in the lower Volga. In: International Conference on Industrial Engineering, Applications and Manufacturing (ICIEAM 2017), pp. 1–5 (2017)
18. Anufriev, D., Petrova, I., Kravets, A., Vasiliev, S.: Big data-driven control technology for the heterarchic system (building cluster case-study). Stud. Syst. Decis. Control **181**, 205–222 (2019)
19. Dronova, O., Smagorinskiy, B.P., Yastrebov, V.: Counteraction to E-commerce crimes committed with the use of online stores. Stud. Syst. Decis. Control **181**, 121–131 (2019)
20. Golubev, A.V., Shcherbakov, M.V., Scherbakova, N.L., Kamaev, V.A.: Automatic multi-steps forecasting method for multi seasonal time series based on symbolic aggregate approximation and grid search approaches. J. Fundam. Appl. Sci. **8**(3S), 2429–2441 (2016)

21. Kravets, A.G., Gurtjakov, A., Kravets, A.: Corporate intellectual capital management: learning environment method. In: Proceedings of the IADIS International Conference ICT, Society and Human Beings 2013, Proceedings of the IADIS International Conference e-Commerce 2013, pp. 3–10 (2013)
22. Novikov, D., Belov, M.: Methodological foundations of the digital economy. Stud. Syst. Decis. Control **181**, 3–14 (2019)
23. Korotkov, A., Kravets, A.G., Voronin, Y.F., Kravets, A.D.: Simulation of the initial stages of software development. Int. J. Appl. Eng. Res. **9**(22), 16957–16964 (2014)
24. Makarov, V., Gaponenko, V., Toropov, B., Kupriyanov, A.: Theoretical and applied aspects of orthogonal coding in computer networks technologies. Stud. Syst. Decis. Control **181**, 47–58 (2019)
25. Burkov, V., Shchepkin, A., Irikov, V., Kondratiev, V.: Methodology and technology of control systems development. Stud. Syst. Decis. Control **181**, 15–27 (2019)
26. Chernyshev, S.L.: Investment management technology with discounting. Stud. Syst. Decis. Control **181**, 231–236 (2019)
27. Cyber-Physical Security Education: https://cybati.org/. Last accessed 28 Mar 2019
28. Trike: http://octotrike.org. Last accessed 28 Mar 2019
29. SDL Threat Modeling Tool: https://www.microsoft.com/en-us/sdl/adopt/threatmodeling.aspx. Last accessed 28 Mar 2019
30. An automated threat modeling platform that secures and scales the enterprise software development life cycle. https://threatmodeler.com/. Last accessed 28 Mar 2019
31. R-Vision Incident Response Platform. https://rvision.pro/. Last accessed 28 Mar 2019

The Problem of Selecting APCS' Information Security Tools

Andrew S. Rimsha and Konstantin S. Rimsha

Abstract This article discusses the problem of selecting information security tools for one of the varieties of cyber-physical systems—automated process control systems. The study of several classes of solutions was conducted, considering their application in automated systems. Based on the formed criteria, a comparison was made; as a result, a tool was determined according to its characteristics that most met the criteria. Requirements were formed for popular existing solutions of this class, considering the specific features of providing information security systems for automated process control systems. None of the solutions fully met the specified requirements, therefore, a methodology was proposed that would solve this problem.

Keywords Cyber-physical systems · Automated process control systems · Threat modeling tools · Risk assessment

1 Introduction

As a result of the development of industrial automation and the use of information technologies in various areas of human life, more and more solutions combine information technologies and physical process control elements. These systems are called cyber-physical systems (CFS) [1]. One of the modern varieties of CFS is an automated process control system (APCS) that interacts in real time with physical and computing processes [2]. The integration of industrial and corporate networks is the main trend in the development of APCS and has a negative impact on their security [3].

Most solutions on the market for providing information security (IS) are both software and hardware tools that require integration into an industrial network to realize their functionality. Considering that the APCS is the implementation of a

A. S. Rimsha (✉) · K. S. Rimsha
Tyumen State University, Perekopskaya st., 15A, Tyumen 625003, Russia
e-mail: RimshaAndrew@gmail.com

K. S. Rimsha
e-mail: RimshaKonstantin@ya.ru

A. G. Kravets et al. (eds.), *Cyber-Physical Systems: Industry 4.0 Challenges*, Studies in Systems, Decision and Control 260,
https://doi.org/10.1007/978-3-030-32648-7_17

continuous process in real time, the introduction of such information security tools in systems in which they were not originally provided for and tested using them may adversely affect part or all of the entire process. Such problems can be caused both by the incompatibility of hardware or software and by the built-in functionality of blocking these funds. Considering that many organizations that use APCS can be hazardous industrial facilities, such an impact can cause equipment failure or even an accident, so when choosing solutions to protect the system, you should be guided primarily by minimal intervention in production processes.

From the classes of solutions offered on the market that have a minimal impact on the system and increase the IS systems, the following can be singled out:

- firewalls;
- one-way interconnection systems;
- intrusion detection systems (IDS);
- duplication and redirection of technological traffic (mirroring);
- correlation of incidents and monitoring of IS events;
- vulnerability scanners;
- threat modeling tools.

2 Analysis of Information Security Tools for APCS

2.1 Firewalls

A firewall is a network security system that monitors and controls incoming and outgoing network traffic based on predefined security rules.

A firewall is usually divided into a network or host. Network firewalls are located on computer gateways for local area networks, wide area networks, and intranets. They are either software solutions running on general-purpose hardware, or separate devices with an embedded OS. A host-based firewall is located on the network node itself and monitors the incoming and outgoing network traffic of these devices. Each type of firewall has its advantages and disadvantages, but in industrial networks, it is more common to use the network type firewall, which is used to set a barrier between a trusted internal network and an unreliable external network, in the case of an APCS—between an industrial network and a corporate one.

Sometimes firewall can be used within the industrial network between the upper and middle levels to monitor network traffic, in which case the access rules are configured in such a way that the list of open ports includes the necessary industrial protocols (for example, access is provided via TCP port 502 (Modbus TCP)).

One type of firewall is the industrial field firewall, used at the controller level. The main difference between the firewall industrial field is the control of the session of industrial protocols at the application level, and not only at the network level, thus the industrial field firewall can conduct a full-fledged inspection of industrial protocols at the application level, thereby allowing more flexible configuration of traffic allowed in the industrial network [4].

When protecting an industrial network between the levels of an APCS, technical characteristics are also important, such as requirements for vibration loads, operating temperature, and moisture resistance. Based on these parameters, you can determine how this device will meet the operational requirements of network equipment at a particular facility.

2.2 One-Way Interconnection Systems

One-way interconnection systems (also known as unidirectional gateways and data diodes) are communication devices that provide secure one-way data transfer between segmented networks. The intelligent design of such systems supports the physical and electrical separation of source and destination networks, establishing a non-routable, completely closed, one-way data transfer protocol between networks. Ensuring the security of all data streams in the network using one-way interconnection systems makes the transfer of unsafe or hostile malware as well as remote access to the industrial system impossible [5].

One-way gateway technology allows you to safely transfer information only in one direction, from protected areas to less secure systems, preventing back access. One example of the use of this tool in the industrial network is a protection against unauthorized access attempts by the corporate network segment, providing only the ability to receive information.

2.3 IDS

IDS is a hardware or software solution that monitors a network or system for malicious actions or breaches of security policy. Any malicious actions or violations are usually reported either to the administrator, or they are collected centrally using security information and event management (SIEM) [6].

The main difference between IDS and firewall is that firewall restricts access between networks to prevent intrusion and does not signal an attack within the network. The IDS, in turn, describes the intended intrusion after it has occurred, and signals the danger. IDS also tracks attacks from the OS. Such opportunities are achieved by studying network interactions, identifying heuristics and signatures of common computer attacks and taking measures to alert operators.

IDS types range from individual computers to large networks. The most common classifications are network intrusion detection systems (NIDS) and host intrusion detection systems (HIDS). A system that tracks important operating system files is an example of HIDS, and a system that analyzes incoming network traffic is an example of NIDS. IDS can also be classified using the detection approach: the most well-known variants are signature detection (recognition of dangerous signatures, such as malware or attacks); and anomaly-based detection (detection of deviations from the allowable traffic model, which often relies on machine learning). An enhanced IDS variant with the ability to respond to detected intrusions is called an intrusion prevention system. In industrial networks, such protection is not recommended due to the potential blocking of the technological process in case of abnormal activity, which can significantly affect the entire industrial network as a whole, therefore using IDS with the inability to respond to detected intrusions is a more acceptable option.

Thus, the formation of the demilitarized zone at the perimeter level of the network using IDS can significantly increase the level of network security without affecting the technological processes themselves.

2.4 Duplication and Forwarding of Network Packets

As a rule, port mirroring is used on a network switch to send copies of all or individual VLANs of incoming and/or outgoing network packets received on one or several ports to a separate interface for monitoring network traffic [7]. From a security point of view, this method can be most effectively used in conjunction with the previous means of protection—IDS. This method is most suitable for use in IDS industrial networks since it ensures that when analyzing traffic there are no threats to the operation of the industrial network.

Another useful use of a port mirroring is analyzing and debugging the information received or diagnosing network errors. This helps administrators closely monitor network performance and alert them when problems occur.

2.5 Correlation of Incidents and Monitoring of Information Security Events

In the field of computer security, SIEM software products and services combine security information management (SIM) and security event management (SEM). They provide real-time analysis of security alerts generated by applications and network hardware.

SIEM can be either a software or hardware solution or service provided [8]. This decision class can perform the following functions and components:

- data aggregation—collecting and processing data from multiple sources, including network devices, servers, databases, applications, security logs, providing the ability to consolidate monitored data so that important events do not go unnoticed;
- correlation—searching for common attributes and combining events into meaningful bundles, which provides the ability to perform various methods of finding

relationships when integrating various sources so that event information is the most complete and does not contain repetitions;

- alert—automated analysis of processed events with customizable incident notification;
- verification of compliance with the requirements—the system can be used to automate the verification of collected data for compliance with security requirements, reporting that adapt to existing processes or a security audit;
- storage—the use of long-term storage of irrelevant data for a more accurate correlation of data over time, as well as for investigating potential incidents;
- forensic analysis—the ability to search through the event logs of available network nodes at different time intervals based on certain criteria.

The focus is on monitoring and managing user and service rights, directory services and other system configuration changes; and incident response.

SIEM combines output from multiple sources and uses alarm filtering techniques to distinguish malicious activity from false alarms, eliminating the need to keep track of various consoles of protection tools and manually analyzing events received from them, automating and simplifying this process as much as possible.

2.6 Vulnerability Scanners

A vulnerability scanner is a computer program designed to evaluate computers, network infrastructure or applications for known vulnerabilities. They are used to detect and detect vulnerabilities due to incorrect configuration or errors in the program code of network devices and programs, such as firewall, a router, a web server, an application server, and so on [9]. Modern vulnerability scanners allow scanning with or without authentication. Most vulnerability scanners have the ability to generate reports on vulnerabilities, as well as installed software, open ports, certificates, and other information about the device that can be requested during the scan.

Authentication scan allows the scanner to directly access network resources using remote administration protocols such as SSH or RDP and authenticate using the provided system credentials. This allows a Vulnerability Scanner to access low-level data, such as certain services and device OS configuration information. After scanning, detailed and accurate information about the OS and installed software is provided, including configuration flaws and missing security patches.

Scanning without authentication is a method that can lead to a large number of false positives and does not allow providing detailed information about certain services and operating system configuration, as well as installed software. This method is commonly used by threat agents or security analysts trying to determine the security status of external available assets.

2.7 *Threat Modeling Tools*

Threat modeling is the process by which potential threats, such as structural vulnerabilities, can be identified, listed and prioritized at the very early design stage—from the perspective of a hypothetical attacker. The purpose of threat modeling is to provide IS professionals with a systematic analysis of the profile of a likely attacker, the most likely attack vectors and assets subject to a more likely attack.

Unlike vulnerability scanners, threat modeling tools do not need to interact with devices to analyze current vulnerabilities.

As a rule, IT threat modeling processes begin with the creation of a visual representation of the application or network infrastructure being analyzed. The application or infrastructure is divided into individual elements in order to analyze the most detailed analysis of the entire system. After the analysis is complete, a visual presentation is used to list potential threats. Further analysis of the model with regard to the risks associated with the identified threats, the prioritization of threats and the listing of relevant mitigating management measures depends on the methodological basis of the threat model process used [10].

It is advisable to model threats at the early stages of the design cycle, when potential problems can be identified at the early stages and eliminated, thus avoiding a more expensive solution to the problem in the future. Using threat modeling to analyze security requirements can lead to proactive architectural solutions that will help reduce threats from the very beginning of the operation of the system.

3 Comparing Information Security Tools

To select the most secure solution for ensuring IS systems of an APCS, it is necessary to create a list of criteria for comparing the considered classes. Thus, the following characteristics will be used as criteria:

- impact on industrial traffic;
- work at all levels of the APCS;
- changing the settings of the used hardware and software;
- using this class of solutions in the projected APCS.

Based on the listed criteria, a comparison was made of the decision classes shown in Table 1.

First of all, it is necessary to use the built-in functionality of the automated system to increase security, including eliminating the vulnerabilities of the current infrastructure. If we take into account the results of comparing decision classes shown in Table 1, where only threat modeling tools do not require any manipulations with the operating system (such as setting up network equipment and introducing additional software or hardware–software solutions into an industrial network), it becomes obvious that threat modeling tools are the most secure tool that allows you to raise

Table 1 Comparison of current classes of IS solutions for APCS

	Impact on industrial traffic	Work at all levels of the APCS	Changing the settings of the used hardware and software	Using this class of solutions in the projected APCS
Firewall	Yes	Yes	No	No
Data diode	Yes	Yes	No	No
IDS	Yes	Yes	Yes	No
Mirroring	No	Yes	Yes	No
SIEM	Yes	No	Yes	No
Vulnerability scanner	Yes	No	Yes	No
Threat modeling tool	No	Yes	No	Yes

the level of information systems of industrial systems [11]. It should also be noted that such tools allow you to effectively design an already existing industrial network, eliminating potential threats in the early stages [12].

The following solutions in the field of threat modeling are the most popular and supported today:

- Attack-Defense Tree Tool (ADTool) is a component of the CybatiWorks training platform developed at the University of Luxembourg as part of the ATREES project to develop a toolkit for building an attack tree and defensive measures. ADTool allows users to model and analyze attack protection scenarios with the help of visual creation of "attack trees" and measures to neutralize them. Since this tool works only with models of threats prepared by the developer in advance, this greatly limits its use.
- Microsoft Threat Modeling Tool 2016 is a free solution for modeling threats from Microsoft, focused on software developers. This product allows you to integrate into systems used in tracking errors and other problems in software. Despite all the advantages of this tool, its use for analyzing IS threats from an APCS is not practical, since this solution is primarily designed to inventory threats inside or between applications, while the vast majority of software products used in APCS are proprietary, only developers have access to the source code.
- OWASP Threat Dragon is a web-based online threat modeling application. This tool is designed to model threats in the early stages of the application development life cycle. The main disadvantage of this solution, like the previous one, is focusing only on vulnerabilities and threats in applications; this tool does not have the ability to simulate threats to network equipment or OS, which makes it inapplicable when used in a distributed infrastructure.
- ThreatModeler is an automated threat modeling solution from MyAppSecurity that identifies, predicts, and identifies threats throughout the application development life cycle, as well as the ability to simulate threats to cloud solutions and IoT

Table 2 Compliance with the requirements of threat modeling tools

	ADTool	Threat Modeling Tool 2016	Threat Dragon	ThreatModeler
Visual network view	Yes	No	No	No
Quantitative threat and risk assessment	No	Yes	No	Yes
Modeling threats for hardware and OS	Yes	No	No	Yes
Database replenishment	No	No	Yes	No

devices. This product does not have the ability to supplement the database of threats and vulnerabilities, thus, the modeling of threats to non-standard equipment is not possible, limiting the use of the solution.

Taking into account the specific features of providing IS systems for APCS [13], it is necessary to determine the main requirements for the considered tools for modeling threats to APCS:

- visual representation of the industrial network infrastructure;
- quantitative assessment of threats and risks;
- modeling of threats to hardware and operating systems;
- replenishment of the database with current and own vulnerabilities and threats.

Based on the listed requirements, an analysis was conducted for compliance with the requirements of the considered threat modeling tools shown in Table 2.

Many of the solutions considered are focused solely on the level of software development, not intended for the correct modeling of threats to industrial networks. Also, many of the products do not allow to model threats with their own parameters and do not support the risk assessment of found threats in the system. In this connection, the development of a threat modeling tool and the risk assessment of an APCS is an urgent task.

4 Risk Assessment Methodology

Before starting the development of a threat modeling tool, it is necessary to determine the methodology used to assess the risks of IS in simulated threats [14]. In the absence of an adequate risk assessment, it is difficult to answer questions about where to start building an information protection system, which resources and which threats should be protected and which countermeasures are the highest priorities [15]. It is also difficult to resolve the issue of the need and sufficiency of one or another set of countermeasures and their adequacy to existing risks [16]. Thus, the issue of assessing the risks associated with the implementation of network attacks is of prime importance [17].

It is worth paying attention to the fact that most of the parameters used in the methodologies considered in [18] are based on qualitative assessments, which in itself reduces their effectiveness. Also, the proposed methods allow assessing risks for individual sections of the system and do not always take into account the consequences for the system as a whole, that is, when assessing risks, negative consequences for various classes of attacks that use a combination of vulnerabilities are not considered, which is quite critical for industrial networks [19], therefore using existing open-access methodologies for assessing IS risks of an APCS is inefficient and it is necessary to use a new approach to threat modeling and risk assessment of IS.

The first thing with which threat modeling begins is the identification of the used components of the APCS that make up the industrial network—C. Following the inventory of the components, it is necessary to determine the related software (software and hardware) tools—S^C. In addition to the set of associated software, each component also has a value parameter, C^W, which determines the cost of an organization's costs incurred in the event of exploiting the vulnerability associated with the software component.

The repository where the vulnerability search for software components will be performed is the Data Security Threat Database (the Russian Federation's national vulnerability database—BDU), since the BDU contains detailed information about vulnerabilities from various public sources (CVE, OSVDB, NVD, CISCO-SA and others), adapted to a single format and focused on industrial facilities. As a vulnerability assessment, the common Common Vulnerability Scoring System version 3.0 (CVSS) [20] is used, the result of which takes a value from 0 to 10. The advantage of this metric is including the ability to self-assess vulnerabilities, using the described criteria. The vulnerability vector looks like this:

$$v_i = \mathrm{CVSS:3.0/AV}:X/\mathrm{AC}:X/\mathrm{PR}:X/\mathrm{UI}:X/\mathrm{S}:X/\mathrm{C}:X/\mathrm{I}:X/\mathrm{A}:X/\mathrm{E}:X/\mathrm{RL}:X/\mathrm{RC}:X, \tag{1}$$

where X—criteria values.

One of the characteristics of vulnerabilities is its type, which combines vulnerabilities similar in parameters to a single set. The directory of types of vulnerabilities is Common Weakness Enumeration (CWE), and the identifier set in accordance with the general list of CWE errors is called the error type identifier (ETI). Thus, each vulnerability will correspond to one or several ETI, that is,

$$v_i^{cwe} = \{CWE\}, \tag{2}$$

where CWE—a set of ETI in the CWE database.

Using the statistics of third-party resources associated with the identification of vulnerabilities of various types, each ETI cwe_j corresponds to the total share of incidents associated with it. In the absence of any of the ETIs in the statistics, they will be assigned a residual percentage, usually referred to as "others." The share of incidents ETI cwe_j has the following designation—cwe_j^p. Thus, the proportion of incidents related to a particular vulnerability of one or more types will be determined as follows:

$$v_i^p = \frac{\sum_{j=1}^{n} v_i^{cwe_j^p}}{n}, \tag{3}$$

where n—the number of ETI related to the vulnerability v_i.

According to ISO/IEC 27000: 2014, vulnerability is a weakness of an asset or control that can be exploited by one or more threats. Therefore, to assess the impact of a threat on a component of a system, we will use chains of vulnerabilities that combine the exploitation characteristics of all vulnerabilities of (1) component, that is, each threat consists of several vulnerabilities (4):

$$T_x = \{T_x^{v_1}, \ldots, T_x^{v_m}\}, \tag{4}$$

where m—the number of vulnerabilities that are included in the T_x threat.

The threat vector is the maximum parameter from all its multiple vulnerabilities:

$$T_x = CVSS : 3.0/\max(AV)/\max(AC)/\max(PR)/$$
$$\max(UI)/\max(S)/\max(C)/\max(I)/\max(A) \tag{5}$$

Since a threat can affect several software/hardware components of a component, to determine the damage, it is necessary to determine the matrix of the interrelation of vulnerabilities (Table 3) of the threat T_i with software components that it affects.

In order to determine which components were affected when the threat T_x was realized for each of the components of the c_y matrix shown in Table 3, it is necessary to calculate the degree of influence of the threat T_x on the component c_y:

$$T_x^{c_y} = \sum_{i=1}^{m} \sum_{j=1}^{k} w_{ij}^{c_y}, \tag{6}$$

where $w_{ij}^{c_y} \in \{0, 1\}$—the coefficient of the impact of the vulnerability v_i on the s_j of the c_y component.

Thus, the potential damage caused by the threat of T_x is calculated as follows:

Table 3 Matrix of the interconnection of threat vulnerabilities and software

	c_1			…	c_n		
	$c_1^{s_1}$	…	$c_1^{s_{k_1}}$		$c_n^{s_1}$	…	$c_n^{s_{k_n}}$
$T_x^{v_1}$	$w_{11}^{c_1}$	…	$w_{1k_1}^{c_1}$	v_1	$w_{11}^{c_1}$	…	$w_{1k_1}^{c_1}$
…	…	…	…	…	…	…	…
$T_x^{v_m}$	$w_{m1}^{c_1}$	…	$w_{mk_1}^{c_1}$	v_m	$w_{m1}^{c_1}$	…	$w_{mk_1}^{c_1}$

$$T_x^Q = \sum_{y=1}^{n} c_y^Q, \quad (7)$$

where $T_x^{c_y} \neq 0$,
c_y^Q—the cost of the component.

After calculating the potential damage of all threats, it is necessary to assess this damage, for this the maximum value of all threats $\max(T^Q)$ is determined and the damage of the remaining threats is estimated relative to it:

$$E_x = \frac{T_x^Q}{\max(T^Q)}, \quad (8)$$

where $E_x \in \{0, \ldots, 1\}$—the damage assessment.

Each software corresponds to a certain type, which determines what type this tool belongs to. Using the statistics of the BDU, each type of software will correspond to the number of associated vulnerabilities, and the percentage of related vulnerabilities will have the following form—s_j^p, so the likelihood of a threat T_x per component c_y according to the matrix of the interconnection of threat vulnerabilities and software (Table 3) will look like:

$$c_y^{T_x} = \frac{\sum_{i=1}^{m} \sum_{j=1}^{k} w_{ij}^{c_y} \times T_x^{v_j^p} \times c_y^{s_j^p}}{T_x^{c_y}} \quad (9)$$

The following formula will be applied to determine the likelihood of a certain T_x threat being implemented to all system components:

$$p_x = \frac{\sum_{y=1}^{n} c_y^{T_x}}{n}, \quad (10)$$

where $p_x \in \{0,\ldots,1\}$—the probability of realization of the threat T_x.

By risk we mean the possibility that when a certain threat is realized with an assessment of T_x, an adverse event will occur that has an estimated damage E_x with a probability of occurrence of this event p_x [21]. Thus, the overall risk assessment of the system R_x will be the product of the assessment of the damage E_x, the assessment of the threat T_x and the probability of occurrence of this event p_x and will take a value from 0 to 10:

$$R_x = E_x \times T_x \times p_x. \quad (11)$$

5 Conclusion

As a result of the analysis of the existing IS security tools for the APCS, it was decided that the most secure in terms of the potential impact on the process is the use of threat modeling tools. Considering that none of the most popular solutions offered on the market is fully suitable for modeling threats of an APCS [22], the authors of the article proposed their own methodology for modeling threats and risk assessment.

The proposed methodology has several features:

- risk assessment is a quantity that allows you to rank risks according to their significance, eliminating primarily the most dangerous ones;
- risks can be assessed not only for known vulnerabilities but also for undeclared vulnerabilities in open sources;
- threat is a combination of several vulnerabilities with the ability to add and remove them.

References

1. Zegzhda, D., Poltavtseva, M., Lavrova, D.: Systematization and security assessment of cyber-physical systems. Autom. Control Comput. Sci. **51**(8), 835–843 (2017). https://doi.org/10.3103/S0146411617080272
2. Alguliyev, R., Imamverdiyev, Y., Sukhostat, L.: Cyber-physical systems and their security issues. Comput. Ind. **100**, 212–223 (2018). https://doi.org/10.1016/j.compind.2018.04.017
3. Zegzhda, D., Stepanova, T.: Approach to APCS protection from cyber threats. Autom. Control Comput. Sci. **49**(8), 659–664 (2015). https://doi.org/10.3103/S0146411615080179
4. Huang, S., Zhou, C.J., Yang, S.H., Qin, Y.Q.: Cyber-physical system security for networked industrial processes. Int. J. Autom. Comput. **12**(6), 567–578. https://doi.org/10.1007/s11633-015-0923-9 (2015)
5. Okhravi, H., Sheldon, F., Haines, J.: Data diodes in support of trustworthy cyber infrastructure and net-centric cyber decision support. In: Pappu, V., Carvalho, M., Pardalos, P. (eds.) Optimization and Security Challenges in Smart Power Grids. Energy Systems. Springer, Berlin, Heidelberg (2013). https://doi.org/10.1007/978-3-642-38134-8_10
6. Oman, P., Phillips, M.: Intrusion detection and event monitoring in SCADA networks. In: Goetz, E., Shenoi, S. (eds.) Critical Infrastructure Protection. ICCIP 2007. IFIP International Federation for Information Processing, vol. 253. Springer, Boston, MA (2008). https://doi.org/10.1007/978-0-387-75462-8_12
7. Parry, J., Hunter, D., Radke, K., Fidge, C.: A network forensics tool for precise data packet capture and replay in cyber-physical systems. In: Proceedings of the Australasian Computer Science Week Multiconference, No. 22 (2016). https://doi.org/10.1145/2843043.2843047
8. Coppolino, L., D'Antonio, S., Formicola, V., Romano, L.: Enhancing SIEM technology to protect critical infrastructures. In: Hämmerli, B.M., Kalstad Svendsen, N., Lopez, J. (eds.) Critical Information Infrastructures Security. Lecture Notes in Computer Science, vol. 7722. Springer, Berlin, Heidelberg (2013). https://doi.org/10.1007/978-3-642-41485-5_2
9. Samtani, S., Yu, S., Zhu, H., Patton, M., Chen, H.: Identifying SCADA vulnerabilities using passive and active vulnerability assessment techniques. In: IEEE Conference on Intelligence and Security Informatics (ISI), pp. 25–30 (2016). https://doi.org/10.1109/ISI.2016.7745438

10. Paté-Cornell, M.E., Kuypers, M., Smith, M., Keller, P.: Cyber risk management for critical infrastructure: a risk analysis model and three case studies. Risk Anal. **38**(2), 226–241 (2017). https://doi.org/10.1111/risa.12844
11. Kerzhner, A.A., Tan, K., Fosse, E.: Analyzing cyber security threats on cyber-physical systems using Model-Based Systems Engineering. In: AIAA SPACE 2015 Conference and Exposition (2015). https://doi.org/10.2514/6.2015-4575
12. Al-Mohannadi, H., Mirza, Q., Namanya, A., Awan, I., Cullen, A., Disso, J., Cyber-attack modeling analysis techniques: an overview. In: 2016 IEEE 4th International Conference on Future Internet of Things and Cloud Workshops (FiCloudW), pp. 69–76 (2016). https://doi.org/10.1109/W-FiCloud.2016.29
13. Yang, Y., Lu, J., Choo, K.K.R., Liu, J.K.: On lightweight security enforcement in cyber-physical systems. In: Güneysu, T., Leander, G., Moradi, A. (eds.) Lightweight Cryptography for Security and Privacy. LightSec 2015. Lecture Notes in Computer Science, vol. 9542. Springer, Cham (2016). https://doi.org/10.1007/978-3-319-29078-2_6
14. Xie, F., Peng, Y., Zhao, W., Gao, Y., Han, X.: Evaluating industrial control devices security: standards, technologies and challenges. In: Saeed, K., Snášel, V. (eds.) Computer Information Systems and Industrial Management. CISIM 2015. Lecture Notes in Computer Science, vol. 8838. Springer, Berlin, Heidelberg (2014). https://doi.org/10.1007/978-3-662-45237-0_57
15. Cherdantseva, Y., Burnap, P., Blyth, A., Eden, P., Jones, K., Soulsby, H., Stoddart, K.: A review of cyber security risk assessment methods for SCADA systems. Comput. Secur. **56**, 1–27 (2016). https://doi.org/10.1016/j.cose.2015.09.009
16. Kalashnikov, A., Sakrutina, E.: The model of evaluating the risk potential for critical infrastructure plants of nuclear power plants. In: Eleventh International Conference "Management of Largescale System Development", Moscow (2018). https://doi.org/10.1109/MLSD.2018.8551910
17. Singhal, A., Ou, X.: Security risk analysis of enterprise networks using probabilistic attack graphs. In: Network Security Metrics. Springer, Cham (2017). https://doi.org/10.1007/978-3-319-66505-4_3
18. Rimsha, A., Zakharov, A.: Method for risk assessment of industrial networks' information security of gas producing enterprise. In: Global Smart Industry Conference (GloSIC), Chelyabinsk (2018). https://doi.org/10.1109/GloSIC.2018.8570079
19. Sadeghi, A., Wachsmann, C., Waidner, M.: Security and privacy challenges in industrial internet of things. In: 2015, 52nd ACM/EDAC/IEEE Design Automation Conference (DAC), pp. 1-6 (2015). https://doi.org/10.1145/2744769.2747942
20. Doynikova, E., Chechulin, A., Kotenko, I.: Analytical at-tack modeling and security assessment based on the common vulnerability scoring system. In: 20th Conference of Open Innovations Association (FRUCT), St. Petersburg, pp. 53–61 (2017). https://doi.org/10.23919/FRUCT.2017.8071292
21. McCormac, A., Zwaans, T., Parsons, K., Calic, D., Bu-tavicius, M., Pattinson, M.: Individual differences and information security awareness. Comput. Hum. Behav. **69**, 151–156 (2017). https://doi.org/10.1016/j.chb.2016.11.065
22. Colombo, A.W., Karnouskos, S., Bangemann, T.: Towards the next generation of industrial cyber-physical systems. In: Colombo, A., et al. (eds.) Industrial Cloud-Based Cyber-Physical Systems. Springer, Cham (2014). https://doi.org/10.1007/978-3-319-05624-1_1

Determining the Parameters of the Mathematical Model of the Process of Searching for Harmful Information

Igor Kotenko and Igor Parashchuk

Abstract The research object is the process of detecting harmful information in social networks and the global network. The chapter proposes an approach for determining (verifying) the parameters of a mathematical model of a stochastic process for detecting harmful information with unreliable, inaccurate (contradictory) specified initial data. The approach is based on the use of stochastic equations of state and observation based on controlled Markov chains with finite differences. Moreover, the determination (verification) of key parameters of the mathematical model of this type (the elements of the matrix of one-step transition probabilities) is carried out by using an extrapolating neural network. This allows one to take into account and compensate for the inaccuracy of the original data inherent stochastic processes of searching and detecting harmful information. In addition, this approach allows one to increase the reliability of decision-making in evaluating and categorizing the digital network content for detecting and counteracting information of this class.

Keywords Harmful information protection · Mathematical model · Activation function · Neural network · Matrix · Transition probability · State · Estimation

1 Introduction

Mathematical models of complex processes, including the process of detecting harmful information (HI), require a large number of a priori (initial) data about these processes. Such data is on the basis of the parameters of adequate mathematical models of this class.

I. Kotenko (✉) · I. Parashchuk
St. Petersburg Institute for Informatics and Automation of Russian Academy of Sciences (SPIIRAS), 39, 14 Liniya, St. Petersburg 199178, Russia
e-mail: ivkote@comsec.spb.ru

I. Parashchuk
e-mail: parashchuk@comsec.spb.ru

A. G. Kravets et al. (eds.), *Cyber-Physical Systems: Industry 4.0 Challenges*, Studies in Systems, Decision and Control 260,
https://doi.org/10.1007/978-3-030-32648-7_18

However, in the overwhelming number of cases, this data is poorly formalized, qualitative, expert, and inaccurate (contradictory). This is due to the diversity of HI circulating in social networks and the Internet [1–3]. In recent years, along with other methods, extrapolating neural networks (ENS) are increasingly being used to process such data. These networks are able to accurately approximate poorly formalized, often unreliable, inaccurate (contradictory) data, bringing them to a form suitable for reliable and adequate modeling of complex processes [4].

The modeling of the HI detection process is carried out in the interests of building intelligent systems for analytical processing of digital network content, designed to search for and resist undesirable, illegal, and in general harmful information. The use of ENS to determine (verify) the parameters of the mathematical model of the HI detection process is expedient and relevant. It will allow one to specify mathematically these parameters in terms of qualitative and quantitative subjective assessments. At the same time, weight coefficients (synaptic weights) and weight ratios must be determined, and the inaccurate (contradictory) data must be formalized and determined by means of ENS verification.

The quality of determining (verifying) the parameters of the mathematical model of the HI detection process increases, as the volume of the accumulated data and knowledge increases [4, 5]. Using ENS allows one to automatically accumulate empirical knowledge about the properties of the HI detection process and make decisions based on accumulated data and knowledge.

2 Related Works

ENS applications are used in such problems of determining model parameters (verification) [4, 5], where traditional probabilistic and other methods are ineffective. This is due to uncertainty, for example, due to the incompleteness and inconsistency of these parameters. But the differences in the levels of uncertainty do not allow to unambiguously determining the real HI features, which makes it difficult to directly apply these methods for this class of problems.

In [6] an adaptive extrapolation of process states using recurrent neural networks is considered. But such an approach requires consideration of auxiliary extrapolation parameters, and this is not always possible. In the interests of identifying and verifying the state of nonlinear modeling objects, multilayer neural networks are sometimes used [7, 8]. But they do not guarantee the high accuracy of determining the parameters and the accuracy of their identification and verification.

In [9], an extended approach to neural network identification and verification is presented. But it is applicable for modeling quasi-stationary processes, which limits its scope. The papers [10, 11] are devoted to a method that allows modeling complex processes using neuro-fuzzy analysis schemes with fuzzy links. But this method is very difficult for mathematical description and laborious.

For our case, when it is necessary to verify the parameters of the mathematical model of the HI search process, the definition is unreliable, inaccurate (contradictory) given initial data for modeling this stochastic process, reduces to a trivial classification problem. This is the task of classifying the parameters of the mathematical model using an extrapolating neural network algorithm. This approach is known, simple and is used in the tasks of recognition and estimation of the parameters of telecommunication networks [12].

3 Theoretical Part

The procedure of the extrapolating neural network verification will be examined by an example of determining the key parameters of the mathematical model of the HI search process, i.e. the values of the elements of the matrix of one-step transition probabilities (OTP) $\tilde{\beta}(t, t+1, c)$ for different states of the parameter of this process. The known papers [13–18] devoted to the synthesis of mathematical models for discrete in state and continuous in time stochastic processes allow one to formulate the unified models of the HI detection process on the basis of controlled Markov chains.

These models are usually specified in the form of differential or difference stochastic equations [19–21]. In this case, the model of the state change dynamics of any j-th unreliable or inaccurate (contradictory) given parameter of the HI detection process (for example, for nine states) can be specified as a system of difference equations of state and observation:

$$\vec{q}_j(t+1) = C^{\mathrm{T}}_{\vec{q}_j}(t+1)\, \vec{\mathfrak{R}}_{\vec{q}_j}(t+1); \tag{1}$$

$$\vec{\mathfrak{R}}_{\vec{q}_j}(t+1) = \tilde{\beta}^{\mathrm{T}}_{q_j}(t, t+1, c)\, \vec{\mathfrak{R}}_{\vec{q}_j}(t) + \Delta\vec{\mathfrak{R}}_{\vec{q}_j}(t+1); \tag{2}$$

$$\vec{z}_{\vec{q}_j}(t+1) = H_{\vec{q}_j}(\vec{q}_j(t+1))\, \vec{\mathfrak{R}}_{\vec{q}_j}(t+1) + \vec{\omega}_{\vec{q}_j}(t+1), \tag{3}$$

where

$\vec{q}_j(t+1)$—a vector of unreliable or inaccurate (contradictory) given values of the j-th parameter of the HI detection process;
$C^{\mathrm{T}}_{\vec{q}_j}(t+1)$—a N-dimensional (in our case N = 9) transposed matrix-row of possible unreliable, inaccurate (contradictory) given values of the j-th parameter of the HI detection process;
$\vec{\mathfrak{R}}_{\vec{q}_j}(t+1)$—an auxiliary vector indicator of the state of the j-th unreliable, inaccurate (contradictory) given parameter of the HI detection process;
$H_{\vec{q}_j}(\vec{q}_j(t+1))$—a nine-dimensional matrix of unreliable (inaccurate, controversial) observations of the state change dynamics of the HI detection process;
$\Delta\vec{\mathfrak{R}}_{\vec{q}_j}(t+1)$—a vector of increment values of the state indicators of the HI detection process;

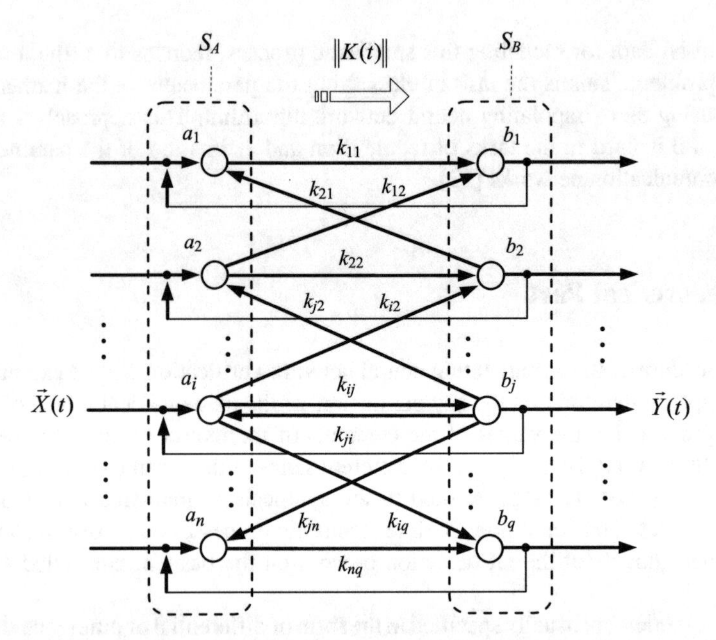

Fig. 1 An example of the structure of a two-layer ENS to determine (verify) the elements of the OTP matrix

$\vec{\Re}_{\vec{q}_j}(t)$—a value of the vector of the auxiliary state indicators of the j-th unreliable, inaccurate (contradictory) given parameter at the previous step (step) t of the HI detection process;
$\vec{z}_{\vec{q}_j}(t+1)$—a vector of the state observation of the j-th unreliable, inaccurate (contradictory) given parameter of the HI detection process;
$\vec{\omega}_{\vec{q}_j}(t+1)$—a vector of observation noise of the state change dynamics of the HI detection process;
$\tilde{\beta}^{\mathrm{T}}_{q_j}(t, t+1, c)$—a transposed matrix of the unreliable, inaccurately (inconsistently) given one-step transition probabilities of the j-th parameter of the HI detection process from the t-th to (t + 1)-th state, taking into account external and control actions c(t).

The definition (verification) of the elements of the OTP matrix $\tilde{\beta}_{q_j}(t, t+1, c)$ is the main feature to develop the model of the HI detection process. This process is carried out using an extrapolating neural network algorithm.

The algorithm is designed to convert poorly formalized, often unreliable, inaccurate (contradictory) given source data to a form suitable for reliable and adequate modeling of the HI detection process.

It can be implemented on the basis of a typical ENS with two or more layers. An example of building an ENS with two layers is shown in Fig. 1.

In a two-layer network, the neurons of the input layer SA (a_1, …, a_n) are responsible for entering the source data vector, and the neurons of the output layer SB (b_1, …, b_q) specify a verified correlation dependence of the probability of transition from one state to another, i.e. elements of the vector $\vec{Y}(t)$, taking into account the matrix of relations (weights) $\|K(t)\|$.

The process of determining (verifying) the elements of the OTP matrix $\tilde{\beta}_{q_j}(t, t+1, c)$, using the extrapolating neural network algorithm, for example, for nine states of the j-th parameter of the HI detection process, has several peculiarities. A typical ENS uses cognitive maps for each of the possible states.

These states are traditionally defined by special connection matrices that have (for example, for nine states of the parameter of the HI detection process) the following form:

$$K(t) = \begin{vmatrix} k_{11}(t) & k_{12}(t) & k_{13}(t) & \dots & k_{18}(t) & k_{19}(t) \\ k_{21}(t) & k_{22}(t) & k_{23}(t) & \dots & k_{28}(t) & k_{29}(t) \\ k_{31}(t) & k_{32}(t) & k_{33}(t) & \dots & k_{38}(t) & k_{39}(t) \\ k_{41}(t) & k_{42}(t) & k_{43}(t) & \dots & k_{48}(t) & k_{49}(t) \\ \vdots & \vdots & \vdots & \ddots & \vdots & \vdots \\ k_{91}(t) & k_{92}(t) & k_{93}(t) & \dots & k_{98}(t) & k_{99}(t) \end{vmatrix}. \tag{4}$$

According to the input vectors, characterizing the dependence of the transition probabilities in the row of the OTP matrix, the elements of these connection matrices are formed. The elements of the connection matrices are formed at the first stage of ENS learning based on the expert opinions, arriving at the input layer of the neural network. At the same time, the ENS is a typical two-layer neural network with direct distribution of information.

In accordance with the steps of determining (verifying) the elements of the OTP matrix $\tilde{\beta}_{q_j}(t, t+1, c)$, using the extrapolating neural network algorithm, each element kji(t) characterizes one row of the OTP matrix. This element describes the correlation dependence of the probability of transition from one state to another, the relationship of the j-th and i-th transition probabilities on the t-th cycle (step) of the HI detection process.

4 Experiments and Discussion

Let us consider a specific example of the procedure for determining (verifying) the key parameter of the mathematical model—the elements of the OTP matrix.

The first stage is the formation of a connection matrix. For nine states of the parameter of the HI detection process, nine weights matrices should be specified. During the formation of the first row of the OTP matrix, the probabilities of transitions from the first state to the second, third, fourth, fifth, sixth, seventh, eighth, and ninth state, and the probability to remain in the first state are determined (verified). In this case, the vector characterizing the opinion of the first expert on the dependence of the transition probabilities in the first row of the OTP matrix is written as

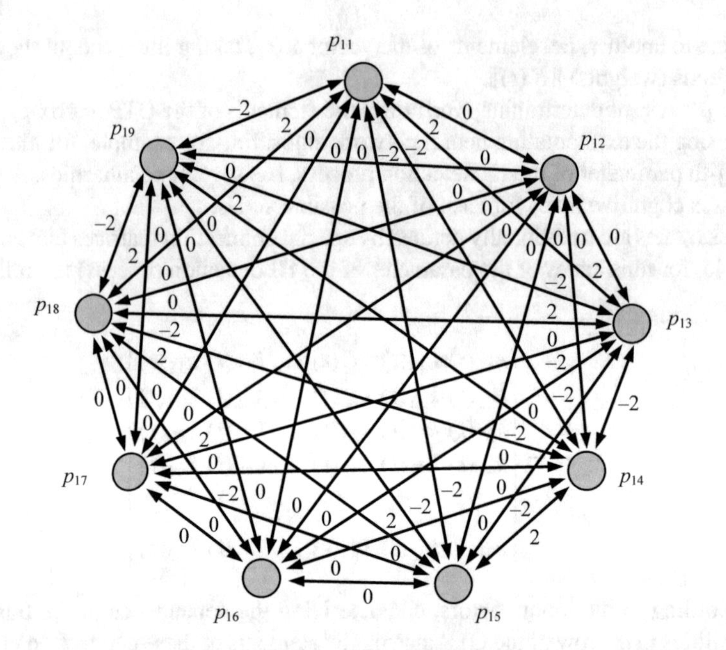

Fig. 2 Cognitive map for the example of the ENS in the interests of determining (verifying) the nine elements of the first row of the OTP matrix

$$\vec{X}_1 = (1, 0, -1, -1, 0, 0, 1, -1). \tag{5}$$

The expression (5) is a formal notation of a mathematical statement: an increase in the probability of being in the first state and the probability of remaining in it $p_{11}(t)$ increases the probability of transition from the first to the third state $p_{13}(t)$ and increases the probability of transition from the first to the eighth state $p_{18}(t)$.

This leads to a decrease in the probability of transition from the first to the fourth $p_{14}(t)$, the fifth $p_{15}(t)$ and from the first to the ninth state $p_{19}(t)$.

And on the impact of changes in the probability of transition from the first to the second $p_{12}(t)$, the sixth $p_{16}(t)$ and the seventh $p_{17}(t)$ states, the expert's opinion is missing. If the positive value of the element of the vector $\vec{X}_1$ corresponds to the positive value of another element of the vector $\vec{X}_1$, this relationship between them is designated as positive.

The relationship is denoted as negative if a positive value of an element of the vector $\vec{X}_1$ corresponds to a negative value of another element of this vector. The quantitative characterization of the relationship is the sum modulo for all the values that describe the relationship of transition probabilities in the OTP matrix. Taking expression (5) as the initial condition for the work of the extrapolating neural network algorithm, one can obtain a cognitive map for the ENS (Fig. 2) and the corresponding connection matrix.

The connection matrix, for example, for the first row of the OTP matrix, looks like this

$$K_1(t) = \begin{vmatrix} 0 & 0 & 2 & -2 & -2 & 0 & 0 & 2 & -2 \\ 0 & 0 & 0 & 0 & 0 & 0 & 0 & 0 & 0 \\ 2 & 0 & 0 & -2 & -2 & 0 & 2 & 0 & 0 \\ -2 & 0 & -2 & 0 & 2 & 0 & 0 & -2 & 2 \\ -2 & 0 & -2 & 2 & 0 & 0 & -2 & 2 & 0 \\ 0 & 0 & 0 & 0 & 0 & 0 & 0 & 0 & 0 \\ 0 & 0 & 0 & 0 & 0 & 0 & 0 & 0 & 0 \\ 0 & 0 & 2 & -2 & -2 & 0 & 0 & 2 & -2 \\ -2 & 0 & -2 & 0 & 2 & 0 & 0 & -2 & 2 \end{vmatrix}. \quad (6)$$

Cognitive maps and connection matrices $K_2(t)$, $K_3(t)$, …, $K_9(t)$ for the eight remaining rows of the OTP matrix are determined (verified) in a similar way. The work of the ENS provides that cognitive expert cards are combined into a cognitive summary map.

The cognitive map and the matrix of weights combine the collective opinion of all involved experts on the correlation dependences of the probabilities of the transition of the parameter of the HI detection process from state to state. The total matrix is described by the expression

$$G_1(t) = \sum_{j=1}^{J} K_j(t), \quad (7)$$

where $J = 9$ is the number of experts (individual connection matrices) participating in the definition (verification) of the elements of the OTP matrix and the nature of the connections between them.

The United matrix reflects the conflicting opinions of all experts on the correlation dependences of the probabilities of the transition of a HI detection process parameter from state to state. Therefore, it contains not only elements 1 and –1, but also 0.

This allows one to more completely reflect the causality relationships between the states of the parameter of the HI detection process. The united cognitive maps and weights matrices for the remaining rows of the OTP matrix $\tilde{\beta}_{q_j}(t, t+1, c)$ are defined similarly.

Let us consider in detail the steps of determining (verifying) the key parameter of the mathematical model—the elements of the OTP matrix using the usual two-layer ENS [6–8].

This extrapolating neural network algorithm is designed to determine the values of the modeling parameters and their extrapolation in the interests of evaluating the effectiveness of the HI detection. The neural network verification based on ENS includes a number of stages. The block diagram of the neural network algorithm for determining (verifying) the values of the transition probabilities of the parameter of the HI detection process from state to state is proposed in Fig. 3.

Initially, the activation of the input ENS layer is performed, i.e. bringing the neurons of the input layer to the initial state. Then the initial initialization of the

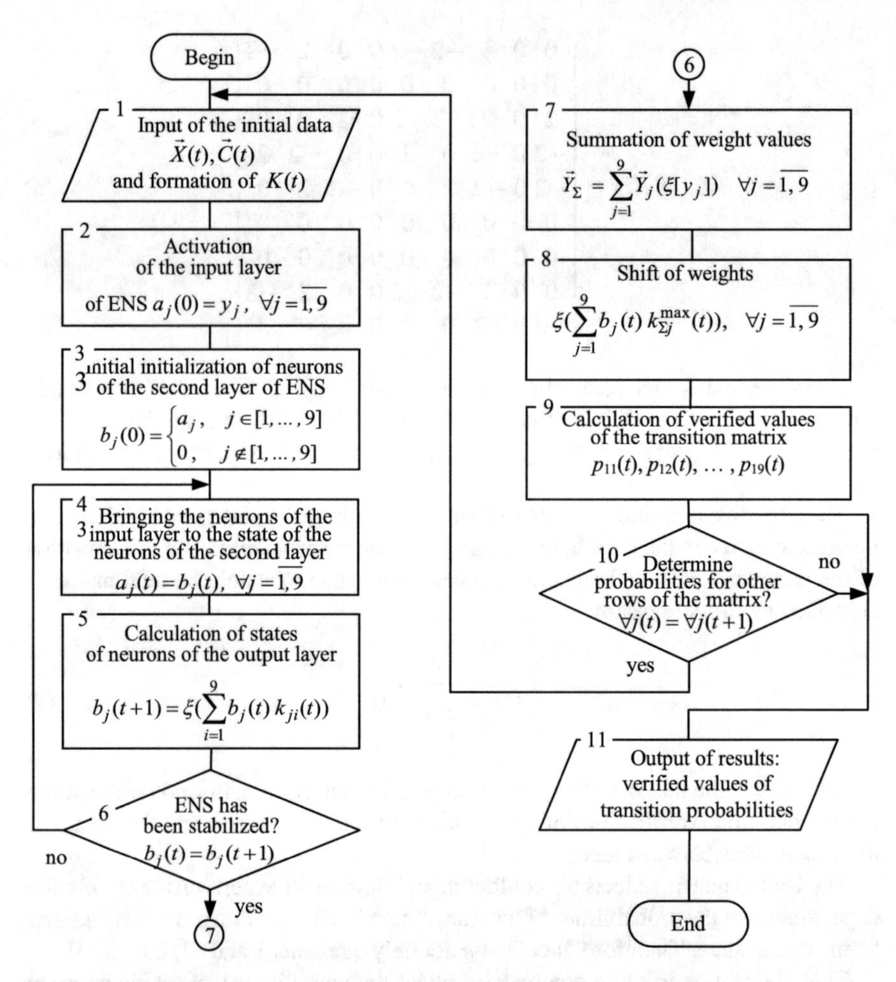

Fig. 3 The block diagram of the neural network algorithm for sequential determination (verification) of the values of the elements of nine rows of the OTP matrix

neurons of the second layer occurs, after this the neurons of the input ENS layer are brought to the state of the neurons of the second layer: $x_j = y_j, \forall j = \overline{1,9}$.

Then the new states of the neurons of the second layer for all $j \in [1, \dots, 9]$ are calculated according to the formula:

$$y_j(t+1) = \xi\left(\sum_{i=1}^{5} y_j(t)\, k_{ji}(t)\right), \forall j = \overline{1,9}, \tag{8}$$

where ξ is the step activation function. The calculations (8) are repeated until the ENS reaches a stable state. Checking (if the ENS has reached the stable state) is carried out by comparing the states of the neurons of the second layer at the previous t-th and next (t + 1)-th cycle.

When the ENS reaches a stable state, the weight coefficients are summed.

$$\vec{Y}_{\Sigma} = \sum_{j=1}^{9} \vec{Y}_j(\xi[y_j]). \tag{9}$$

The elements of the vector (9) characterize certain (verified) weight coefficients of the correlation connection of the values of the transition probabilities of the HI detection process parameter for one row of the OTP matrix. The total correlation matrix of weights is formed on their basis.

The next step of the neural network verification is the calculation of the probabilities of transition of the HI detection process parameter from the first state to the second $p_{12}(t)$, the third $p_{13}(t)$, the fourth $p_{14}(t)$, the fifth $p_{15}(t)$, the sixth $p_{16}(t)$, the seventh $p_{17}(t)$, the eighth $p_{18}(t)$, the ninth state $p_{19}(t)$, and the probabilities of remaining in the first state $p_{11}(t)$.

To do this, the quadratic metric, applied to the offset values of the total weights, is used [12].

The final step of the neural network verification is the reiteration of the steps of the algorithm until the elements in all the remaining *j*-th rows of the OTP matrix are determined.

Within our example and taking into account the stages described in Fig. 3, the output vector of the second layer at each *t*-th cycle of the ENS operation successively takes a series of state values. These states are determined on the basis of expression (8) and for our example will be equal:

$$\vec{Y}_1(1) = \xi([0, 0, 2, -6, 2, 0, 0, 2, -2]) = [1, 0, 1, -1, 1, 0, 0, 1, -1];$$
$$\vec{Y}_1(2) = \xi([10, -4, 2, -4, -4, 4, 2, 4, -4]) = [1, -1, 1, -1, -1, 1, 1, 1, -1];$$
$$\vdots \qquad \vdots \qquad \vdots$$
$$\vec{Y}_1(5) = \xi([6, 10, 6, -20, -12, 10, 6, 10, -10]) = [1, -1, 1, -1, -1, 1, 1, 1, -1].$$

The example shows that the ENS reaches a stable state already in the fourth step.

Similarly, we obtain the output vectors of the second ENS layer for input vectors, which characterize the correlation dependences between the probability of being in the first state and remaining in it and the probabilities of transition from the first to the third, to the fourth, to the fifth, sixth, seventh, eighth and ninth states.

The output vectors of the second layer of the ENS contain the following elements:

$$\vec{Y}_2(4) = \xi([6, 10, 6, -20, -12, -10, -12, 10, 6]) = [1, 1, 1, -1, -1, -1, -1, 1, 1];$$
$$\vec{Y}_3(5) = \xi([6, 10, 6, -20, -12, -10, -12, 10, 6]) = [1, 1, 1, -1, -1, 1, -1, 1, 1];$$
$$\vec{Y}_4(3) = \xi([-6, -10, -6, 20, 12, -10, 10, 6, -12]) = [-1, -1, -1, 1, 1, -1, 1, 1, -1];$$
$$\vec{Y}_5(4) = \xi([-6, -10, -6, 20, 12, -10, 10, 6, -12]) = [-1, -1, -1, 1, 1, -1, 1, 1, -1];$$
$$\vec{Y}_6(4) = \xi([6, 10, 6, -20, -12, -10, -12, 10, 6]) = [1, 1, 1, -1, -1, -1, -1, 1, 1];$$

$$\vec{Y}_7(5) = \xi([-6, -10, -6, 20, 12, -10, 10, 6, -12]) = [-1, -1, -1, 1, 1, -1, 1, 1, -1];$$
$$\vec{Y}_7(5) = \xi([-6, -10, -6, 20, 12, -10, 10, 6, -12]) = [-1, -1, -1, 1, 1, -1, 1, 1, -1];$$
$$\vec{Y}_8(4) = \xi([6, 10, 6, -20, -12, -10, -12, 10, 6]) = [1, 1, 1, -1, -1, 1, -1, 1, 1];$$
$$\vec{Y}_9(3) = \xi([6, 10, 6, -20, -12, -10, -12, 10, 6]) = [1, 1, 1, -1, -1, -1, -1, 1, 1].$$

These results characterize the overall preference for the predominance of the values of one probability of transition of the HI detection process parameter from state to state, relative to another.

Then the values of the elements of the total weight vector are equal

$$Y^1_\Sigma = ([12, 20, 12, -40, -24, 20, -30, 68, -6]).$$

The normalization stage in the problem of determining (verifying) a key parameter of a mathematical model, i.e. the elements of the OTP matrix using ENS, is mandatory.

It allows one to get rid of the negative values of weights while maintaining their proportional dependence.

As a result, we obtain the values of probabilities of the transition of a given j-th parameter (q_j) for detecting the HI from state to state.

These are probabilities of transition of the HI detection process parameter q_j from the first state to the second $p_{12}(t)$, the third $p_{13}(t)$, the fourth $p_{14}(t)$, the fifth $p_{15}(t)$, the sixth $p_{16}(t)$, the seventh $p_{17}(t)$, the eighth $p_{18}(t)$, the ninth state $p_{19}(t)$ and the probability of remaining in the first state $p_{11}(t)$, i.e. the elements of the first row of the OTP matrix:

$$p_{11}(t) = y^2_{11}/(y^2_{11} + y^2_{21} + y^2_{31} + y^2_{41} + y^2_{51} + y^2_{61} + y^2_{71} + y^2_{81} + y^2_{91}) = 0.074;$$
$$p_{12}(t) = y^2_{22}/(y^2_{12} + y^2_{22} + y^2_{32} + y^2_{42} + y^2_{52} + y^2_{62} + y^2_{72} + y^2_{82} + y^2_{92}) = 0.114;$$
$$p_{13}(t) = y^2_{33}/(y^2_{13} + y^2_{23} + y^2_{33} + y^2_{43} + y^2_{53} + y^2_{63} + y^2_{73} + y^2_{83} + y^2_{93}) = 0.074;$$
$$p_{14}(t) = y^2_{44}/(y^2_{14} + y^2_{24} + y^2_{34} + y^2_{44} + y^2_{54} + y^2_{64} + y^2_{74} + y^2_{84} + y^2_{94}) = 0.015;$$
$$p_{15}(t) = y^2_{55}/(y^2_{15} + y^2_{25} + y^2_{35} + y^2_{45} + y^2_{55} + y^2_{65} + y^2_{75} + y^2_{85} + y^2_{95}) = 0.041;$$
$$p_{16}(t) = y^2_{66}/(y^2_{16} + y^2_{26} + y^2_{36} + y^2_{46} + y^2_{56} + y^2_{66} + y^2_{76} + y^2_{86} + y^2_{96}) = 0.114;$$
$$p_{17}(t) = y^2_{77}/(y^2_{17} + y^2_{27} + y^2_{37} + y^2_{47} + y^2_{57} + y^2_{67} + y^2_{77} + y^2_{87} + y^2_{97}) = 0.034;$$
$$p_{18}(t) = y^2_{88}/(y^2_{18} + y^2_{28} + y^2_{38} + y^2_{48} + y^2_{58} + y^2_{68} + y^2_{78} + y^2_{88} + y^2_{98}) = 0.44;$$
$$p_{19}(t) = y^2_{99}/(y^2_{19} + y^2_{29} + y^2_{39} + y^2_{49} + y^2_{59} + y^2_{69} + y^2_{79} + y^2_{89} + y^2_{99}) = 0.094.$$

The remaining values of the key parameter of the mathematical model can be determined (verified) in a similar way. These are the probabilities of the transition of unreliable, inaccurately (contradictory) given parameter q_j of the HI detection process (i.e. the elements of the second, third, fourth, fifth, sixth, seventh, eighth and ninth rows of the OTP matrix $\tilde{\beta}_{q_j}(t, t + 1, c)$).

5 Conclusion

Thus, the approach for determining the parameters of the mathematical model of the stochastic HI detection process with unreliable, inaccurate (contradictory) given initial data has been proposed. This approach is based on the use of stochastic equations of state and observation based on controlled Markov chains in finite differences. In this case, the determination (verification) of key parameters of the mathematical model of this type, i.e. the elements of the matrix of one-step transition probabilities, is carried out by using an extrapolating neural network. It allows one to take into account and compensate the inaccuracy of the original data, inherent to stochastic processes for HI detection. It stipulates the increase of the reliability of decision-making for evaluation and categorization of the digital network content to detect and counteract the information of this class.

Acknowledgements The work is performed by the grant of RSF #18-11-00302 in SPIIRAS.

References

1. Elahi, G., Yu, E., Zannone, N.: A modeling ontology for integrating vulnerabilities into security requirements conceptual foundations. In: Proceedings of ER-2009, pp. 99–114. Springer (2009)
2. Kotenko, I., Saenko, I., Chechulin, A., Desnitsky, V., Vitkova, L., Pronoza, A.: Monitoring and counteraction to malicious influences in the information space of social networks. In: Proceedings of 10th Social Informatics Conference (SocInfo2018), September 25–28, 2018, Saint Petersburg, Russia. Part II. Lecture Notes in Computer Science, vol. 11186, pp. 159–167. Springer (2018)
3. Kotenko, I., Saenko, I., Chechulin, A.: Protection against information in eSociety: using Data Mining methods to counteract unwanted and malicious data. In: Proc. of Digital Transformation and Global Society. In: Second International Conference DTGS. Communications in Computer and Information Science (CCIS), vol. 745, pp. 170–184 (2017)
4. Kosko, B.: Neural Networks and Fuzzy Systems. A Dynamical Systems Approach to Machine Intelligence. Prentice-Hall, Englewood Cliffs (1992)
5. Kriesel, D.: A Brief Introduction to Neural Networks. Cambridge Press, Cambridge (2010)
6. Parlos, A., Menon, S., Atiya, A.: An algorithmic approach to adaptive state filtering using recurrent neural networks. IEEE Trans. Neural Netw. **12**(6), 1411–1432 (2001)
7. Rojas, R.: Neural Networks. Springer, Berlin (1996)
8. Muller, B., Reinhardt, J., Strickland, M.: Neural Networks: An Introduction. Springer (1995)
9. Anderson, J., Rosenfeld, E.: Neurocomputing: Foundation of Research. MIT Press, Cambridge (1988)
10. Nesteruk, G., Kupriyanov, M.: Neural-fuzzy systems with fuzzy links. In: Proceedings of the VI-th International Conference SCM'2003. St.Pb, StPSETU «LETI», vol. 1, pp. 341–344 (2003)
11. Kotenko, I., Parashchuk, I., Omar, T.: Neuro-fuzzy models in tasks of intelligent data processing for detection and counteraction of inappropriate, dubious and harmful information. In: Proceedings of the 2nd International Scientific-Practical Conference Fuzzy Technologies in the Industry (FTI-2018). CEUR Workshop Proceedings (CEUR-WS), vol. 2258, pp. 116–125 (2018)

12. Parashchuk, I.: System formation algorithm of communication network quality factors using artificial neural networks. In: Proceedings of 1st IEEE International Conference on Circuits and System for Communications (ICCSC'02), St.Pb, SPbGTU, pp. 263–266 (2002)
13. Stewart, N., Thomas, G.: Markov processes. In: Proceedings of Probability and Mathematical Statistics, pp. 214–234. Wiley, New York (1986)
14. Bini, D., Latouche, G., Meini, B.: Numerical Methods for Structured Markov Chains. Oxford University Press, New York (2005)
15. Yuksel, S.: Control of Stochastic Systems. Queen's University Mathematics and Engineering and Mathematics and Statistics (2017)
16. Van Handel, R.: Stochastic Calculus, Filtering and Stochastic Control. Springer (2007)
17. Dobre, T.: In: T.G. Dobre, J.G. Sanchez Marcano (eds.) Chemical Engineering: Modeling, Simulation and Similitude. Wiley, Weinheim (2007)
18. Oliver, D., Kelliher, T., Keegan, J.: Engineering Complex Systems With Models and Objects. McGraw-Hill, New York (2007)
19. Oksendal, B.: Stochastic Differential Equations. An Introduction with Applications. Springer, Berlin (2007)
20. Iacus, S.: Simulation and Inference for Stochastic Differential Equations. With R Examples. Springer, New York (2008)
21. Quarteroni, A.: Mathematical Models in Science and Engineering. Proc. Not. AMS **56**(1), 9–19 (2009)

Decision Support Algorithm for Parrying the Threat of an Accident

Alexander A. Bolshakov, Aleksey Kulik, Igor Sergushov and Evgeniy Scripal

Abstract It is proposed to increase the degree of flight safety of aircraft based on the use of an onboard control system using approaches to build cyber-physical systems. For this purpose, an algorithm for countering the threat of an accident has been developed, which is implemented in a decision support device. This device is the main element of the flight safety control system of the aircraft and represents a dynamic expert system. The device provides the formation of recommendations for the crew to parry the accident. For this purpose, information on changes in the values of the input variables affecting the flight safety of the aircraft in time, as well as the psychophysical condition of the crew members, the technical condition of the control object, external factors, as well as the forecast of changes in flight conditions is used. The set of decision support rules is evaluated for completeness and absence of data inconsistency. Computer simulation of the algorithm, combined with the evaluation of a set of rules for decision support, allowed to confirm its performance. The research results should be used in the development of flight safety control systems for aircraft.

Keywords Flight safety · Expert system · Decision support

1 Introduction

In recent years, the active development of aviation technology in a significant degree contributed to the increase in the level of flight safety of aircraft of various types. According to the statistics presented in [1], a sufficiently large number of accidents is associated with the human factor (87%). Therefore, various decision support systems are used by the crew at different stages of the flight [2–7]. An example of such a system is an intelligent decision support system, which is part of an integrated management

A. A. Bolshakov
Peter the Great St. Petersburg Polytechnic University (SPbPU), St. Petersburg 195251, Russian Federation

A. Kulik (✉) · I. Sergushov · E. Scripal
JSC "Design Bureau of Industrial Automation", Saratov 410005, Russian Federation
e-mail: kulikalekse@yandex.ru

A. G. Kravets et al. (eds.), *Cyber-Physical Systems: Industry 4.0 Challenges*, Studies in Systems, Decision and Control 260,
https://doi.org/10.1007/978-3-030-32648-7_19

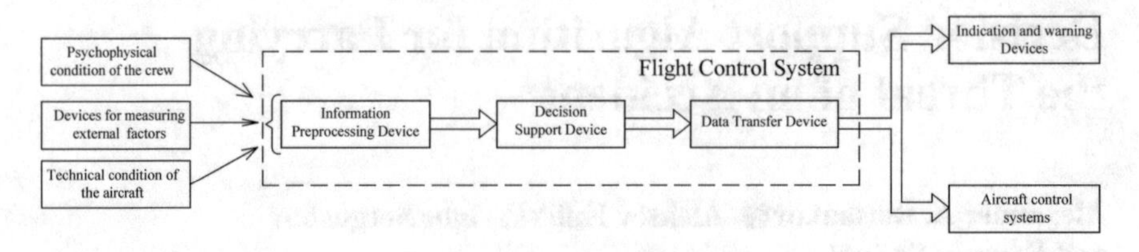

Fig. 1 A block diagram of a control system of safety of flight aircraft

system, intelligent support for the actions of the flight crew—an expert system [2], the principle of operation of which is based on the receipt and evaluation of data of the equipment used. The expert system allows you to recognize the presence of an emergency in the process of piloting the device, as well as to predict the options for its development and prevention. Another example of flight safety devices of an aircraft are systems aimed at eliminating accidents on the runway and supporting the operator in dangerous situations of managing complex technical objects [3–7].

The disadvantage of the presented systems is the lack of a comprehensive assessment of the flight conditions of the aircraft on the set of external and internal factors affecting the safety of aircraft flight, taking into account the prediction of their changes. The use of accident prediction tools at their early stages will allow to timely determine the presence of an accident threat with its subsequent parrying.

Also in recent years, cyber-physical systems have been actively developed, which constantly receive data from the environment and use them to further optimize management processes. Cyber-physical systems were developed on the basis of real-time systems, distributed computing systems, automated control systems of technical processes. The peculiarity of such systems is digital integration, digital distributed communication, parallel numerical processes, the presence of descending, ascending and regulating flows, the combination of synchronous and asynchronous control [8–14].

Therefore, increasing the flight safety of the aircraft can be achieved by the use of on Board control systems flight safety using the approaches of building cyber-physical systems. Figure 1 shows the block diagram of the aircraft flight safety control system [15].

The proposed system has two levels of recognition of changes in the operating conditions of aircraft at the stage of pre-processing of information and support of decision-making, which eliminates the false formation of data on the flight accident and its consequences. At the same time, the computational core of the system is software and logic integrated circuits, as well as converting devices, the type of which depends on the type of interfaces of data exchange between the system and the peripheral devices of the onboard aircraft equipment complex.

It should be noted that the main element of the considered systems are decision support devices. The main functions of the software and algorithmic support of these devices are to issue recommendations to the crew about the presence of aircraft accident threats, methods of their parrying and interaction with on-Board equipment.

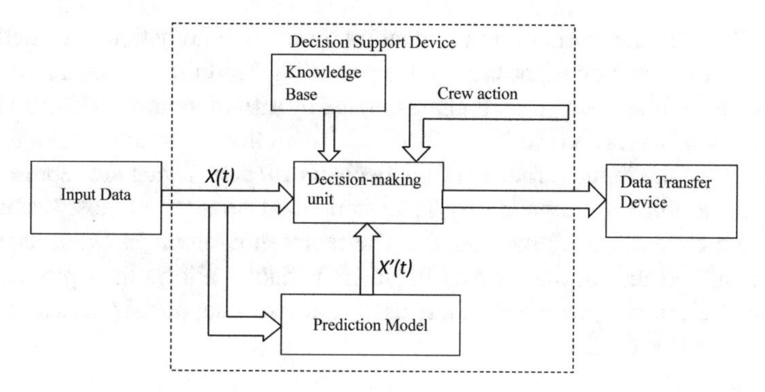

Fig. 2 A block diagram of a device decision support

2 Problem Statement

The purpose of this work is to develop an algorithm to support the decision-making of the crew, which allows forming recommendations to the pilot and signals to the aircraft control system to parry the threat of aviation accident. The algorithm should ensure the issuance of signals to parry the threat of aviation accident, taking into account the pre-processing of external and internal factors affecting the safety of the flight apparatus, as well as forecasting the development of the threat of the incident.

To achieve this goal it is necessary to: (a) formalize the input variables of the algorithm; (b) create a set of rules for decision support; (c) develop an algorithm to support decision-making by the crew.

3 The Formalization of Input Variables

The proposed decision support device is a dynamic expert system, the peculiarity of which is to form recommendations to the crew on parrying an accident when the values of the input variables change over time. The block diagram of the decision support device is shown in Fig. 2 [15].

Here $X(t)$—an array of input data after preprocessing; $X'(t)$—the results of predicting the threat of an accident; $Y(t)$—the output values of the decision support device, characterizing the recommendations to the pilot to parry the threat of an accident or signals for parrying automation.

The input variables of the device are external and internal factors affecting the safety of the aircraft flight, such as the technical condition of the control object, the psychophysical characteristics of the crew, the weather conditions of the flight. Also, the input of the decision support unit receives the values of the forecast of changes in the controlled variables and the results of the evaluation of the aircraft flight conditions. At the output of the system, recommendations are formed to the pilot to neutralize the threat of an accident or signals of its parrying by means of automatic control. Parrying the threat of an accident by means of automation is carried out in the absence of a reaction of the crew to the prevailing conditions of the aircraft flight.

At the same time, each group is characterized by a set of input variables that assess the state of the factors and their impact on the aircraft flight. These factors are weakly formalized, so the input variables of the decision support device are represented as linguistic variables. A linguistic variable is defined on a set of fuzzy values that belong to a certain space-time domain. The representation of the input data of the decision support unit in the form of linguistic variables allows their processing by the fuzzy logic device, which is widely used in decision support devices and aircraft control systems [16–22].

4 Formation of a Set of Rules for Decision Support

According to the block diagram (Fig. 2) the main element of the flight safety management system is the decision-making unit and the knowledge base. Software implementation of the algorithm for parrying the threat of aviation accident is carried out in the decision-making block based on the input information and a set of rules of the knowledge base of the system. The structure of the rules to support the decision to parry the threat of aviation accident is as follows:

$$\text{RULE} <\#> : \text{IF “}Z = \{k_i\}\text{” \& “}X_{1j} = \{f_i, k_i\}\text{” \& “}X_{2j} = \{f_i, k_i\}\text{” \& “}X_{3j} = \{f_i, k_i\}\text{” \&}$$
$$\text{“}Z' = \{k_i\}\text{” \& “}X'_{1j} = \{f_i, k_i\}\text{” \& “}X'_{2j} = \{f_i, k_i\}\text{” \& “}X'_{3j} = \{f_i, k_i\}\text{” THEN “}Y = \{g_i\}\text{”} \quad (1)$$

where Z, X1j, X2j, X3j, X4j, Z′, X1j′, X2j′, X3j′, X4j′—input variables of the decision block; fi, ki—values of the input variable; Y—output variable of the decision block (action to parry the threat of aviation accident); gi—value of the output variable.

The expression (1) shows that the decision support rule has a rather complex structure, the implementation of which can lead to high computational costs. Therefore, the set of rules for decision support, it is advisable to structure into groups of flight conditions of the aircraft. The output variable of the set of rules characterizes the actions of the crew, the signals of its informing, as well as the automatic reconfiguration of the control systems of the apparatus. It should be noted that the composition of the set of rules depends on the control object, its on-Board equipment, the functions performed and is determined in the process of developing the aircraft flight safety control system.

In the process of creating a set of rules of the decision support system, their validation, which consists in assessing the completeness and consistency of the conclusions formed by the system, is quite important. So the completeness of the output data characterizes the share of coverage of the output coordinate to the General range of its changes. As a rule, the completeness of the set of rules of the expert system is estimated using the completeness index [23, 24], which has the following form:

$$\mathrm{IP} = \int_u z5(u) \cup z1(u) / \int_u 1(u)du \tag{2}$$

where $1(u)$ is the value of the output variable corresponding to the full range of its change; $z5(u)$, $z1(u)$—the values of the output coordinate under different conditions for the input variables.

The absence of inconsistency of the set of rules is characterized by the presence of the same value of the output data for the same values of the input data and is estimated by the index of inconsistency [25], which is represented by the formula:

$$\mathrm{IPR} = 1 - \left(\int_u z5(u) \cap z1(u) / \min\left(\int_u z5(u); \int_u z1(u) \right) \right) \tag{3}$$

Taking into account the division of the spacecraft flight conditions into classes and applying the precedent matrix presented in [22], we obtain the following set of decision support rules:

1. Flight conditions are accident-free Z = k1

   RULE<1> : IF "X1j = {f1, k1}" & "X2j = {f1, k1}" &"X3j = {f1, k1}" THEN "Y = {g1}". (4)

   where g1—threat of aviation accident is absent, parrying is not required.
2. Flight conditions are difficult Z = k2

   RULE <2.1 >: IF "X1j = {f3, k3}" & "X2j = {f1, k1}" & "X3j = {f1, k2}" THEN "Y = {g2}",
   RULE <2.2 >: IF "X1j = {f3, k4}" & "X2j = {f1, k1}" & "X3j = {f1, k2}" THEN "Y = {g2}"
   RULE <2.3 >: IF "X1j = {f2, k2}" & "X2j = {f1, k1}" & "X3j = {f1, k2}" THEN "Y = {g2}", (5)

   where g2 is the threat Countered by automation, is improving the controllability of an object by signals from automatic control systems, improvement of stability and controllability, etc.

Thus, the set of rules of the decision support algorithm consists of two main groups: the current and projected flight conditions, each of which is divided into four subgroups depending on the type of threat aviation accident.

5 Development of Decision Support Algorithm by the Crew

On the basis of the obtained set of rules and input information for the decision-making block, an algorithm for parrying the threat of an accident is proposed. The output

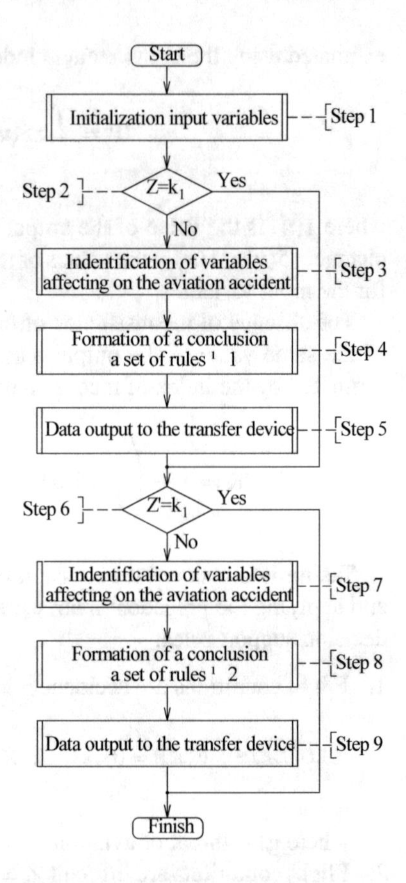

Fig. 3 Block diagram of decision support algorithm for parrying the threat of aviation accident

data of the algorithm in the process of its software implementation are “light” and “speech” information about changes in flight conditions and the threat of an accident. Moreover, speech information on parrying the threat of aviation accident, as well as control signals are issued to the systems of on-Board equipment of the aircraft. For Fig. 3 the block diagram of the decision support algorithm for parrying the threat of aviation accident is presented.

According to the presented scheme, the algorithm of decision support on parrying the threat of aviation accident works as follows:

Step 1. Initialization of the input variables characterizing the state of the flight conditions of the aircraft by the results of pretreatment of external and internal factors.
Step 2. Determination of the absence of a threat of an accident. If there is no threat of an accident, then the transition to the analysis of the forecast results of changes in the controlled variables and flight conditions of the device is carried out.
Step 3. If the condition for step 2 is not met, the procedure is carried out to determine the variables affecting the accident, compare them with the reference values, as well as with the values of the variable aircraft flight conditions.

Step 4. The formation of a method of parrying the threat of an accident in accordance with a set of rules for the current threat (rule set 1), followed by the conclusion of recommendations and management teams in the system of the onboard equipment complex.
Step 5. Data transmission is carried out according to the method of parrying the threat of an accident under the influence of the current values of the variables affecting flight safety, presented in the form of a command coming to the input of the warning and control systems of the aircraft onboard equipment complex.
Step 6. Determination of the absence of the predicted threat of an accident. If the condition is met, then the transition to the end of the program.
Step 7. If the condition is not met, the procedure is implemented to determine the variables that affect the threat of aviation accident based on the results of forecasting.
Step 8. A method of parrying the potential threat of an accident is being formed.
Step 9. Data transmission is performed according to the method of countering the threat of aviation accidents with the forecasted values of the variables that affect flight safety, presented in the form of commands issued to the input of the systems of notification and management of onboard equipment of the aircraft.

Based on the results of numerical modeling, it is possible to determine the compliance of the set of rules of the decision support device with the criteria of completeness and absence of inconsistency. So the situation with $IP = 1$ corresponds to the account of all possible States of the input variables and changes in flight conditions. In its turn, the fairness of $YPI \leq 0.4$ characterizes the absence of inconsistency between the output variables of the decision support rule set at the same values of the input variables.

The proposed algorithm makes it possible to form recommendations to the crew and control signals to neutralize the threat of an accident, taking into account the projected changes in external and internal factors affecting the flight conditions of the aircraft.

6 Experiment

The simulation of the threat of an accident with the issuance of recommendations to the crew for countering it was performed using the proposed decision support algorithm a set of rules (2)–(5) and the values of the input variables of the aircraft flight safety system. The characteristics of the results of the accident threat modeling are shown in Figs. 4, 5 and 6.

In the process of numerical simulation, the following states of the aircraft flight conditions are obtained:

- for linguistic variables equal to one, the value of the flight conditions is 1.0, which corresponds to the accident-free flight mode, consequently, the threat of an accident is absent and parrying the threat is not required when the value of $Y = 0.8$ (Fig. 4);

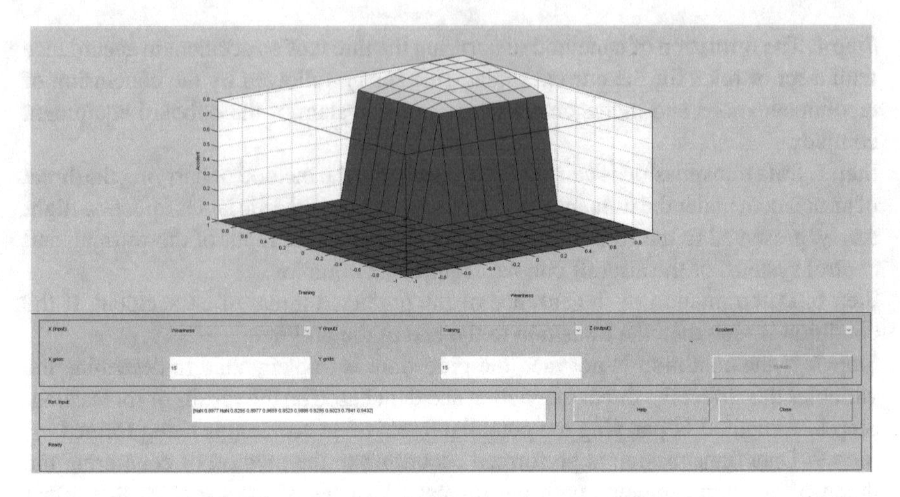

Fig. 4 Characteristics of the results of numerical simulation of aircraft flight conditions for accident-free situation

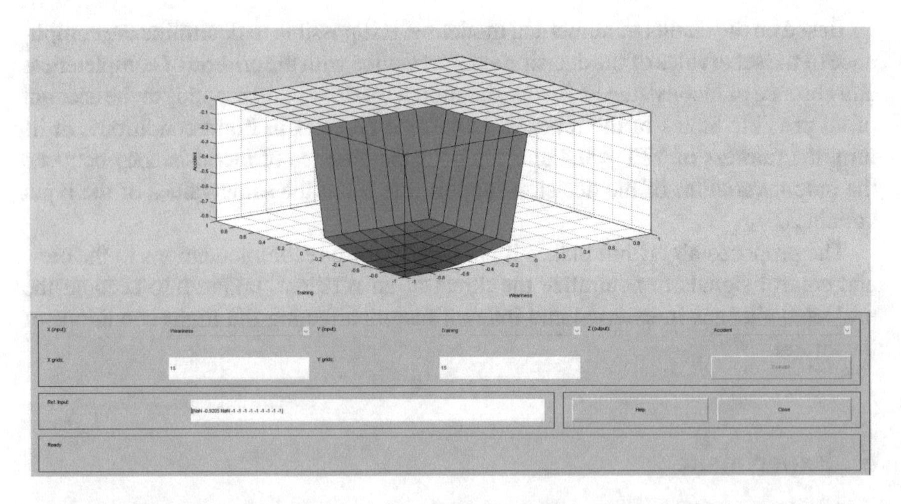

Fig. 5 Characteristics of the results of numerical simulation of aircraft flight conditions for a catastrophic situation

- with linguistic variables of input variables equal to -1, the value of the flight conditions is -1.0, which corresponds to the catastrophic flight condition, i.e. under the influence of a set of influencing factors, a threat of a catastrophic accident is created and the crew evacuation is required at a value of $Y= -0.8$ (Fig. 5);
- at linguistic input variables corresponding to difficult flight conditions (low level of aircraft controllability and increase of ECI-page fatigue from monotonous load), it is required to improve flight conditions by means of automation, which corresponds to $Y = 0.5$ (Fig. 6).

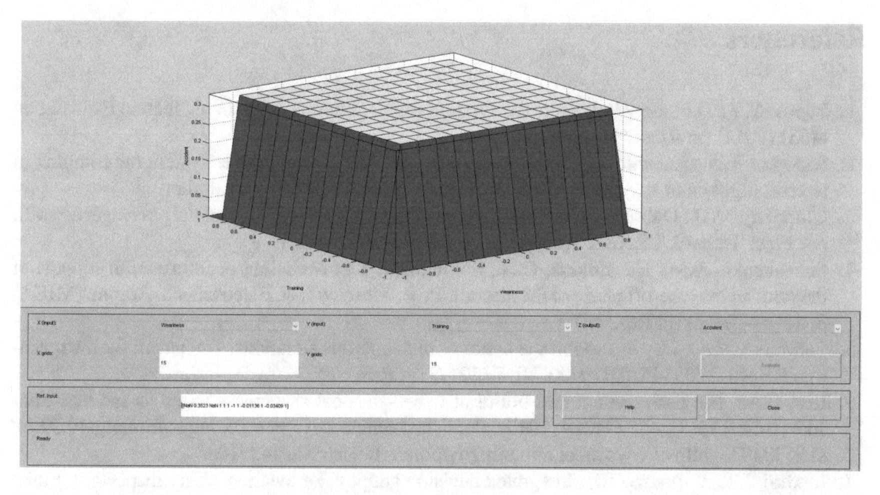

Fig. 6 Characteristics of the results of numerical simulation of aircraft flight conditions for complex flight conditions

Based on the results of numerical modeling, it is possible to determine the compliance of the set of rules of the decision support device with the criteria of completeness and lack of inconsistency. So the situation with IP = 1 corresponds to the account of all possible States of the input variables and changes in flight conditions. In turn, equity and PR = 0.4 characterizes the absence of inconsistency between the output variables of the set of decision support rules for the same values of the input variables.

The proposed algorithm makes it possible to form recommendations to the crew and control signals to neutralize the threat of an accident, taking into account the projected changes in external and internal factors affecting the flight conditions of the aircraft.

7 Conclusion

In the course of the work, the analysis of input variables of the decision support unit is carried out, a set of rules of the knowledge base is proposed, on the basis of which an algorithm for parrying the threat of aviation accident is developed. A distinctive feature of the algorithm is the planning of the current and projected threats of an accident. Also the operation of the flight safety management system involves the formation of the aircraft control signal in automatic mode in the absence of a positive reaction of the pilot to parry the threat of an accident.

Further work on the establishment of safety control systems flight aircraft focus on its software and hardware implementation of subsequent ground-based testing and testing in the composition of the flying laboratories.

References

1. Popov, Y.V.: The safety indicators for aviation flights. Internet-J. Technol. Techno Field Secur. **6**(58) (2014) (in Russian)
2. Sapogov, V.A., Anisimov, K.S., Novozhilov, A.V.: Fail-safe computer system for complex of aircraft flight control systems. Electron. J. Proc. MAI **45** (2008) (in Russian)
3. Glubokay, M.D.: Onboard decision support system at the take-off stage of the passenger aircraft. Air Fleet Technol. LXXXII **1**(690), 21–30 (2008) (in Russian)
4. Shevchenko, A.M., Nachinkina, G.N., Solonnikov, Y.I.: Modeling of information support of the pilot on the take-off phase of the aircraft. Proc. Moscow Inst. Electromech. Autom. (MIEA) **5**, 54–64 (2012) (in Russian)
5. Suholitko, V.A.: Way to support the operator in dangerous situations. The patent for the invention RF No. 220544 G05D 1/00. 2017 (2017) (in Russian)
6. Lee, E.A.: The past, present and future of cyber-physical systems: a focus on models. Sensors (Basel) **15**(3), 4837–4869 (2015). Published online 2015 Feb 26. https://doi.org/10.3390/s150304837. https://www.ncbi.nlm.nih.gov/pmc/articles/PMC4435108/
7. Luxhoj, J.T., Williams, T.P.: Integrated decision support for aviation safety inspectors. Finite Elem. Anal. Des. **23**, 381–403 (1996)
8. Khaitan, et al.: Design techniques and application of cyber physical systems: a survey. IEEE Syst. J. (2014)
9. Lee, E.A., Seshia, S.A.: Introduction to Embedded Systems—A Cyber-Physical Systems Approach. LeeSeshia.org 2011 (2011)
10. Fawzi, H., Tabuada, P., Diggavi, S.: Secure estimation and control for cyber-physical systems under adversarial attacks. IEEE Trans. Autom. Control (2014)
11. Wolf, W.: Cyber-physical system. Computer **3**, 88–89 (2009)
12. Crop: a cyber-physical system architecture model in the field of precision agriculture. In: Conference Agriculture for Life, Life for Agriculture, vol. 6, pp. 73–79
13. Lee, J., Bagheri, B., Kao, H.-A.: A cyber-physical systems architecture for Industry 4.0-based manufacturing systems. Manuf. Lett. **3**, 18–23 (2015). https://doi.org/10.1016/j.mfglet.2014.12.001
14. Chuprynousky, V.P., Namiot, D.E., Sinyakov, S.A.: A cyber-physical system as the basis for the digital economy. Int. J. Open Inf. Technol. **4**(2), 18–25 (2016). ISSN: 2307-8162 (in Russian)
15. Bolshakov, A.A., Kulik, A.A., Sergushov, I.V., Skripal, E.N.: Method of aircraft accident prediction. Mechatron. Autom. Control **19**(6), 416–423 (2018) (in Russian)
16. Fedunov, B.E., Prokhorov, M.D.: Conclusion on precedent in knowledge bases of onboard intelligent systems. Artif. Intell. Decis.-Making **3**, 63–72 (2010) (in Russian)
17. Kosko, B.: Fuzzy systems as universal approximators. IEEE Trans. Comput. **43**(11), 1329–1333 (1994)
18. Cordon, O., Herrera, F.A.: General study on genetic fuzzy systems. Genetic Algorithms in Engineering and Computer Science, pp. 33–57 (1995)
19. Veshneva, I.V., Chistyakova, T.B., Bolshakov, A.A.: The status functions method for processing and interpretation of the measurement data of interactions in the educational environment. SPIIRAS Proc. **6**(49), 144–166 (2016)
20. Averchenkov, V., Miroshnikov, V., Podvesovsky, A., Korostelyov, D.: Fuzzy and hierarchical models for decision support in software systems implementation, pp. 410–422. Knowledge-Based Software Engineering. Springer (2014)
21. Veshneva, I.V., Chistjakova, T.B., Bol'shakov, A.A., Singatulin, R.A.: Model of formation of the feedback channel within ergatic systems for monitoring of quality of processes of formation of personnel competences. Int. J. Qual. Res. **9**(3), 495–512 (2015)
22. Denisov, M., Kozin A., Kamaev, V., Davydova, S.: Solution on decision support in determining of repair action using fuzzy logic and agent system. Knowledge-Based Software Engineering, pp. 433–541. Springer (2014)

23. Veshneva, I., Bolshakov, A., Kulik, A.: Increasing the safety of flights with the use of mathematical model based on status functions. In: Dolinina, O., et al. (eds.) ICIT 2019, SSDC 199, pp. 608–621. Springer (2019). https://doi.org/10.1007/978-3-030-12072-6_49
24. Protalinski, O.M.: Application of Artificial Intelligence Techniques in the Automation of Technological Processes: Monograph, 183 p. Publishing House of ASTU, Astrakhan (2004) (in Russian)
25. Bolshakov, A.A., Sergushov, I.V., Skripal, E.N.: Intelligent method for assessing the threat of an accident. Bull. Comput. Inf. Technol. **5**(167), 3–9 (2018) (in Russian)

Predictive Model for Calculating Abnormal Functioning Power Equipment

Alexandra A. Korshikova and Alexander G. Trofimov

Abstract A method of early detection of defects in technological equipment of energy facilities is proposed. A brief analysis of the Russian market of cyber-physical industrial equipment monitoring systems was carried out. Special attention is paid to the problems of preparing initial data for training a model, in particular, the problem of obtaining adequate data on accidents that have occurred. A mathematical problem is formulated for modeling the anomaly index, which takes values from 0 (normal operation) to 1 (high probability of an accident). The model is based on well-known statistical methods. A method for dividing the periods of operation of technological equipment into "normal" and "anomalous" is proposed. The method of binary classification AUC ROC allows you to limit the number of signs involved in the formation of the anomaly indicator, signs that have a good "separation" ability. Using the Spearman's rank correlation criterion, signs are selected that are most sensitive to the development of process equipment malfunctions. As an anomalous indicator, it is proposed to consider the ratio of the densities of distribution of the final signs, estimated in the anomalous and normal areas of operation of the process equipment. A method is proposed for generating an alarm for detecting the anomalous operation of the technological equipment of power units. It is shown that the proposed model made it possible to identify the beginning of the development of an emergency, while individual measurements did not detect any features of the operation of equipment of energy facilities in the pre-emergency time interval.

Keywords Technological equipment · Cyber-physical systems · Defect detection · Predictive analytics · Linear regression · AUC ROC

A. A. Korshikova (✉)
OOO Inkontrol, Moscow, Russia
e-mail: aakorshikova@gmail.com

A. G. Trofimov
National Research Nuclear University "MEPhI", Moscow, Russia

A. G. Kravets et al. (eds.), *Cyber-Physical Systems: Industry 4.0 Challenges*, Studies in Systems, Decision and Control 260,
https://doi.org/10.1007/978-3-030-32648-7_20

1 Introduction

Predictive analytics is a combination of methods for detecting faults in the operation of process equipment in the early stages of their occurrence [1, 2].

At the current stage of technology development, cyber-physical systems that solve the tasks of predictive analytics or predictive services are becoming increasingly important. Predictive maintenance [3] allows specialists from different areas to prevent unwanted events occurring in any process, to predict the timing of repair work or error correction, and, as a result, reduce the cost of equipment maintenance due to preventive troubleshooting. World-famous companies such as Siemens [4], IBM, Baker Hughes, General Electric, have already appreciated the advantages of using automatic monitoring and forecasting systems [5]. According to experts, by 2020 the money turnover in the market of predictive services will increase by one and a half to two times compared with the current state [6].

To date, the Russian market of cyber-physical systems for monitoring industrial equipment amounts to 55 billion rubles. There are more than a dozen companies that are engaged in predictive analytics at a professional level. For example, Datadvance develops, implements and maintains the pSeven software product [7] already used in the fields of transport and aerospace technology. The Russian company Clover Group is the creator of the platform of intellectual analytics and predictive analysis Clover, designed to solve the problem of predicting the state of equipment on the basis of continuously improved mathematical models for industrial companies [8].

However, in power engineering (both Russian and foreign) there are much less successes in the early diagnostics of the technical condition of technological equipment. It can be said that at the moment there are no systems that allow to reliably predict future malfunctions and accidents of the process equipment of power units. This is a simple explanation.

For a nearly century-long history of the Russian energy sector, a volume of measurements of the parameters of the state-of-the-art equipment has been created, which allows, when all the necessary conditions (regulations) are met, to operate the equipment without incident. In the overwhelming majority of cases, the prevention of malfunction/ accident does not require special methods, but is carried out according to existing measurements (permissible, warning, alarm signaling). The accidents, however, are caused primarily by either non-compliance with the rules and regulations of equipment operation, or external influences that cannot be foreseen (accidents in the power system, natural phenomena, the human factor, etc.).

Nevertheless, a certain percentage of malfunctions/ accidents remains, which do not appear clearly at the level of individual measurements but the development of which can be diagnosed by the methods of predictive analytics on the behavior of a certain set of measured parameters [9].

In Russia, one of the main players is AO Rotek: the Prana remote monitoring and prognostics system created by it [10] over the past three years has been equipped with gas turbine units at a number of heat and power plants. However, a significant drawback of this system is the complexity of its operation, which requires constant attention from the specialists of the development company, the presence of a significant number of setting parameters of the methods used, which are not obvious at the choice of their values, which greatly complicates and lengthens the adjustment period (the so-called your learning process).

Since 2017, AO "Interavtomatika" and OOO "Incontrol" are actively conducting their own developments in the field of creating cyber-physical systems with the hope of their more successful application in predictive analytics in power engineering.

2 Software and Analytical Predictive Analytics Module (PAM PA)

2.1 Principle of Operation

The software-analytical module of predictive analytics (PAM PA) represents a system of software and hardware tools that allow real-time recording of changes in the technical state and forecasting with a certain probability the output of the monitored technological equipment to a faulty state.

The work of the predictive analytics module contains statistical and intellectual methods for constructing a model for predicting incipient defects.

The essence of the method is as follows. It is assumed that a number of technological parameters and/or their derivatives (average values, variance, …), characterizing the work of the diagnosed equipment and instituted in the software technical complex of the automatic control system, starting from a certain period before the detected fault, react in a certain way to nascent defect. This reaction can be determined by statistical methods, a criterion has been formed—an indicator of anomalous equipment operation, which should vary in the range from 0 (no fault is foreseen) to 1 (the probability of a malfunction/ accident is extremely high).

2.2 Algorithm for Building a Model

To build the model, historical (archival) data on the functioning of the diagnosed equipment is used (Figs. 1 and 2):

- matrix of technological parameters;
- vector moments of detection of faults/ accidents.

After specifying the structure of the model, it should go through the training stage. This is an extremely important stage in the development of a model, on the quality of which its future predictive ability substantially depends. For learning, basic data are needed. In this capacity, there are archival data on equipment performance indicators

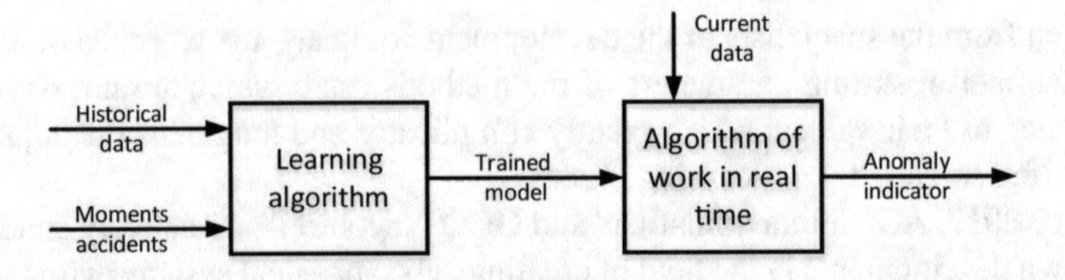

Fig. 1 The scheme of the algorithm of the model

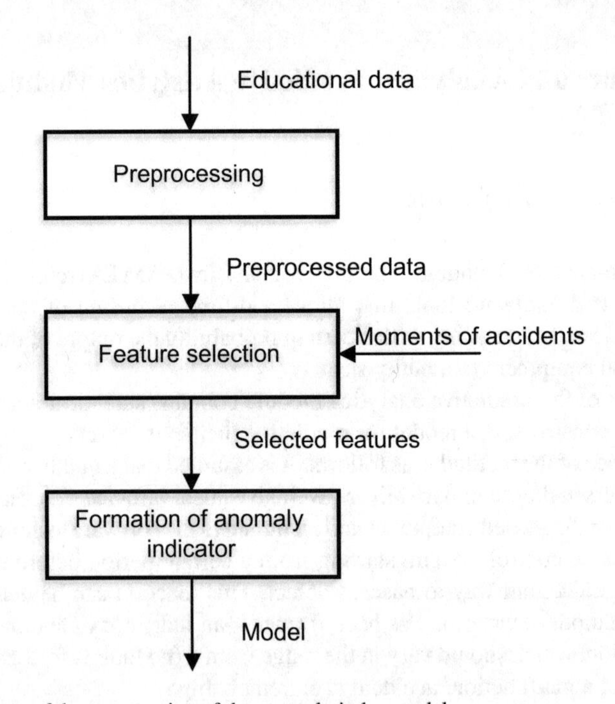

Fig. 2 Diagram of the construction of the anomaly index model

and information on accidents that occurred (type of accident, time of its detection, time to restore the correct operation).

The time depth of the archived data must be at least 1–3 months past from the earliest accident. Model data is trained on this data, that is, all necessary parameters are determined and its operation is configured. Adequate data on defects in the equipment being diagnosed are crucial for building a qualitative predictive model.

The fact that building a model based on retrospective data on the work of the diagnosed process equipment over a sufficiently long period of time when its properties may change should not be confused, since, when and if these changes are "harmless" and lead to emergencies, they will be taken into account by the model in the field of "normal" functioning; if changes in the operation of equipment are such that their

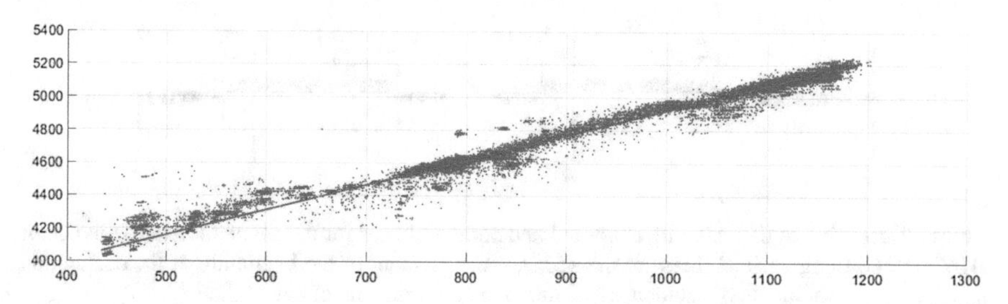

Fig. 3 Graph of the simplest linear regression model

tendency to increase can lead to an accident, then the task of building a model is to identify them and refer them to the area of "anomalous" functioning.

Preprocessing Data

At the preprocessing stage of the initial data, all the observed performance indicators of the diagnosed equipment are corrected to the base parameter x_b, on which the values of the other parameters depend (for example, power).

For each remaining indicator from $x_1, \ldots, x_m$, we build the simplest linear regression model on the indicator x_b (Fig. 3):

$$x_i = \beta_{0i} + \beta_{1i} x_b + e_i \tag{1}$$

The calculation of the parameters of the ith regression model is carried out according to the method of least squares [11].

It is necessary to note the validity of the application in this case of a linear regression model:

1. As is known, in the heat and power processes, there is an almost linear dependence of the heat engineering parameters on the load or on any parameter determining it (changes in the flow rates of fuel, water, steam, air, gases, chemical reagents, etc.) proportional to changes in power, other parameters—temperature, pressure, electrical parameters—under the action of the corresponding automatic control systems retain their constant value, which is also a special case of a linear relationship with the prop factor rationality 0.
2. As the experiments show, the residues e_i in dependence (1) have a close to a normal distribution with zero mean value.
3. The residuals e_i have the same variability for all dependent ("predicted") values x_i.

The fulfillment of conditions (1–3) fully substantiates the linear regression dependence [12].

Feature Selection

Next is the selection of informative features, that is, the parameters and their derivatives are the most "sensitive" to the development of the fault. It is proposed to use the

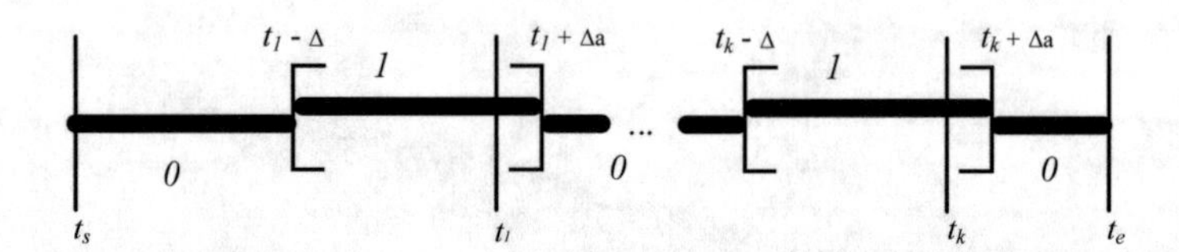

Fig. 4 Illustration of data markup abnormal and normal classes for the calculation of the indicator $AUC\ ROC$. Long vertical lines are marked: t_s—the moment of the beginning of the calculation a_i; $t_1, \ldots, t_k$—moments of accidents; t_e—the moment of the end of calculation a_i

measure of its predictive ability to predict accidents [13] as a criterion for the informativeness of a feature.

The selection of signs with the greatest predictive ability is carried out in two stages.

At the first stage, signs are identified that have the best "separability" in "normal" and pre-emergency ("abnormal") areas (Fig. 4).

As a measure of separability, we choose the indicator $AUC\ ROC$. The $AUC\ ROC$ indicator, which makes it possible to assess the quality of a binary classification, displays the ratio between the parameters of the total number of carriers of a feature (in this case, "accident rate") correctly classified as carrying a feature, and the proportion of parameters of the total number of parameters, not carrying a feature, mistakenly classified as carrying a feature when varying the threshold of a decision rule [14].

A certain value of Δ (prediction interval) is specified, which determines the maximum time ago relative to the accident when the process of development of the anomalous situation that led to the accident could presumably begin. As data of the "anomalous" class, we use the values of this indicator in the time intervals $[t_1 - \Delta,\ t_1 + \Delta_a], \ldots, [t_k - \Delta,\ t_k + \Delta_a]$; as data of a "normal" class, at all times that do not belong to specified intervals, where Δ_a is the time delay associated with possible inaccuracy in detecting the accident moments $t_1, \ldots, t_k$ and the time required for their elimination (Fig. 4).

The data of the anomalous class are formed from the intervals labeled "1", the normal class—labeled "0".

As a result of the first stage, features with the greatest "separation" ability are distinguished.

At the second stage, the selection of parameters with the highest predictive ability is performed (Fig. 5).

The concept of a predictive ability of a trait is that a trait that has a predictive ability demonstrates a positive dynamic in the degree to which its values differ from normal values as it approaches the time of the accident, i.e. the dependence of the $AUC\ ROC$ values on time for the ith parameter $a_i(\tau)$ is increasing.

As a measure of the monotony of the increase of the function $a_i(\tau)$, the Spearman rank correlation coefficient is used. The Spearman coefficient allows us to statistically establish the existence of a relationship between phenomena. Its calculation assumes the establishment of a serial number—rank for each attribute. To calculate

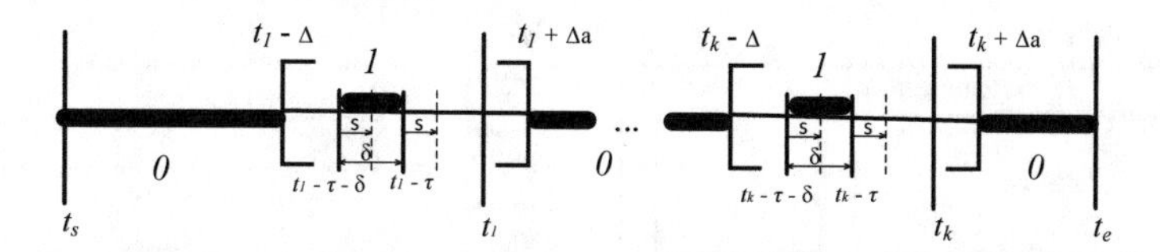

Fig. 5 Illustration of data marking of anomalous and normal classes for determining the dynamics $a_i(\tau)$: t_s—the moment of the beginning of the calculation $a_i(\tau)$; $t_1, \ldots, t_k$—moments of accidents; t_e is the moment when the calculation is finished $a_i(\tau)$; s—calculation step

the Spearman ratio, use the formula:

$$\rho_{xy} = 1 - \frac{6\sum d^2}{n\left(n^2 - 1\right)}, \tag{2}$$

where: n—displays the number of ranked signs; d—difference between ranks in two variables; $\sum d^2$ is the sum of squared difference of ranks. The ranks, in this case, will be: the first rank is the ordinal number of the calculation time interval $a_i(\tau)$; the second rank is the number of $a_i(\tau)$ value calculated for a given interval, in the series of $a_i(\tau)$ values calculated at all other time intervals.

The feature selection algorithm has a single parameter—the number of features selected. As a result, of the algorithm, a vector of indices of indicators is formed, sorted in decreasing order of predictive ability.

Calculation of Anomaly Indicator

The anomaly index is calculated based on the analysis of the density functions of the distribution of selected signs in the "anomalous" and "normal" intervals of operation of the diagnosed process equipment (in terms of machine learning, the data distribution of the "anomalous" and "normal" classes).

To estimate distribution densities, the kernel density estimation method (Parzen window) is used.

For the characteristic value z_i, the ratio of probability densities for the "anomalous" and "normal" time intervals is calculated:

$$\lambda_i(z_i) = \ln\left(\frac{f_{1i}(z_i)}{f_{0i}(z_i)}\right), \quad \text{where } i = \overline{1, M} \rightarrowtail \bar{\lambda}(z) = \frac{1}{M}\sum_{i=1}^{M}[\nu_i \cdot \lambda_i(z_i)]. \tag{3}$$

For the anomaly indicator, a logistic transformation is used:

$$p(z) = \frac{1}{1 + e^{-\alpha\bar{\lambda}(z)}}, \tag{4}$$

where $\alpha > 0$ parameter that determines the degree of steepness of the logistic curve.

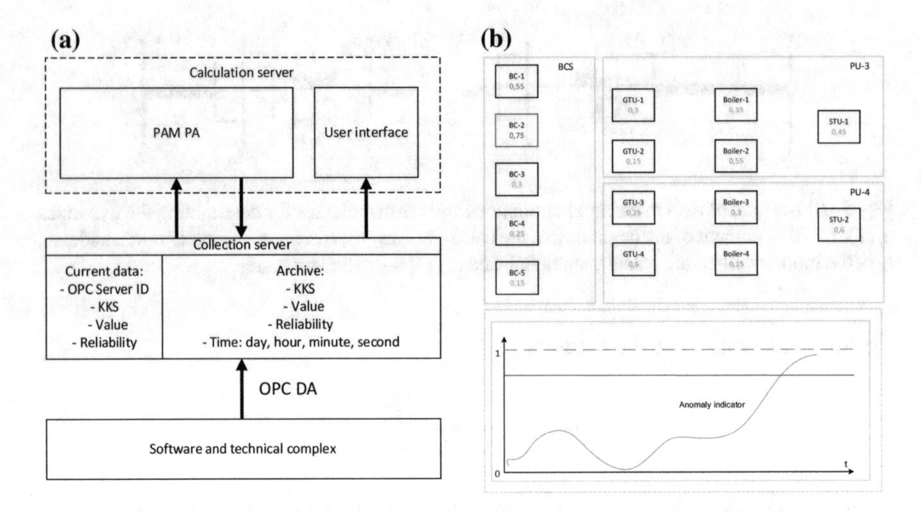

Fig. 6 Software and hardware implementation: **a** block diagram of a complex of software and hardware; **b** user interface

2.3 Software and Hardware Implementation

From the point of view of architecture, PAM PA consists of the following functional parts (Fig. 6):

- collection server, primary preprocessing (data is "cleared", emissions, noise are removed) and storage (buffering) of the obtained data;
- a calculation server on which calculations are made and the results of calculations are displayed for the maintenance personnel of the energy facility.

2.4 Testing the Model

The training of the model took place on one actually detected gas turbine unit malfunction (damage to the flow section of the compressor).

In Fig. 7 shows a graph of the predictive values of indicators, sorted in descending order.

From the graph, it is clear that many indicators have a good predictive ability, and for some indicators, the values of the predictions are equal, which is probably due to their strong correlation.

The trained model was tested on the of the automatic control system gas turbine unit model. The fault was modeled several times in different operating modes of the gas turbine unit (Fig. 8). The anomaly index, calculated in real-time, has a good predictive ability.

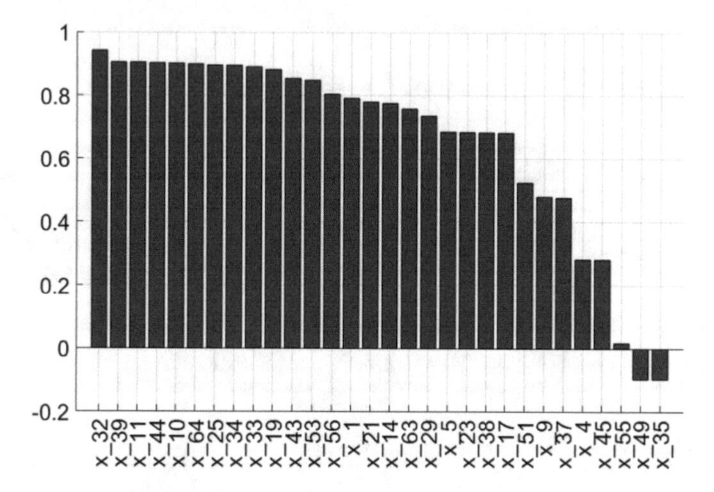

Fig. 7 Graph of values of predictors of indicators

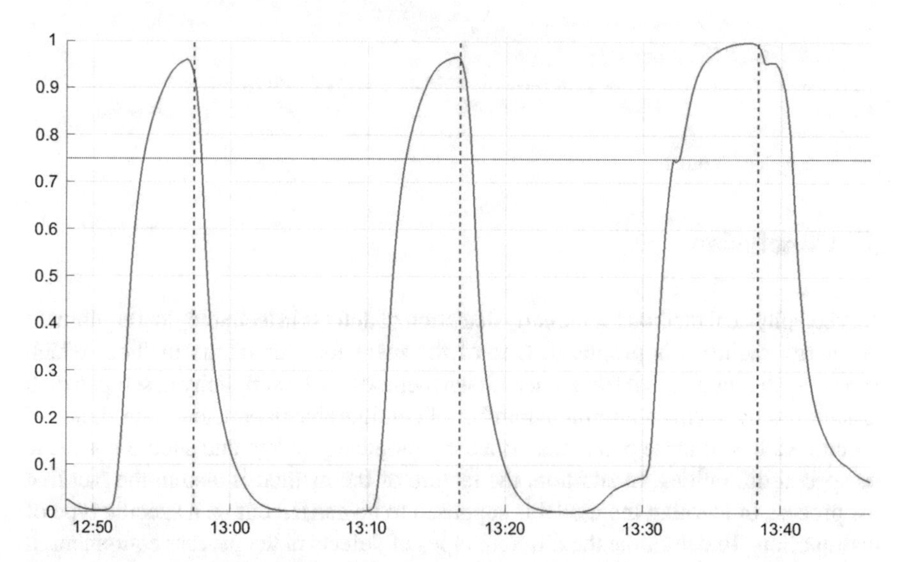

Fig. 8 Real-time anomaly indicator

When the $p(z)$ value exceeds the set limit, the PAM PA issues an alarm (Fig. 9). The appropriate specialists should make the decision on the possibility of further operation of the equipment.

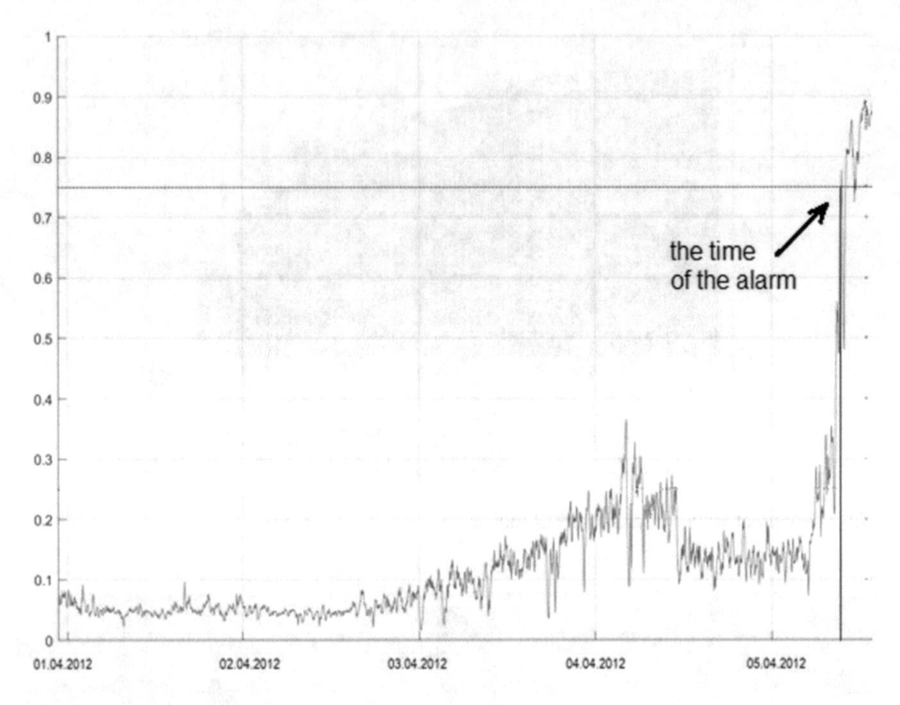

Fig. 9 Graph of anomaly

3 Conclusion

A cyber-physical method for the early detection of defects in technological equipment of energy facilities is proposed. One of the main features of the method, which favorably distinguishes it from other known methods, is the simplicity in setup, which is achieved by having a minimum number of configurable parameters. The choice of specific values of these parameters is a technological problem and should not cause its special difficulties. In addition, the feature of the method is also in the fact that the process of learning the model is supposed to be carried out on a specific type of malfunction. To determine the different types of defects in the process equipment, it is necessary to include defects in the process of training different types and obtain the corresponding ensemble of models. This allows you to move from the diagnosis of "something went wrong" to an indication of the likelihood of a very specific defect.

The model was tested on the stand of the automatic control system of a gas turbine installation and showed good results.

The main problem hindering the widespread introduction of PAM PAs to energy facilities in Russia is the lack of adequate data on faults in the past with this equipment. However, this circumstance does not detract from the suitability of the model for industrial operation, but only affects the duration of commissioning works. However, the lack of reliable data on faults is the "bottleneck" of any kind of development of such systems.

Our expectations from the introduction of the predictive analytics module:

- transfer as many failures as possible from the category of sudden to the category of predictable ones;
- prediction of the residual life of parts and equipment;
- the transition from scheduled repairs to repairs on-state.

References

1. Siegel, E.: Predictive Analytics: The Power to Predict Who Will Click, Buy, Lie, Or Die. Wiley, New York (2016)
2. Waller, M.A., Fawcett, S.E.: Data science, predictive analytics, and big data: a revolution that will transform supply chain design and management. J. Bus. Logistics **34**(2), 77–84 (2013)
3. Preventive and Predictive Maintenance 700ZB00102, https://www.lce.com/pdfs/The-PMPdM-Program-124.pdf. Last accessed 20 May 2019
4. He, L., Levine, R.A., Fan, J., Beemer, J., Stronach, J.: Random forest as a predictive analytics alternative to regression in institutional research. Practical Assess. Res. Eval. **23**(1), 1–16 (2018)
5. Siemens Automation Technology Homepage, https://www.energy.siemens.com/ru/pool/hq/automation/automation-control-pg. Last accessed 26 May 2019
6. Coleman, C., Damodaran, S., Deuel, E.: Predictive-maintenance and the smart factory. Deloitte: [2017]. URL: https://predictive-maintenanc20Deloitte.pdf. Last accessed 25 April 2019
7. DATADVANCE Homepage, https://www.datadvance.net/ru/. Last accessed 24 May 2019
8. COVER GROUP Homepage, https://clover.global/contacts/. Last accessed 24 May 2019
9. Cheng, S., Pecht, M.: Multivariate state estimation technique for remaining useful life prediction of electronic products. In: AAAI Fall Symposium on Artificial Intelligence for Prognostics, Arlington (pp. 26–32) (2007)
10. PRANA Homepage, https://prana-system.ru. Last accessed 23 May 2019
11. Draper, N.R., Smith, H.: Applied Regression Analysis. Wiley, New York (2014)
12. Abbott, D.: Applied Predictive Analytics: Principles and Techniques for the Professional Data Analyst, p. 427. Wiley, Indianapolis (2014)
13. Korshikova, A.A., Trofimov, A.G.: Model for early detection of emergency conditions in power plant equipment based on machine learning methods. Therm. Eng. **66**(3), 189–195 (2019)
14. Bowman, A.W., Azzalini, A.: Applied Smoothing Techniques for Data Analysis. Oxford University Press Inc., Oxford (1997)

The System for Operational Monitoring and Analytics of Industry Cyber-Physical Systems Security in Fuel and Energy Domains Based on Anomaly Detection and Prediction Methods

N. V. Nashivochnikov, Alexander A. Bolshakov, A. A. Lukashin and M. Popov

Abstract The distinctive features and key areas of analytical tools application for the operational monitoring and security analysis of cyber-physical systems of critical information infrastructure are highlighted. Problems of applying data analytics methods and technologies to ensure the security of cyber-physical systems at enterprises of the fuel and energy industries are described. The architectural solutions of the advanced security analytics platform are proposed. The chapter discusses the use of data analysis methods and technologies to ensure the security of cyber-physical systems at enterprises of the fuel and energy complex. Identified the distinctive features and highlighted the key areas of application of analytical tools of the system of operational monitoring and security analysis of cyber-physical systems of critical information infrastructure. Problem questions are formulated, the possibilities and limitations of advanced analytics tools in solving the tasks of ensuring the security of cyber-physical systems at the enterprises of the fuel and energy complex are defined. Architectural solutions of the advanced cybersecurity analytics platform are described. The presented results are based on the analysis of information from open sources: materials of scientific-practical conferences, analytical and technical reviews on the subject of security of industrial systems, and generalization of practical experience in the development, implementation, and support of integrated security systems at enterprises of the fuel and energy complex.

N. V. Nashivochnikov · A. A. Bolshakov
Gazinformservice Ltd., St. Petersburg, Russia
e-mail: nashivochnikov-n@gaz-is.ru

A. A. Bolshakov
e-mail: bolshakov-a@gaz-is.ru

A. A. Lukashin (✉) · M. Popov
Peter the Great St. Petersburg Polytechnic University, St. Petersburg, Russia
e-mail: alexey.lukashin@spbstu.ru

M. Popov
e-mail: popov_m@spbstu.ru

A. G. Kravets et al. (eds.), *Cyber-Physical Systems: Industry 4.0 Challenges*, Studies in Systems, Decision and Control 260,
https://doi.org/10.1007/978-3-030-32648-7_21

Keywords Data analysis practices · Anomaly detection · SIEM · Advanced analytics platform · Machine learning · Import substitution · Operational monitoring and analysis system

1 Introduction

Cyber-physical systems (CPS) are systems that comprise sensors, actuators affecting the external physical world and a computer (cyber control section) providing control of the entire system. Enterprises of Fuel and Energy Complex (FEC) have long been equipped with CPS, the logic of which is completely determined by the software used. The fourth industrial revolution (Industry 4.0) is characterized by the development of more complex network-intensive systems through CPS integration, usually produced by different manufacturers. CPS operating as part of a single gas transmission network within a large enterprise (corporation) can be taken as an example of a network system.

The increasing complexity of interacting industrial CPS and the importance of ensuring their security while operating, especially at Sensitive Information Infrastructure (SII) facilities [16], determine the relevance of an effective In-process Monitoring and Analysis System (IMAS) designed to identify and analyze security events and incidents in FEC primary and supporting processes. Moreover, IMAS should be designed as a component of situation centers at an enterprise, corporation or industry level. Situation centers solve the problems of integrated industrial, environmental and information security of FEC facilities to minimize the risks of accidents and emergencies, as well as to eliminate the consequences of such extremal situations. One of the most relevant areas of IMAS Analytical Tools (AT) use is the proactive monitoring automation: detection and forecasting of processing facilities and Industrial Control Systems (ICS) abnormal operation.

AT refer to data analysis practices: appropriate scientific and methodological apparatus (disciplines) and technology (tools) supporting its implementation [2]. To assess the CPS operation security, a comprehensive analysis using advanced data analysis practices is required. The complexity allows taking into account the degree of mutual interference of various components and systems, to identify hidden cause-and-effect relationship and to achieve a synergistic effect in ensuring the stability of engineering processes (EP). Advanced analytics refers to the exploration of data or content using complex methods and tools, usually beyond traditional business analysis methods, providing deeper insights, forecasts and recommendations. Advanced analytics practices include such disciplines as machine learning, pattern matching, forecasting, visualization, data/text mining, network and cluster analysis, multivariate statistics, graph analysis, complex event processing, and neural networks.

2 CPS Security as a Subject of Monitoring and Analysis

It is significant that IMAS does not replace EP monitoring in the dispatching control loop of FEC facilities. In addition, IMAS and ICS/SCADA (Supervisory Control and Data Acquisition) have different functional purposes. In particular, the distinctive features of IMAS are listed below:

1. The CPS operation security at enterprises should be analyzed in three allied lines simultaneously, as well as in an integrated form: analysis of sensory data, IT metrics and information security events (Fig. 1).
2. Maximum IMAS independence from ICS/SCADA software, i.e. invariance with respect to EP control algorithms implemented in industrial applications of specific manufacturers.
3. Minimal impact of IMAS on SII: the ICS/SCADA operation security control loop should be independent of the dispatching control loop and automatic EP control systems. Separate computing and telecommunication resources should be allocated for the monitoring functions of situation centers.
4. To ensure the interoperability of ICS/SCADA and IMAS, the sensory data and low-level signals of engineering process control (direct access (DA)) received from the SCADA OPC server according to the OPC DA standard should be taken as the initial data for assessing the stability of engineering processes.

Traditional information security tools are not enough for industrial CPS due to their cyber-physical nature. Moreover, well-known information security tools require

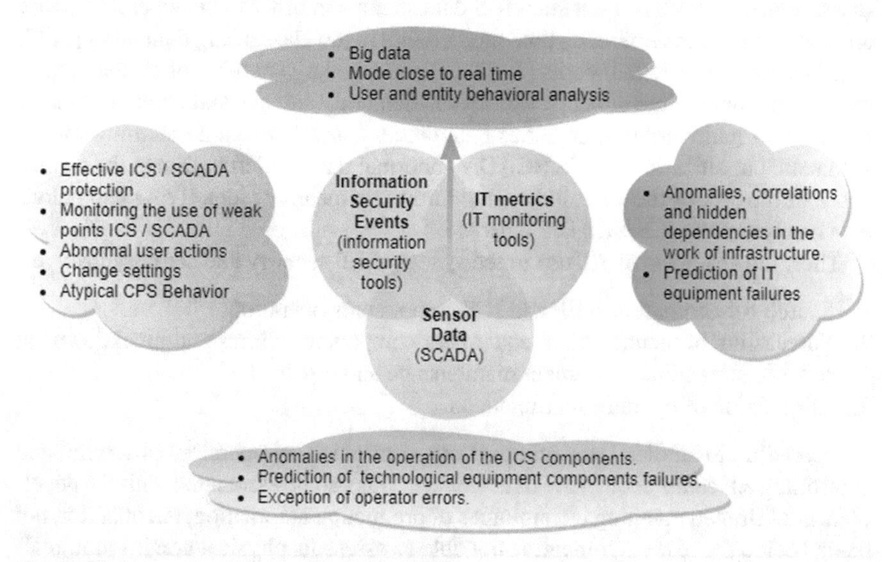

Fig. 1 Comprehensive CPS security analysis

adaptation due to system constraints and historical experience of ICS/SCADA segment creation and development in isolation from the corporate network and the Internet. Detailed issues of architectural solutions for ICS/SCADA information security provision are considered in [1, 3, 9, 14, 21], for example.

The key areas of AT use in the sphere of ICS/SCADA information security involve:

1. Improving the ICS security provision efficiency.
2. Monitoring the SCADA security vulnerability:
   - SCADA vulnerability exploit;
   - authentication/authorization fail, built-in passwords, hidden accounts;
   - file resolutions (Read/Execute for all users);
   - memory corruption (buffer overflow);
   - openness to malicious code injections.
3. Analysis of interaction with SCADA systems—detection of anomal user actions, changes in settings; aberrant CPS behavior.

The key areas of AT use in the sphere of IT monitoring involve:

1. Search for anomalies, correlations and hidden dependencies in the infrastructure equipment operation.
2. Forecasting of equipment failures and breakdowns in order to streamline its maintenance and repair.

The number of monitored ICS/SCADA signals in IMAS of FEC facilities reaches tens and hundreds of thousands, which generate a daily stream of up to several million measurements. Processing of such ICS data streams in order to detect and visualise anomal values in the near-real-time mode refers to the class of big data tasks [7, 12, 15]. The abovementioned works [7, 12, 15] review a large number of modern publications announcing the effect of various mathematical models and applied statistics methods on multivariate time-series data processing and machine learning for the automatic classification of ICS/SCADA abnormal data, which indicates the significance of the task of the EP stability monitoring by means of identifying anomalous data received from ICS/SCADA sensors.

Thus, the key areas of AT use in the sphere of EP security and stability involve:

1. Search for anomalies in EP and ICS components operation.
2. Forecasting of technological equipment component failures and breakdowns in order to streamline equipment maintenance and repair.
3. Elimination of operator malfunctions.

According to studies [10], malfunctions committed by operators observing and controlling EP cause about 42% of accidents. This can be associated with the development of limited manning technologies of production automation. An operator, not being beside the real equipment, is not able to assess its physical condition in real-time and in the current operating conditions. The identification of aberrant behavior of operators, EP or ICS/SCADA processes determine the relevance of AT of User and Entity Behavior Analytics (UEBA) class [13, 17] designed to remove the restrictions inherent in formal rules of traditional Security Information and Event Management (SIEM) correlation [8].

3 Practical Experience in AT Use

In the light of features listed in the previous section, the pilot IMAS project was integrated in the segment of the main gas compressor station based on the integration of IT monitoring platforms (HP Operations Bridge and HP ArcSight SIEM) and Qlik Sense Analytics Platform. Various types of anomalies were detected during the experimental operation using AT:

- EP deviations associated with changeover periods;
- transfer of control loops into manual mode;
- situations caused by incorrect sensor readings.

Complex security incidents were revealed:

- prevention of EP emergency stop associated with the SCADA server UPS shutdown;
- identification of the cause for frequent SCADA alarms within the primary EP associated with power supply disruptions.

Statistical methods of anomaly detection, fault tree models, visual analysis (exploration) methods in interactive mode based on the associative in-memory data model were recognized as the most useful for the early warning of dangerous situations, anomaly detection and interpretation by the operating organization. These methods are not supported by SCADA.

It is observed that AT are most in demand during testing and configuration of CPS joint operation within the shakedown period, commissioning period and start of new and modernized facilities operation. Promising directions of AT use can involve the tasks of aberrant behavior identification in operators, ICS control objects and SCADA processes.

Along with successful results, the identified problems of technological and methodological availability of AT for industrial use should be also pointed out.

1. The source data problem. CPS data features involve:
   - a large number of different data sources, large data volumes and a large variety of data: tens of thousands of signal tags, high tag updating rate up to 10 times per second, noisy data and data skipping;
   - absence of any engineering data association providing reference data and configuration information;
   - absence of a single standard (mechanism) for data retrieval from SCADA produced by different manufacturers;
   - absence of representative data sampling and labeled data sets.

These features complicate the direct application of machine learning methods, in particular, neural network models [5, 19, 20]. The data must be pre-processed to be used with advanced analytics methods. The development of appropriate pre-processing procedures (Extract, Transform, Load (ETL)) takes 80% of design time of a data analyst (engineer) [16].

2. The problem associated with the complexity of monitoring and analysis objects. There are no mathematical models of objects for reliable forecasting of security events and incidents. Namely, there are no formalized models of physical processes that would adequately describe the signs of CPS normal operation. The complexity and the cost of mathematical model building can be compared to ICS/SCADA development.
3. The problem associated with the AT feasibility analysis. The economic interest of potential customers in AT is low because engineering and economic evaluation is quite a complex process. There are no appropriate methods to do that. At the same time, the negative factor of additional responsibility and burden for operators resulting from the introduction of new AT is obvious. A customer must have in mind a clearly expressed (preferably in rubles, hours or in physical terms of saved materials) and significant economic or technical effect.
4. The problem of state regulation. Statutory instruments of regulators ensuring the SII facilities security do not provide specific requirements for a comprehensive analysis of ICS/SCADA operation in terms of anomaly detection.
5. The problem of import substitution. There are no Russian platforms for research and analysis of big data received from industrial CPS, which are essential for the successful IMAS AT application on industrial scale. Characteristics of an industrial platforms supporting the advanced AT analytics practices and processes of AT full life cycle are reviewed in Gartner publications [4, 6, 17, 18].

4 Advanced Security Analytics Platform Architecture

Due to the diversity of data and analysis aspects, a set of machine learning models is required. In this regard, it is important to create a universal Advanced Security Analytics Platform (ASAP) to automate the processes of optimal model selection, their learning cycles and delivery to the operating environment in order to process data flows in near-real-time mode. The architectural scheme of such a platform is shown in Fig. 2.

The platform consists of three levels for data handling:

1. The data collection level involves the processes of collecting, receiving and pre-processing data from systems such as SIEM or ICS/SCADA, network traffic analysis and other procedures for processing contextual, reference and configuration information.
2. The data storage level involves the processes of consolidation and long-term storage of data obtained at the data collection level.

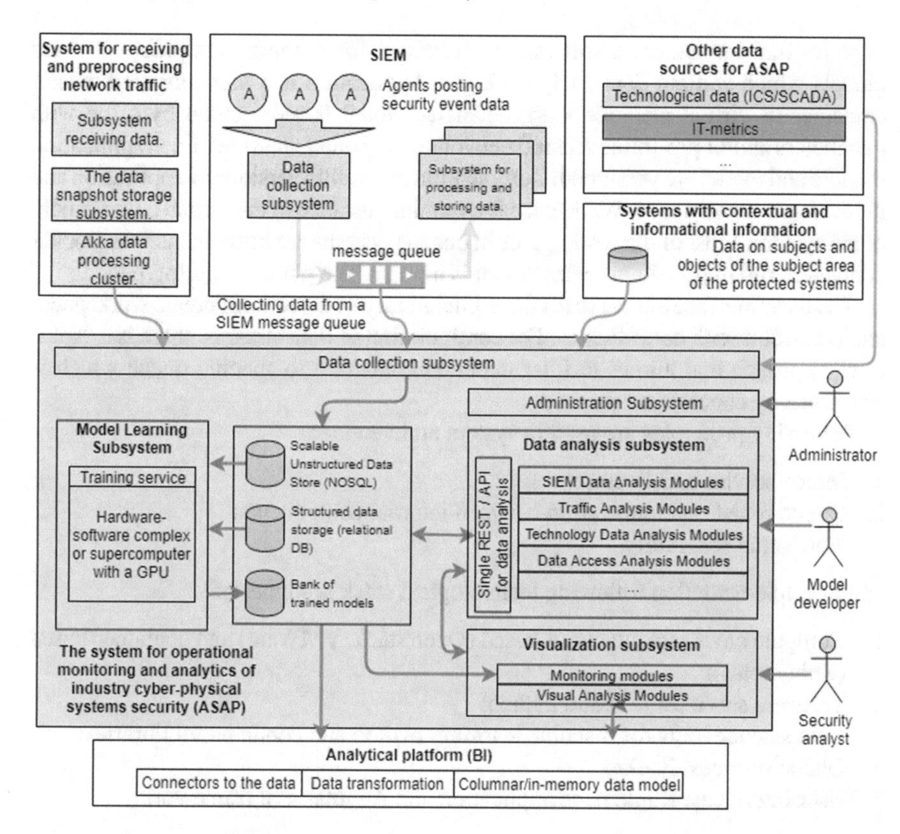

Fig. 2 The architecture ASAP

3. The data visualization level involves the processes of building interactive forms and visualizations that allows the platform user to carry out basic analysis of consolidated data from various sources.

The ASAP platform consists of the following subsystems:

1. Data gathering system;
2. Data processing system;
3. User interface for analytics and control of data flows.

The data processing layer is a set of services that are gathering data from SIEM system. The platform supports a messages stream from agents in CEF format. But it can be integrated with other CPS systems. The integration platform is implemented using Kafka queues. Such an approach allows scaling the data gathering system according to the load in term of events per second. Processed messages are saved in local storage for offline analytics and also buffered and sent to online processors for online anomaly detection and other data processing methods.

Data processing system is represented as a set of analyzers templates which is called the digital library of detectives. This library consists of prepared application modules that implement a software environment for running data processing and models which contain data analytics logic. Application is a preconfigured docker container or virtual machine with metadata. Model is represented by a file with parameters and/or pre-trained neural network or any other classifier. Each application module and model are versioned. Combination of specific versions of application and model form a workspace. Workspace is a running instance in the virtual environment of the system. Most of the workspaces in our subsystems are implemented as Docker containers running in Kubernetes which is a part of the platform prototype.

The user interface allows to fill the digital library of detections, create workspaces and connect it with data streams. For configuration of data streams, there is a "split-rule" approach that allows to filter data and connect it to specific queue which is processed by concrete workspace.

Following principles are used in system architecture:

1. Micro-service architecture;
2. Queue based communication between internal components;
3. Horizontal scalability.

For implementation following technological stack is suggested:

1. Compute environment: cloud-based (OpenStack, VMWare) and container-based (Kubernetes);
2. Software services: java and python;
3. Data science tools for machine learning: python and conda based libraries;
4. Queue services: Kafka;
5. User interfaces: single page application and Angular as a framework.

The prototype of the platform was implemented in the OpenStack [11] environment for running experiments and tests on real SIEM data.

5 Anomaly Detection Method for Semistructured Data

SIEM systems produce a lot of events with different contact. The method of extracting feature vectors from messages with different data fields is introduced in this section. One of the standards in SIEM systems for the format of messages is CEF (common event format). This format is introduced by ArcSight, and this standard contains mandatory fields like timestamp, type, product name, and version, etc. Message specific information is not fixed in CEF standard and placed as a set of key-value fields to the extension section. So, message flow can be considered as time-series data with a different set of fields in a key-value format.

Formulation of the problem: there is a stream of events from different SIEM sources, and it is required to establish whether a particular event (or a multitude of events) is an anomaly compared to the previous observation.

The unsupervised method based on Isolation Forest was chosen to identify anomalies. Feature engineering for analysis was developed drawing on statistics collection, which consists of collecting information on the frequency of occurring values for the observed keys of the analyzed fields in events.

Collecting statistics is the process of obtaining weights for each value of each key from the interesting fields in the event. The weights obtained allow us to form vectors for further analysis and search for anomalies. The parametrization of the statistics collection allows flexible adjustment of the feature vector correlation, which was obtained.

In the software implementation, the statistics are presented in the form of map M, containing such values as, $m[asa\_dest\_port|1000] = 0.1$, where $asa\_dest\_port|1000$ is a component key consisting of polar events and a specific value.

Statistics are calculated periodically by time or by a number of events. At each step, the statistics merge with the previous statistics through a forgetting ratio. This approach allows the system to implement memory and gradual adaptation to the current situation, as well as conduct an analysis of events on the stream, close to real time.

Basic steps to get feature vectors:

1. Getting the occurrence frequency of value in the chunk of events.
2. Obtaining weights for each key value based on the previous and current statistics using the "averaging" algorithm and using the forgetting coefficient (taking into account the current and previous values using the coefficients). This allows us to implement a system with memory and flexible adaptation to changes in input data.
3. Normalization of weights and refusal from weights by a threshold value.
   Statistics are calculated periodically by a number of events.
4. Getting the frequency of occurrence of the value in the chunk of events is according to the formula (1).

   (1)

   where is the number of events with the same value for a specific key, N is the number of events in a chunk in total.
5. Obtaining weights for each key value based on the previous and current statistics is according to the formula (2).

   (2)

   where is the value of the normalized weight at the previous step (previous chunk), is the coefficient of forgetting the weight for the previous step, is the occurrence frequency of value for the current step.
6. The normalization of weights and refusal from weights by a threshold value is according to the formula (3).

   (3)

where is the weight of the value for a particular key k, is weight before normalization.

If the , then this value is removed from the list of weights, where is the threshold of the value of the statistics.

Qualitative assessment of statistics calculated periodically by the number of events is the following:

Positive:

- The normalized values at each step allow each element of the vector to be obtained in one range from 0 to 1, which makes it possible to use different models of machine learning without reconfiguration when the intensity and heterogeneity of the input stream changes in the future.
- The correlation between the values in each step enables us to interpret the flow regardless of the flow intensity.

Negative:

- The use of statistics by the number does not allow an adequate assessment of the temporal nature of the events arrival, for example, the intensity of the events at night and during the day can be dramatically different and the transition between inactive and active stages can be interpreted as an anomaly.

Statistics are calculated periodically by time.

1. Getting the frequency of occurrence of the value in the chunk of events is according to the formula (4).

(4)

where is the number of events with the same value for a specific key, T is the window period for which events are collected.

2. Obtaining weights for each key-value based on the previous, current statistics and normalization of weights and refusal from weights by a threshold value is according to the formula (5).

(5)

where is the value of the normalized weight at the previous step (previous chunk), is the coefficient of forgetting the weight for the previous step, is the frequency of occurrence of the value for the current step, is the number of unqueue value for a specific key.

Qualitative assessment of statistics, which is calculated periodically by the time is the following:

Positive:

- Time specificity of events and correlation of events by time window are taken into account.

– It allows us to get a generalized characteristic of the flow, which in the future makes it possible to catch not only anomalies in events but also anomalies in the behavior of their sources.

Negative:

– Due to the data collection specificity, the events are not sorted. Because of that, statistics validity is distorted with a large data stream. To bring the analysis closer to real-time, it was decided not to waste time on sorting the data. If desired, two approaches to windows are used: a fixed window in time above and below. In this case, events that have a timestamp below the lower threshold of the window can be opened. It also becomes possible to skip events due to their delay from certain sources due to their temporary unavailability. The second approach is to use only the upper limit for events. There may be the possibility of incorrect calculation of statistics due to the presence of events not from the current time window.

6 Conclusion

In this chapter the system architecture for operational monitoring and analytics of industry cyber-physical systems is proposed. The proposed system can solve described challenges by using new approaches based on intelligent data analytics methods including anomaly detection We implemented the architecture and performed experiments with anomaly detection based on the method of extracting features from the unstructured flow of events coming from SIEM system. Experiments with network devices (agent Cisco ASA), windows domain events, and Checkpoint firewall are demonstrated good results of detecting anomaly behaviour. In the future work, we will extend a set of anomaly detection and anomaly prediction methods in our system and experiment with different machine learning methods including:

- Research of seq2seq methods to predict attacks (LSTM networks and others);
- Research of possibility to use LSTM method for anomaly detection;
- Research of survival analysis methods for evaluating the probability of performing network attacks;
- Extending presented method by adding the running window of events and introducing several statistics for long and current memory.

The proposed architecture allows extending the prototype by adding new analysers to the digital library and perform the different experiments on the same data. This approach also allows having historical data sets with annotated time when an attack or malicious behaviour is performed.

References

Ackerman, P.: Industrial Cybersecurity: Efficiently Secure Critical Infrastructure Systems. Packt Publishing Ltd., Birmingham (2017)

Anderson, K.: Analytical Culture: From the Data Collection to the Business Results. Mann, Ivanov and Ferber, Russia (1999)

Chernov, D., Sychugov, A.: Modern approaches to ensuring information security of automated process control systems. Tekhnicheskiye nauki **10**, 58–64 (2018)

D'yakonov, A., Golovina, A.: Identification of anomalies in the mechanisms of machine learning methods. Analytics and data management in data intensive areas. In: Proceedings of XIX International Conference DAMDID/RCDL 2017 (pp. 469–476) (2017)

Filonov, P., Lavrentyev, A., Vorontsov, A.: Multivariate industrial time series with cyber-attack simulation: fault detection using an lstm-based predictive data model. arXiv preprint arXiv: 1612.06676 (2016)

Idoine, C., Krensky, P., Linden, A., Brethenoux, E.: Magic quadrant for data science and machine-learning platforms (2019). https://www.gartner.com/en/documents/3899464/magic-quadrant-for-data-science-and-machine-learning-pla. Last accessed 29 April 2019

Islam, R.U., Hossain, M.S., Andersson, K.: A novel anomaly detection algorithm for sensor data under uncertainty. Soft. Comput. **22**(5), 1623–1639 (2018)

Kavanagh, K., Sadowski, G., Bussa, T.: Magic quadrant for security information and event management (2018). https://www.gartner.com/en/documents/3894573/magic-quadrant-for-security-information-and-event-manage. Last accessed 29 April 2019

Kotenko, I., Levshun, D., Chechulin, A., Ushakov, I., Krasov, A.: An integrated approach to ensuring the security of cyber-physical systems based on microcontrollers. Voprosy kiberbezopasnosti **3**(27), 29–38 (2018)

Kraevski, J., Ivanov, A.: Situational perception. A new approach to the design of human machine interfaces. Avtomatizatsiya v promyshlennosti **12**, 26–30 (2014)

Lukashin, A., Lukashin, A.: Resource scheduler based on multi-agent model and intelligent control system for openstack. In: International Conference on Next Generation Wired/Wireless Networking (pp. 556–566). Springer, Berlin (2014)

Mart´ı, L., Sanchez-Pi, N., Molina, J., Garcia, A.: Anomaly detection based on sensor data in petroleum industry applications. Sensors **15**(2), (2015)

Matveev, A.: Behavioral analysis systems market review —user and entity behavioral analytics (UBA/UEBA). https://www.anti-malware.ru/analytics/Market_Analysis/user-and-entity-behavioral-analytics-ubaueba. Last accessed 29 April 2019

McLaughlin, S., Konstantinou, C., Wang, X., Davi, L., Sadeghi, A.R., Maniatakos, M., Karri, R.: The cybersecurity landscape in industrial control systems. Proc. IEEE **104**(5), 1039–1057 (2016)

Mehrotra, K.G., Mohan, C.K., Huang, H.: Anomaly Detection Principles and Algorithms. Springer, Berlin (2017)

Rieger, C., Manic, M.: On critical infrastructures, their security and resilience—trends and vision. arXiv preprint arXiv:1812.02710 (2018)

Sadowski, G., Bussa, T., Litan, A., Phillips, T.: Market guide for user and entity behavior analytics (2018). https://www.gartner.com/en/documents/3872885/market-guide-for-user-and-entity-behavior-analytics. Last accessed 29 April 2019

Sallam, R., Richardson, J., Howson, C., Kronz, A.: Magic quadrant for analytics and business intelligence platforms (2019). https://www.gartner.com/en/documents/3900992/magic-quadrant-for-analytics-and-business-intelligence-p. Last accessed 29 April 2019

Utkin, L.V.: A framework for imprecise robust one-class classification models. Int. J. Mach. Learn. Cybernet. **5**(3), 379–393 (2014)

Utkin, L., Zhuk, J.: Robust anomaly detection model using clogging model. Vestnik komp'juternyh i informacionnyh tehnologij **7**, 47–51 (2013)

Zegzhda, D., Vasiliev, Y., Poltavtseva, M., Kefeli, I., Borovkov, A.: Cybersecurity advanced production technologies in the era of digital transformation. Voprosy kiberbezopasnosti **2**(26), 2–15 (2018)

Engineering Education for Cyber-Physical Systems Development

Organization of Engineering Education for the Development of Cyber-Physical Systems Based on the Assessment of Competences Using Status Functions

I. V. Veshneva, Alexander A. Bolshakov and A. E. Fedorova

Abstract The causal graph of the social relationships of the organization is based on the structure of a balanced scorecard. The example of the structural unit of the university is considered. The values of the state estimates in the nodes of the tree are presented on the basis of the status function method. Each value of the status function is formed from two estimates of the intuitive simulation including the present and the desired state. The integral moments are calculated for individual branches of the event graph tree. For integral moments, an analogue of Shewhart control charts is constructed. The applicability of the status function method to obtain an integrative assessment of the organization's activities as a complex social network is shown.

Keywords Status functions · Balanced scorecard · Product quality control charts · Event graph · Quality control

1 Introduction

The development of the digital economy requires the creation of systems in which data in digital form is a key factor of production in all areas of social and economic activity [1]. Horizontal and vertical integration between various functions and operations in a complex social system will allow creating data exchange platforms for embedded cyber-physical systems. To solve problems of collecting and evaluating data from different sources, decision-making in real-time requires the use and design

I. V. Veshneva (✉)
Saratov State University, 83 Astrakhanskaya Street, Saratov 410012, Russia
e-mail: veshnevaiv@gmail.com

A. A. Bolshakov
Peter the Great St. Petersburg Polytechnic University, Polytechnicheskaya, 29, Saint-Petersburg 195251, Russia
e-mail: aabolshakov57@gmail.com

A. E. Fedorova
Ca' Foscari, University of Venice, Dorsoduro 3246, 30123 Venice, Italy
e-mail: anna.fedorova@unive.it

A. G. Kravets et al. (eds.), *Cyber-Physical Systems: Industry 4.0 Challenges*, Studies in Systems, Decision and Control 260,
https://doi.org/10.1007/978-3-030-32648-7_22

of mathematical models [2, 3], the development of technical systems and technologies that allow realizing human interaction with the technical system [4]. To obtain a guaranteed result associated with improving the quality of the activities of complex socio-economic systems based on developed and implemented cyber-physical systems, it is necessary to build a system for monitoring and managing the quality of business processes in educational institutions and enterprises [5, 6].

Imagine the possibility of using modern mathematical models for collecting and analyzing heterogeneous data based on the use of technologies of a balanced scorecard, quality control of six sigma, status functions and cause-effect graphs [7, 8].

To control the quality of business processes, six sigma technology has been successfully used [9]. It allows you to reduce the variability of the process, measure and control processes, anticipate user requirements and expectations, support decisions based on measurement data, and limit deviations [10]. However, the prospects for the application of this technology may be limited due to the inconsistency of monitored indicators, the lack of normalization of values. In addition, it is difficult to achieve comparable results for various aspects of the organization's activities. There may be a mismatch of objectives for individual components and indicators of controlled activities [11]. Combining disparate performance indicators of the organization and ensuring their consistency is possible through the use of a balanced scorecard (BSC) [12, 13].

To develop a mathematical model for controlling disparate performance indicators of an organization, let us show the possibility of applying the technology of six sigma on the structure of the organization's performance indicators. To do this, we construct a cause-and-effect graph in the form of a tree of goals for an educational institution [14]. We will measure the attainability of goals in disparate activities based on the method of status functions (SF) [15, 16].

The proposed SFs represent the orthonormal basis of the system for which the model base has been developed [17]. Their use allows us to solve the problem of comparability of indicators for various activities. The integral moments of the values of estimates based on the SF will be used to apply the technology of six sigma.

2 Methodology

BSC translates the organization's operational strategy into a system of clearly defined objectives and indicators. This indicator determines the degree of achievement of these installations within 4 aspects: finance, market, management, resources.

We will combine the capabilities of information technology, decision-making support, six sigma and SSP on the example of an educational institution, primarily for the training of specialists in the field of cyber-physical systems. First, we will describe the use of MTP for an educational institution. This requires the adaptation of the components of the 4 aspects to the educational institution. Transform these aspects.

2.1 Aspects of Organization's Operational Strategy

Finance. It is important to understand that the educational institution fulfills the social order of society in the state regulation of activities. Therefore, financial profit is not its key goal. Therefore, it is necessary to find a replacement for the aspect of financial activity. In financial activities, the results of the past are mainly used, and in other aspects planning for future activities is carried out. The financial aspect is replaceable by the *aspect of internal activity*. This is justified by the fact that the results of the internal activities of the organization in the past are informative in the financial performance indicators.

Market (*customers*). Clients of the education system in general and of the educational institution, in particular, are the whole of society as a whole [18]. Replace the projection on the market (or customer) – *aspect of the relationship*.

The projection on *management* (or internal business processes) is called the *aspect on the types of activities*, taking into account the complexity and intersection of the educational processes. We will assume that as a result of the implementation of business processes, products of the following groups are being formed:

- educational services,
- scientific and technical products,
- integrated products based on scientific and technical products and educational services,
- educational and methodical products.

For example, it is possible, but difficult, to link educational services and the results of a student design office. In this case, the participation of teachers in the implementation of research and development work may well "hinder" the development and publication of educational and methodical products that are required for the implementation of the educational process. At the same time, these processes have internal unity and crossover.

Training and development (*intellectual potential of the organization*). This aspect is one of the main activities of the educational institution. The implementation of the training and development processes of the organization's staff provides the organization with a strategic advantage in the future. Transform this group of goals into an *aspect for the future*.

When replacing the semantic meaning of the components of the BSC indicators (finance, market, management, resources) on the transformed aspects (internal activities, relationships, activities, future), the logical time structure "yesterday - today - tomorrow" is used. Thus, we will assume that using "yesterday's achievements" in the aspect of internal activity, the organization implements "today" activities and relationships. Development in an organization allows shaping its "future".

As a result, we obtain four groups of indicators for combining in the BSC. A typical project for the implementation of the BSC begins with building a hierarchy of goals—defining the main and auxiliary goals, and the corresponding selection of management levels. Let's carry out this construction in the form of hierarchy. Let's

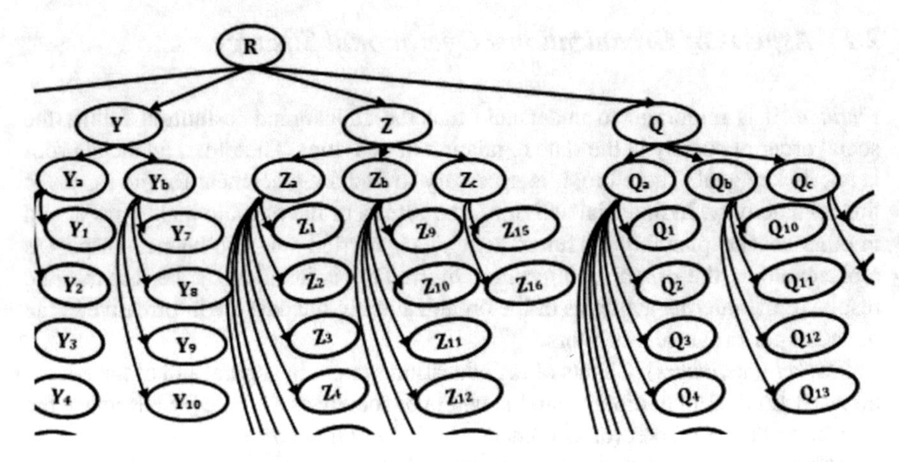

Fig. 1 A fragment of the hierarchy of goals of an educational institution based on the SSP

create a tree graph in the form of a tree of goals. Let us denote the root vertex R. Use certain aspects of the BSC indicators as 4 branches of this causal graph. Let us single out four intermediate outputs of the output tree, which participate in the formation of the resulting assessment of the effectiveness of the activities of the university (university departments).

We use the four aspects of the balanced scorecard that are transformed for a university to form R, obtained as a result of collecting expert information and analyzing expert information and interpreting the results obtained (Fig. 1):

X—The Aspect of Learning and Development. Includes groups of indicators "PPS", "Students", "Administration", "Educational support staff".
Y—Aspect on relationships. Includes groups of indicators "External consumers", "Internal consumers".
Z—Aspect on internal activity. Includes groups of indicators "Financial indicators", "Educational and scientific activities", "Management".
Q—Aspect on activities. Includes groups of indicators "Business processes", "Ensuring processes", "Control processes".

2.2 *Model for the Quality Assessment*

When building a model for assessing the quality of the work of a university department, it is advisable to classify the input variables and, on its basis, build an output tree (Fig. 1), which determines the system of nested statements-knowledge of lower dimension. A total of 64 indicators are monitored. The final assessment of the model, which shows the quality of the unit's work, is denoted by the letter R.

The criteria for evaluating indicators should be clearly regulated. Indicators are evaluated for compliance with the three possible levels (low, medium-high) and target value. For example, the indicator "Passing advanced training courses of the teaching staff of the department" received the following linguistic formulations of possible assessment values: low—1 time in 3 years, with possible financial opportunities of the university and interesting programs may be more often. Medium—1 time in 2 years, with the possibility of passing courses more often than 1 time in 2 years. High—once a year or more. The target value is 1 time in 3 years, established by the standard.

The formation of the level of assessment allows you to bring all the various scales of indicators to a single scale of assessments. However, it is linguistic. For such estimates, it is easy to obtain a mathematical model based on the theory of fuzzy sets.

The analysis of expert information, as well as the interpretation of the results, allows us to form the following system of criteria for evaluating the activities of higher education institutions, based on the system for estimating the parameters of the BSC using the SF [15, 16].

2.3 Analysis of Mutual Influence of Indicators

This is due to the fact that controlled indicators influence each other. For example, resource management processes and learning processes are crossover [19]. Or the processes of scientific development and teaching students. The interrelation of the latter is obvious. However, science can both contribute to better student learning and also interfere. In the first case, students have the opportunity to participate in the most promising scientific research and study modern high technology, including In a different relationship, the main resources of the teacher are directed to the field of research and there may simply not be enough time for active work with students.

It is important to understand that there is a mutual influence of indicators. It is not unique and not linear. Mathematical models of such processes are developed and actively used in the description of physical processes [20–22].

The most striking example is the interference. When adding the characteristics of two interacting processes, it is possible to observe their mutual amplification or weakening (Fig. 2).

2.4 Indicators of the Structural Unit of an Educational Institution

That is why the SF method has good prospects in developing mathematical models of product quality control. This is due to the fact that the resulting models have a

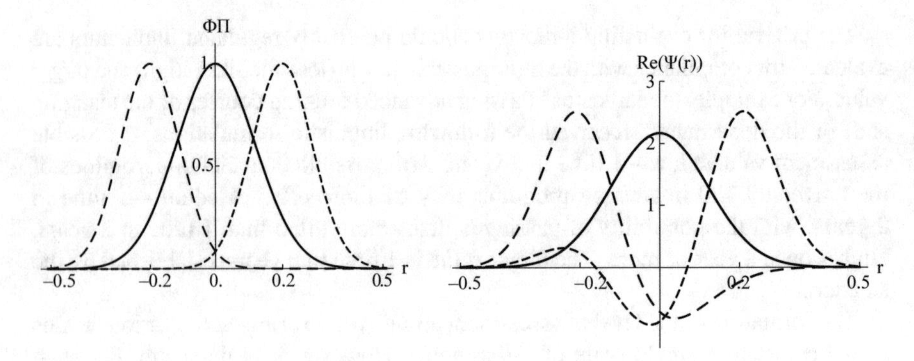

Fig. 2 The kind of membership functions (left) and the amplitude of the status functions obtained as a result of the algorithm orthogonalization Gram-Schmidt (right)

synergistic property to ensure a significant excess of the capabilities of the system as a whole over the sum of the capabilities of its parts. Since the SFs are the semantic analogue of membership functions in the theory of fuzzy sets and the mathematical analogue of wave functions in models of quantum mechanics, they can be used to model the process of interference type in optics or synergy, as is commonly used to denote this phenomenon in humanitarian studies of the interaction of social groups. It is important to note that the basis of such phenomena is the interaction, which is expressed as the intersection of processes in an educational institution. That is why the SF method has good prospects in developing mathematical models of product quality control. This is due to the fact that the resulting models have a synergistic property to ensure a significant excess of the capabilities of the system as a whole over the sum of the capabilities of its parts. Since the SFs are the semantic analogue of membership functions in the theory of fuzzy sets and the mathematical analogue of wave functions in models of quantum mechanics, they can be used to simulate the process of interference type in optics or synergy, as is commonly used to denote this phenomenon in humanitarian studies of the interaction of social groups, including student groups learning cyber-physical systems. It is important to note that the basis of such phenomena is the interaction, which is expressed as the intersection of processes in an educational institution.

The target value thus meets the minimum requirements of compliance with the requirements of the faculty. Many employees participate in advanced training programs.

Examples of indicators are presented in Table 1 as a fragment.

Ratings can be represented by a radar chart. However, for use in the information system, the presented results must be processed and structured to build decision support rules [23]. The most promising is the status function method, described in detail in the work of the authors [15]. For each of the indicators, a status function is built on the basis of the obtained estimates [16]. The real part is determined in accordance with the directly obtained assessment, the imaginary part of the status functions is based on the obtained estimate of the value of the target indicator and has

Table 1 Fragment of estimates of indicators of the structural unit of an educational institution

Partial rate		Target value	Estimate
x_1	Passage of advanced training courses of the teaching staff of the department	Low	Low
x_2	Attending seminars, training of the teaching staff of the department	High	Low
x_3	Participation of faculty members in conferences	High	High
x_4	New teaching methods and technologies	High	Medium
x_5	Participation of the department in obtaining grants	High	low
x_6	Educational programs	High	High
x_7	Graduation of teaching staff ((Number of associate professors + number of professors)/total number of the teaching staff)	Medium	High
…			
q_{12}	Logistics	High	Medium
q_{13}	Information and technical resources	High	High
q_{14}	HR process	High	Medium
q_{15}	Quality Management System	High	Medium

3 possible values: −1, 0, +1 (Fig. 3). This corresponds to the values of the indicator: “below the target value”, “coincides with the target value”, “above the target value”.

3 Results and Discussion

For the fragmentary implementation of the cyber-physical system during training, the verification of the mathematical model and its technical implementation was carried out (Fig. 3). In the information-measuring system (IMS), a non-invasive method of student feedback (remote psychodiagnostics) is widely used with the help of multispectral analysis and a neural network expert system.

The analyzing device takes into account the features of the facial expression, head position, dynamics of the movement of eyes, lips, hands and intensity (saturation) of

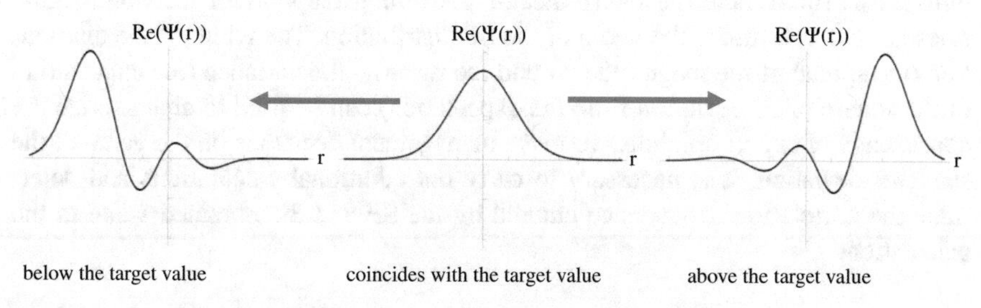

Fig. 3 Illustration of the meaning of the parameter of the expected target value

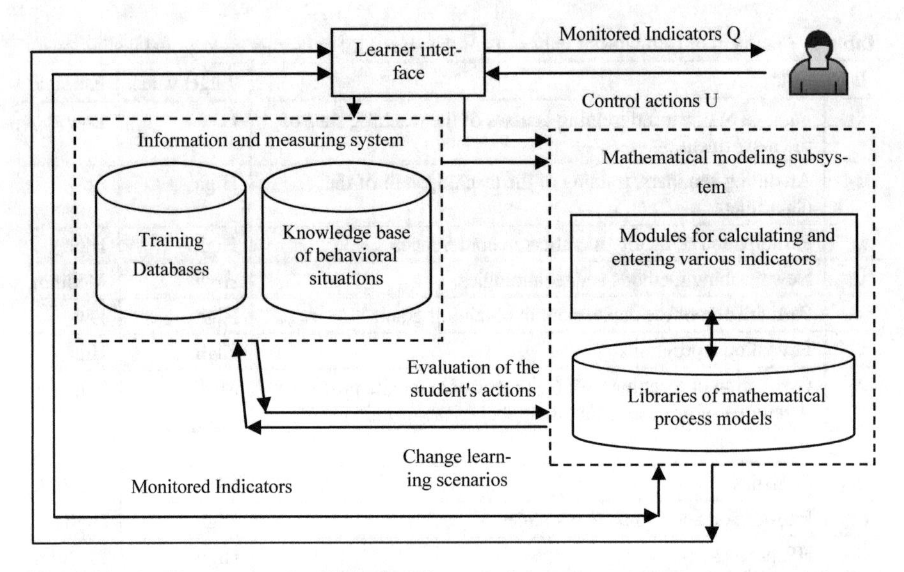

Fig. 4 The functional structure of the e-learning system

the body color (face) of each student. The measurement results are processed statistically using discriminatory analysis, based on new algorithms for optical-geometric analysis, using the method of the Federation Council. The proven technology is associated with a synchronous transition from a collective educational environment to an individual, aimed at the realization of personal needs, formed by students as a result of interaction with the external (collective) educational environment. Due to the inclusion of a feedback channel (non-invasive diagnostics) in IMS, the psychophysical state is constantly monitored and corrected (by adjusting the input information data block), both as an individual student and as a group (training group) as a whole, allowing to flexibly level various features of the program scenario (Fig. 4).

The structure of IMS includes a base unit containing a video system, a structured (infrared) illumination system, a hardware-software unit consisting of a PC and the corresponding software, interface for working with network data, physiological parameters analyzer.

The subsystem of mathematical modeling is based on complex-valued status functions (SF). In this article, the main indicator for forming a support for decision-making rules is used to estimate the width of the SF distribution. The width of the distribution (the spread of the magnitude around the mean)—the variance (the expectation of the square of its deviation from the expectation) can be used to characterize the consistency of expert opinions. To make management decisions on the value of the standard deviation, it is necessary to carry out additional calculations and determine the correlation dependence entered by the SF and the obtained value in the calculation.

$$\sigma = \iint\limits_{-\infty}^{+\infty} (r - m_r)\Psi(r)\Psi^*(r)dr / \iint\limits_{-\infty}^{+\infty} \Psi(r)\Psi^*(r)dr \tag{1}$$

where r is the entered base variable, m_r—expected value Ψ—superposition of basic status functions of assessments, Ψ*—complex conjugation symbol.

The obtained rules for supporting the adoption of solutions have been verified in a number of experiments [23]:

If σ ∈ [0; 0.8[, then “Competency assessments are sufficiently correlated” and “The process does not require management intervention”.

If σ ∈ [0.8; 1[, then “The mismatch of ratings is critical” and “It is necessary to take measures to regulate the state”.

The results are representable in the format of quality checklists. The sample number is plotted on the abscissa—in our case, a controlled group of parameters. The ordinate axis is the value of the mean-square deviation. Mathematical expectation sets the value of the distribution center estimates of indicators. With a normal distribution of the estimated values, the observed deviations do not exceed 3 σ.

In this way, a model was obtained for controlling a large number of disparate indicators and identifying critical values of deviation of estimates (Fig. 5).

For individual branches of the tree, the values of the status functions and their integral moments were calculated. The second integral moment was chosen, which characterizes the width of the distribution (spread of the value around the mean value). The calculation results show that all values are within the boundaries of the statistical controllability of the processes ±3 sigma [−0.18867; 0.94334] (Table 2).

The obtained estimates of disparate indicators were possible to evaluate in a single scale of measurements. Combining the projection “yesterday – today – tomorrow” allowed us to design both the transformation of the structure of the BSC in accordance

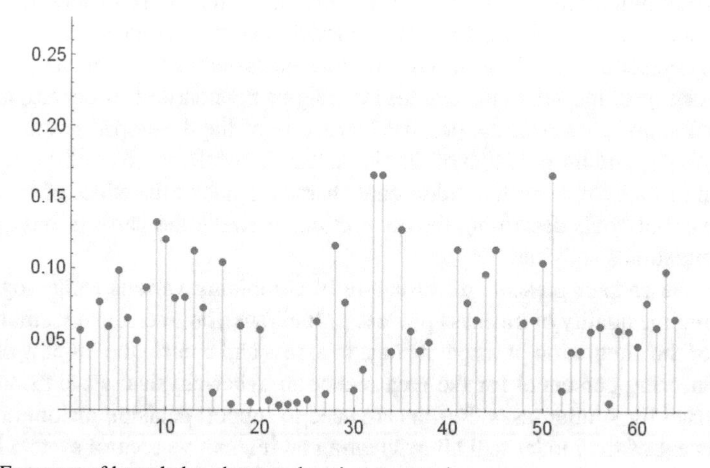

Fig. 5 Fragment of knowledge about students’ competencies

Table 2 Values of integral moments for 4 aspects of the MTP and final characteristics

Fragments of the aspects target tree	Expected value	Dispersion
The relationship aspect	−0.1862	0.0419
Learning and development aspect	0.0569	0.1273
Aspect "Internal Activity"	0.0671	0.0567
Aspect "Activities"	−0.1361	0.0422
Combining aspects in the root vertex of the tree R	−0.0792	0.0486

with the goals of the educational institution, and to apply the gradient estimation method when making assessments based on status functions.

4 Conclusions

The use of the status function method for evaluating the activities of a structural unit of an educational institution can be used in designing mathematical models of the operation of a given social network and for building production rules for an information system.

It is important to note that the use of the values of the integral moments of the status functions for all the estimated indicators allows us to obtain the objective function of the mathematical model of the system. For this, the three-sigma rule is used and the deviations of the values of both individual indicators and selected groups of indicators are monitored. These may be separate branches of the event tree, but the greatest interest from the point of view of studying the properties of the mathematical model of the entire system is the calculation of deviations of groups of indicators of minimum sections from the indicators of the entire system, which is a plan for the development of the proposed model of control of integrative indicators of the organization. In addition, it is the minimal cross-sections, in which the value of the expectation of the values beyond the three sigma boundaries is observed, requires additional study. However, the proposed structure of the developed graph requires the complexity and introduction of the characteristics of the vertices of the graph, as shown in this article, as well as additional characteristics of the edges of the cause-effect graph of goals describing the connections between the vertices and goals of the organization's activities.

Thus, the chapter presents the structure of combining various technologies for controlling the quality of business processes. It allows you to create a mathematical model for the integration of disparate data to assess and control the process of training engineering personnel for the area related to cyber-physical systems, and can also be used for similar tasks. Results are used to support decision making. Further development of the model will allow creating an interactive control system human in the loop.

References

1. Dorosinsky, L.G., Zvereva, O.M.: Information Technology Support the Life Cycle of the Product, p. 243. Zebra, Ulyanovsk (2016)
2. Tomamichel, M.: Quantum Information Processing with Finite Resources: Mathematical Foundations (pp. 1–135). Springer, Berlin. https://doi.org/10.1007/978-3-319-21891-5
3. Jackson, P.: Introduction to Expert Systems, 3rd edn. Addison-Wesley, Harlow (1999)
4. Brovko, A., Dolinina, O., Pechenkin, V.: Method of the management of garbage collection in the "Smart Clean City" project. In: Gaj, P., Kwiecień, A., Sawicki, M. (eds.) CN 2017. CCIS (vol. 718, pp. 432–443). Springer, Cham (2017). https://doi.org/10.1007/978-3-319-59767-6_34
5. Gharib, M., Giorgini, P., Mylopoulos, J.: Analysis of information quality requirements in business processes. Revisited Require. Eng. **23**, 227 (2018). https://doi.org/10.1007/s00766-016-0264-4
6. Skobelev, P.: Towards autonomous AI systems for resource management: applications in industry and lessons learned. In: Proceedings of the XVI International Conference on Practical Applications of Agents and Multi-Agent Systems (PAAMS 2018). LNAI (vol. 10978, pp. 12–25). Springer, Berlin (2018). https://doi.org/10.1007/978-3-319-94580-4_2
7. Bezruchko, B.P., Smirnov, D.A.: Extracting Knowledge from Time Series: An Introduction to Nonlinear Empirical Modeling. Springer Series in Synergetics. Springer, Heidelberg (2010)
8. Ganter, B., Wille, R.: Conceptual scaling. In: Roberts, F. (ed.) Applications of Combinatorics and Graph Theory to the Biological and Social Sciences (pp. 139–167). Springer, New York (1989). https://doi.org/10.1007/978-1-4684-6381-1_6
9. Transactional Six Sigma and Lean Servicing: Leveraging Manufacturing Concepts to Achieve World-Class Service (296p). ByBetsi Harris Ehrlich. https://doi.org/10.1201/9781420000337
10. Fominykh, D., Rezchikov, A., Kushnikov, V., Ivashchenko, V., Bogomolov, A., Filimonyuk, L., Dolinina, O., Kushnikov, O., Shulga, T., Tverdokhlebov, V.: J. Phys. Conf. Ser. **1015**, 032169 (2018)
11. Nikolaychuk, O.A., Yurin, A.Y.: Computer-aided identification of mechanical system's technical state with the aid of case-based reasoning. Expert Syst. Appl. **34**(1), 635–642 (2008)
12. Kaplan, R.S., Mikes, A.: Managing Risks: A New Framework. Harvard Business Review (2012)
13. Olve, N.G., Roy, J., Wetter, M.: Performance Drivers: A Practical Guide to Using the Balanced Scorecard. Wiley, New York (1999)
14. Veshneva, I., Bolshakov, A., Kulik A.: Increasing the safety of flights with the use of mathematical model based on status functions. In: Dolinina, O., Brovko A., Pechenkin V., Lvov A., Zhmud V., Kreinovich V. (eds.) Studies in Systems, Decision and Control (vol. 199, pp. 608–622). Springer Nature Switzerland AG, Berlin (2019). https://doi.org/10.1007/978-3-030-12072-6
15. Veshneva, I., Chistyakova, T., Bolshakov, A.: The status function method for processing and interpretation of the measurement data of interaction in the educational environment. Trudy SPIIRAN **6**(49), 144–166 (2016). (in Russian)
16. Veshneva, I.V.: Quality assessment of a social object based on the construction of a multidimensional "quality field" of a balanced scorecard using the theory of fuzzy sets. Bull. Saratov State Tech. Univ. **3**(1), 227–234
17. Porter, M.E.: Competitive Advantage: Creating and Sustaining Superior Performance. The Free Press, New York (2008)
18. Sengupta, J.: Innovation models. In: Theory of Innovation. Springer, Cham (2014). https://doi.org/10.1007/978-3-319-02183-6_2
19. Xu, M., Sun, M., Wang, G., Huang, S.: Intelligent remote wireless streetlight monitoring system based on GPRS. In: Xiao, T., Zhang, L., Fei, M. (eds.) Communications in Computer and Information Science, vol. 324, pp. 228–237. Springer, Heidelberg (2012)
20. Ruiz, D., Nougues, J.M., Puigjaner, L.: Fault diagnosis support system for complex chemical plants. Comput. Chem. Eng. **25**(1), 151–160 (2001)

21. Venkatasubramanian, V., Rengaswamy, R., Yin, K., Kavuri, S.N.: A review of process fault detection and diagnosis: part I: quantitative model-based methods. Comput. Chem. Eng. **27**(3), 293–311 (2003)
22. Wang, H.C., Wang, H.S.: A hybrid expert system for equipment failure analysis. Expert Syst. Appl. **28**(4), 615–622 (2005)
23. Bolshakov, A.A., Veshneva, I.V., Melnikov L.A.: New Methods of Mathematical Modeling of the Dynamics and Management of the Formation of Competences in the Process of Learning in a Higher School (250p). Hotline–Telecom (2014)

Analysis of Impact Made by the Flagship University on the Efficiency of Petrochemical Complex

Pavel Golovanov, Mikhail Yu. Livshits and Elena Tuponosova

Abstract The chapter deals with the problems of mathematical modeling of engineering education for the development of Cyber-physical systems. The chapter analyzes the impact of labor resources and scientific potential of the flagship university of the Samara Region on the key indicators of production capacity of modern the regional petrochemical complex as a Cyber-physical system. The following statistical characteristics of the university's resources are taken into account: number of graduates, total number of scientific publications, performance of scientific work under grants, and generation of objects of intellectual property. Mathematical models are built in the form of the Cobb-Douglas production function. In identification of the model parameters is done by the least squares method. The output parameters of the model of the regional petrochemical production are as follows: average annual production capacity of oil incoming for processing, and annual production capacity of production of oil products and mineral lubrication oil. The approximate properties of the mathematical model are assessed by the determination factor and the F-statistics factor, and the prognostic parameters of the model with the Durbin-Watson model. Characteristics of efficiency of use of basic resources of the Samara State University are analyzed, qualitative and quantitative correlations are identified. A forecast of the production capacity of the petrochemical complex is made up to the year 2020, and prerequisites for the securing of its labor potential are found.

Keywords Modeling · Cobb-Douglas production function · Mathematical model · Production capacity · Factorial elasticity · University · Scientific publications · Research and development work · Generation of objects of intellectual property · Input and output parameters

---

P. Golovanov · M. Yu. Livshits (✉) · E. Tuponosova
Samara State Technical University, 244 Molodogvardeyskaya Street, Samara 443100, Russia
e-mail: usat@samgtu.ru

A. G. Kravets et al. (eds.), *Cyber-Physical Systems: Industry 4.0 Challenges*, Studies in Systems, Decision and Control 260,
https://doi.org/10.1007/978-3-030-32648-7_23

1 Introduction

Technical universities are a major source of engineering workforce in the region. Besides, innovative programs of regional development are largely formed by the scientific potential of universities. Creation of a network of flagship universities in 2016 ensured the increase in the number of 'growth points' of the regional economy as a prerequisite of technological development, improvement of investment attractiveness including those in the international markets. The chapter will dwell on the impact made by the scientific and workforce potential of a flagship university—Samara State Technical University (SSTU) on the efficiency of the regional petrochemical complex of the Samara Region. The modern petrochemical complex of the Samara Region includes oil production enterprises, large modern oil refineries and petrochemical plants with automated technologies. On the one hand, as a regional flagship university SSTU undertakes the mission of regional development by forming an outlook of the Samara's society, by managing the markets of today and, based on the construction of mechanisms of construction of unique interdisciplinary competences, solution of the challenges of the future [1]. On the other hand, SSTU is the only regional university that has a large oil and technology department in its structure with over 900 young graduates every year. Actually, a large part of engineers engaged in the science-intensive petrochemical complex that determines the economic potential of the Samara Region consists of alumni of SSTU.

2 Problem Statement

We will build mathematical models whereby the efficiency of petrochemical complex Y_{ji} will be related to indicators used to characterize work of a university. Efficiency Y_{ji} will be used to assess mean annual performance that determines maximum production output possible in a year and that is determined from the full usage of stated mode of operation of equipment and production premises. The value of production capacity is determined as ratio of actual production output Bi to mean annual production capacity Di effective in the ith reporting period for the manufacturing of jth products [2, 3].

$$Y_{ji} = \frac{B_i}{D_i}, \quad j = 1, 2; \; i = 1, 2, \ldots, \tag{1}$$

Modeling of functioning of universities is a topic of many papers of various authors [4–14].

Article [4] suggests the method of mathematical modeling of the functioning of a university. Mathematical models of discharge of students are built in the form of the Cobb-Douglas production function. In article [5], the authors dwell on the system of forecasting the market's needs in workforce of various qualifications and calculate workforce demand from the perspective of regional economies, in particular, in the Republic of Karelia.

Article [6] looks at 3 groups of questions related to mathematical modeling of managing quality of educational process at universities: building of basic models of educational process quality control; analysis of factors impacting the system of education and identification of major functions of quality including factors of potential and of delivery of results in the activities of universities. Article [7] proposes a study of basic methods in the building of mathematical models of educational processes using the example of 50 universities of Russia from the web site of "Expert Rating Agency" of the Ministry of Education and Science of the Russian Federation.

The authors of [8] analyzed the statistical data of the average annual numbers of people engaged in economic activities within the Samara Region and built the mathematical models to determine the required number of trained specialists taking into consideration the number of labor force and the gross regional product. Article [9] discusses the results of the analysis of the production function of general education. It is proposed that the qualification of a tutor is the most significant factor for the improvement of quality of general education.

Article [10] analyzes Russian and international ratings of universities, compares key criteria of assessment of Russian and international universities, and identifies the major principles of educational and scientific research activities. Article [11] raises the question of long-term forecasts of labor market demands in specialists with higher professional education. The author provides dependencies of the number of educational institutions of various types, number of students and discharge of specialists starting from 1940s. The uniform appearance of curves of changing of statistical data for every stage allows for creation of invariable analytical models giving an accurate description of the whole process of development; knowing a specific model one might forecast processes in education for a long term with a specified degree of accuracy.

Article [12] proposes a model of performance for 16 state universities in Malaysia for the year 2008. Measurement of efficiency and delivery of results are made using the DEA method. The building of the mathematical model of two universities and its identification using the numerical experiments of the educational systems are analyzed in [13].

In article [14] creation of lifecycle models of educational process, on the basis of CALS-technologies for high school is considered. Appropriate applied programs and the network structure, connecting universities and technological branches, have been developed for the purpose of step-by-step formation and eligibility of all stages of training of the expert for hi-tech branches.

The monograph [15] proposes an analysis of quantitative indicators of national systems of education and results of international ratings in the sphere of education. Role of Russia in the international educational environment is assessed, and a tentative assessment is given of its position in the global ratings of accessibility of higher education and financial possibilities of acquiring it.

However, the problem of quantitative assessment and forecast of impact of activity indicators of flagship universities on the branches of regional economy for the improvement of which those universities were created has hitherto been without an effective solution.

The econometrics widely employ the method of building mathematical models in the form of a non-uniform Cobb-Douglas production function (PF):

$$Y(t) = A \cdot K(t)^{\alpha} \cdot L(t)^{\beta}, \quad \alpha + \beta \neq 1, \tag{2}$$

which also allows to take into consideration the factor of scientific technical progress (STP) [16, 17]:

$$Y(t) = A \cdot K(t)^{\alpha} \cdot L(t)^{\beta} \cdot e^{\tau \cdot t}, \quad \alpha + \beta \neq 1, \tag{3}$$

where Y(t)—yield of end products; K(t)—capital resources, L(t)—labor resources, α, β—characteristics of efficiency of resource usage pecypca—indicator of elasticity, A—scale factor of conversion technology, τ—scientific technical progress (STP) impact factor [17, 18].

3 Mathematical Modeling

The following traditionally considered indicators of universities will be taken as input characteristics of the model K(t) and L(t): discharge of SSTU students—S_i, people; total number of scientific publications—P_i, pcs.; performance of scientific and technical research (R&D) under grants—G_i, units; generation of objects of intellectual property—I_i, units. As output parameters of the model of the regional petrochemical production Y(t) the following will be taken: mean annual production capacity of oil coming to processing—$Y_1(t)$,%; mean annual capacity of production of oil products and lubrication oils—$Y_2(t)$, % [2, 3]. In order to assess sensitivity of model solutions with respect to the respective resources the elasticity factors will be taken: χ_j—for the $S_i$factor; κ_j—for the P_i factor; ϕ_j—for the $G_i$factor; ρ_j—for the I_i factor; and μ_j—for the factor of scientific and technical progress (STP) impact factor.

The identification of the model parameters will be performed by the method of least squares (LS) [19]. The approximate properties of description will be assessed with the determination factor and criterion of F-statistics, and the prognostic properties with the Durbin-Watson model (DW) [4, 20].

We will build a model of annual production capacity of production of oil ($j = 1$)$Y_1(t)$ coming for processing from the year 2008 to 2017. The parameters of models identified using the initial statistical data [2, 3] in the variants of the non-uniform Cobb-Douglas PF (4) and of the model considering the STP (5) are given in Table 1 and in Figs. 1 and 2, where $Y_1(i) = Y_1(t_i)$—statistical data of the mean annual production capacity of the production of oil coming for processing in the ith year; $Y_1(t)$—calculation using model (4) and model (5) considering the STP factor.

$$Y_1(t) = A_1 \cdot S(t)^{\chi_1} \cdot P(t)^{\kappa_1} \cdot G(t)^{\phi_1} \cdot I(t)^{\rho_1}, \tag{4}$$

Table 1 Parameters of models (4), (5)

Parameters of model $Y_1(t)$	STP not considered (4)	STP considered (5)
χ_1	−0.1518	−0.0941
κ_1	0.0186	0.0420
ϕ_1	−0.0411	−0.0353
ρ_1	0.0316	0.0609
μ_1		−0.00887
DW	1.9267	2.0633
R^2	0.8018	0.8476
F	5.0580	6.9500

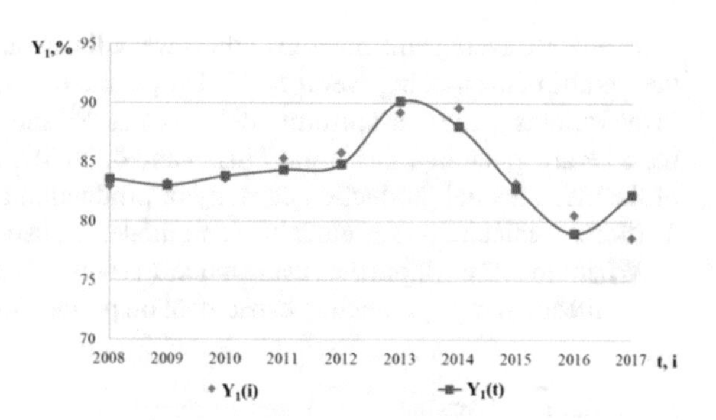

Fig. 1 Calculated data for model (4)

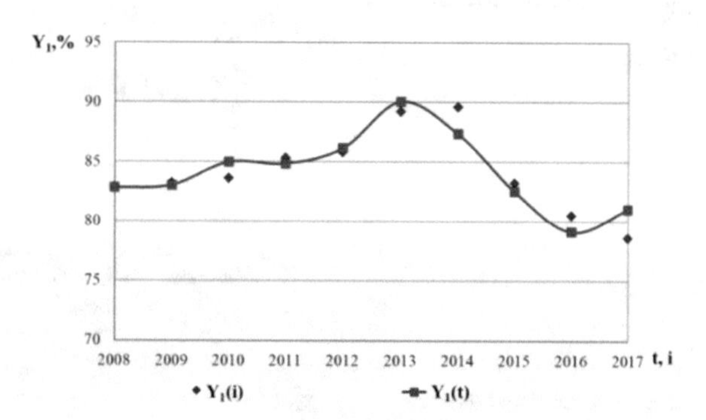

Fig. 2 Calculated data for model (5)

$$Y_1(t) = A_1 \cdot S(t)^{\chi_1} \cdot P(t)^{\kappa_1} \cdot G(t)^{\phi_1} \cdot I(t)^{\rho_1} \cdot e^{\mu_1 t}, \tag{5}$$

It is seen from Table 1 that the biggest component elasticity, ergo, meaning is with the factors P_i—total number of scientific publications and I_i—generation of objects of intellectual property. The model having been augmented with the factor of STP impact, these values increase by 2 times. The approximative and prognostic characteristics of the model are very good ($R^2 \approx 0.8$, $DW \approx 2$).

Table 2 Parameters of models (4), (5) (smoothed data)

Parameters of model $Y_1(t)$сгл	STP not considered (4)	STP considered (5)
χ_1	−0.0793	0.1457
κ_1	0.0535	0.0199
ϕ_1	−0.0892	−0.0512
ρ_1	0.0578	0.2151
μ_1		−0.01757
DW	0.63321	2.6223
R^2	0.78301	0.9763
F	4.51069	51.4504

In order to average the random outliers we will smooth the input characteristics by the method of the moving average [21]. The parameters of models with smoothed data in the variants of the non-uniform Cobb-Douglas PF and non-uniform PF considering the STP are given in Table 2 and Figs. 3 and 4. $Y_1(i)$сгл—averaged statistical data of the mean annual production capacity of production of oil coming to processing; $Y_1(t)$сгл—calculation for model (4) and model (5) using smoothed input data.

When smoothing input data using model (4) without considering the impart of STP on the mean average production capacity of oil production $Y_1(t)$ greatest contribution

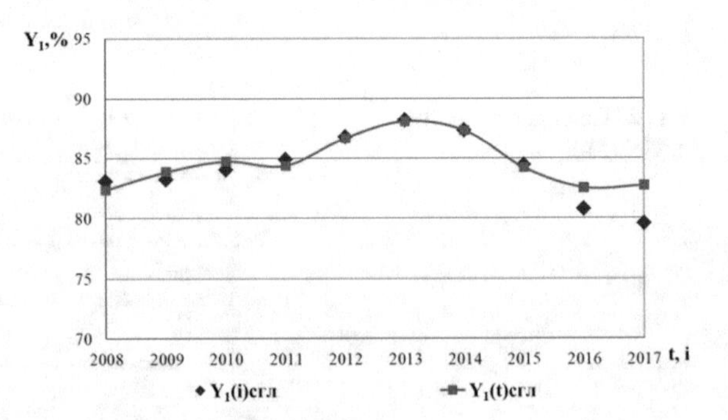

Fig. 3 Calculated data for model (4) (smoothed data)

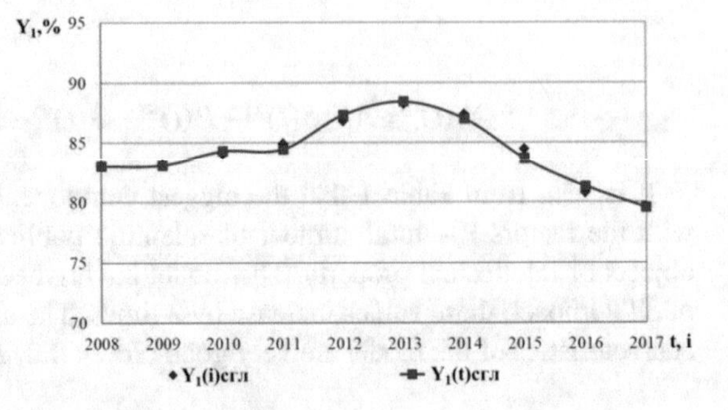

Fig. 4 Calculated data for model (5) (smoothed data)

Table 3 Parameters of models (6), (7)

Parameters of model $Y_2(t)$	STP not considered (6)	STP considered (7)
χ_2	−0.2206	−0.1670
κ_2	−0.0250	−0.0033
ϕ_2	0.0233	0.0286
ρ_2	−0.0684	−0.0412
μ_2		−0.00824
DW	1.6330	1.7472
R^2	0.442	0.4747
F	0.9989	1.1295

is seen from the factors P_i, total number of scientific publications, and I_i, generation of objects of intellectual property. In model (5) considering the impact of the STP, the quality of solution is improved: the determination factor R^2 is increased almost to one, the value of generation of objects of intellectual property I_i and number of students discharged Si having the largest factorial elasticity, i.e. these factors make the most impact on the mean average production capacity of oil production Y_1.

In a similar way, we will build a model of average annual production capacity of production of mineral lubricating oil—$Y_2(t)$, %. The parameters of the model for the input statistical data in the variants of the non-uniform PF (6) and the model considering the STP (7) are given in Table 3 and in Figs. 5 and 6, where Y_2(i)—statistical data of average annual production capacity of production of mineral lubricating oil, and Y_2(t)—calculation for model (6) and model (7) considering the STP.

$$Y_2(t) = A_2 \cdot S(t)^{\chi_2} \cdot P(t)^{\kappa_2} \cdot G(t)^{\phi_2} \cdot I(t)^{\rho_2}, \quad (6)$$

$$Y_2(t) = A_2 \cdot S(t)^{\chi_2} \cdot P(t)^{\kappa_2} \cdot G(t)^{\phi_2} \cdot I(t)^{\rho_2} \cdot e^{\mu_2 t}, \quad (7)$$

In the models (6) and (7), largest factorial elasticity ϕ_2 lies with the parameter G_i—performance of R&D under grants, which is further increased by 22% once the

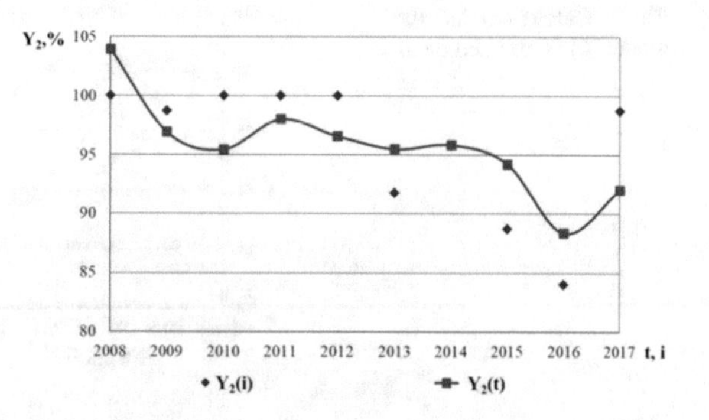

Fig. 5 Calculation data for model (6)

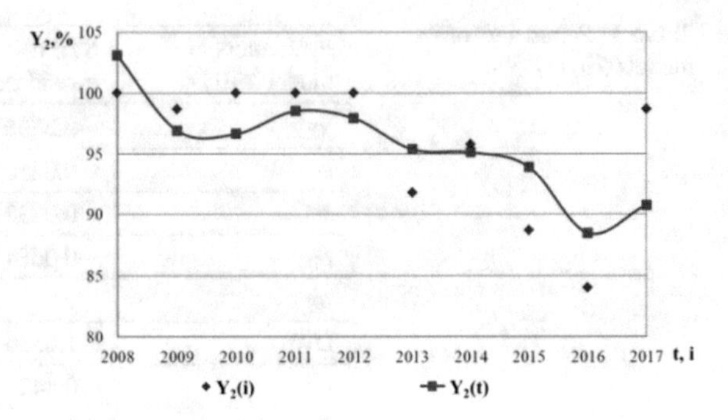

Fig. 6 Calculation data for model (7)

Table 4 Parameters of models (6), (7) (smoothed data)

Parameters of model Y_2(t)сгл	STP not considered (6)	STP considered (7)
χ_2	−0.5063	−0.6709
κ_2	0.1096	0.1106
ϕ_2	0.0136	0.0223
ρ_2	−0.2932	−0.3800
μ_2		0.00837
DW	1.9447	2.3270
R^2	0.9596	0.9614
F	29.6663	31.1145

model is augmented with the external factor of STP. It is to be noted that consideration of the STP factor leads to increase of the factor of determination R^2.

Parameters of models (6), (7) using smoothed data in the variants of the non-uniform Cobb-Douglas PF and of the non-uniform PF considering the STP are given in Table 4 and in Figs. 7 and 8. Y_2(i)сгл—averaged statistical data of average annual production capacity of production of mineral lubricating oil, and Y_2(t)сгл—calculation for the model (6) and model (7) with STP using smoothed input statistics.

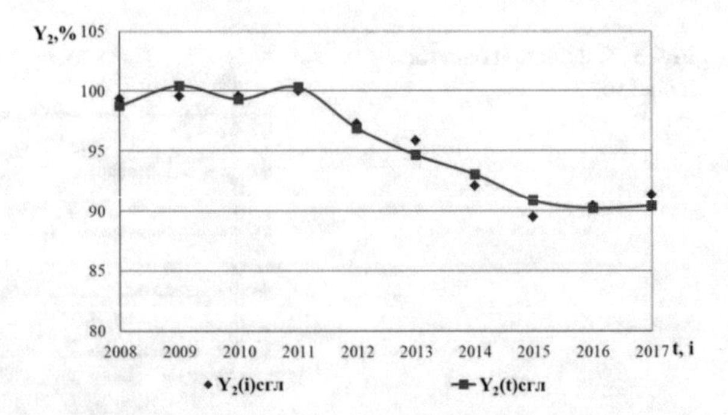

Fig. 7 Calculated data for model (6) (smoothed data)

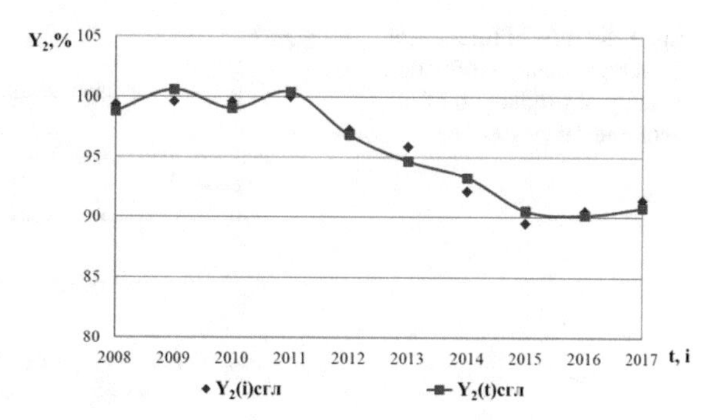

Fig. 8 Calculated data for model (7) (smoothed data)

When modeling the production of mineral lubricating oil (j = 2) greatest factorial elasticity is with the total number of scientific publications P_i and performance of R&D under grants G_i. Once the model is augmented with the impact of STP (model 7), these values grow by 1% and 64% respectively. When input statistical data is smoothed, the approximative properties are improved in a substantial way, the R^2 growing by 2 times.

4 Forecast

Following the results of the studies of the production capacity of the petrochemical complex of the Samara Region one may state adequacy of the produced mathematical models and to use the same for a short-term forecast (3 years) to the year 2020.

To perform the forecast, we will use the LS method to extrapolate the incoming statistical data of the number of discharged students S_i, generation of objects of intellectual property I_i by the quadratic polynomial, performance of R&D under grants G_i by a cubic polynomial, and the total number of scientific publications P_i will be taken in this forecast as compliant with the roadmap of the flagship university [1, 22]. Table 5 summarizes forecasting data of incoming parameters. Based on them and using the produced models (5) and (7) considering the STP we will make a short-term of production capacity to the year 2020.

The assessed value of average annual production capacity of production of oil coming to processing (j = 1) (Fig. 9) $Y_1(t)$ up to the year 2020 will be in the area of 80%, which aligns with the value of the year 2017. The average annual production

Table 5 Forecast values of incoming indicators

Year	2018	2019	2020
S, people	5367	5069	4413
P, pcs	3924	4442	5031
G, units	49	50	51
I, units	48	51	60

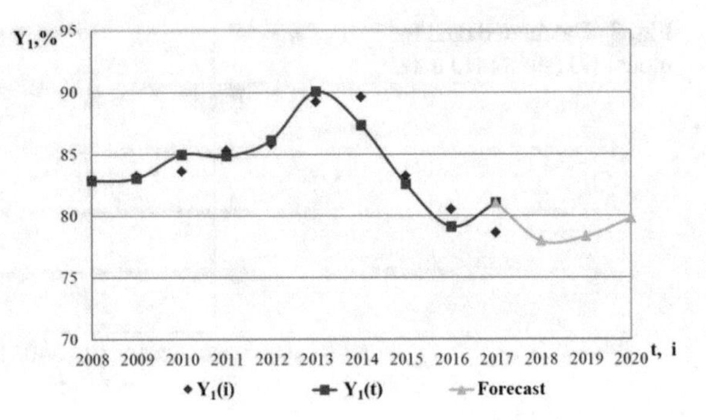

Fig. 9 Results of forecasting of average annual production capacity of production of oil incoming for processing

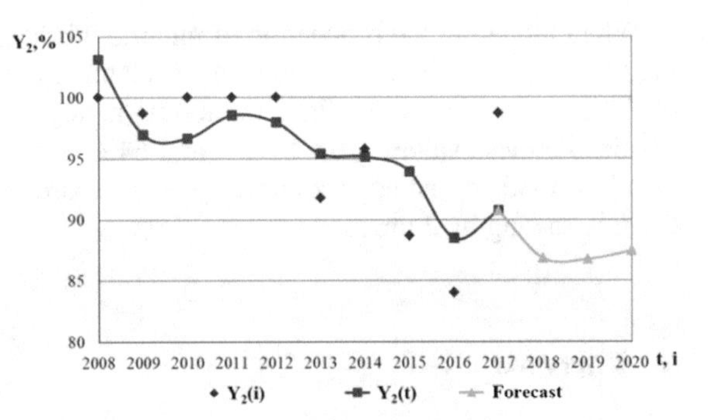

Fig. 10 Results of forecasting of average annual production capacity of production of mineral lubricating oil

capacity of production of mineral lubricating oil (j = 2) $Y_2(t)$, (see Fig. 10), following the forecast, will decrease in 2018 versus 2017 by 5%, and in 2019 it will decrease by 0.5% versus PY, and in the year 2020 it will increase by 1%, which is 3.5% lower than the value of capacity of production capacity in the year 2017. The situation is related to the introduction into the model of incoming values for the factor P_i in the preceding years that are lower than those achieved in the most recent phase of the period in question. The data is taken from the SSTU road map [1, 22] and are likely to be exceeded. Therefore, the forecast will be more optimistic.

5 Conclusion

It may be thus demonstrated that the basic indicators of activity of a flagship university do impact the production capacity of the petrochemical complex of the Samara Region; that impact may be assessed in a quantitative way. Following the results of the modeling we may conclude that the number of specialists discharged from the university determined by the number of students enrolled in state-funded programs and by the demographic situation, impacts the performance of the petrochemical complex to a smaller degree than the quality of their training. In its turn, the quality level of training of graduates is determined by the scientific potential of the university, individual work with students performed by the university tutors, especially in the R&D activities. Engagement of students and tutors alike in the research and development performed by the university has a positive impact on the development of the petrochemical complex of the region.

Acknowledgements The reported study was funded by the Russian Foundation for Basic Research (RFBR), according to research project No. 17-08-00593.

References

1. Samara State Technical University/Mission of the flagship university. URL https://su.samgtu.ru. Last accessed: 14 Mar 2019
2. Official Statistics\Entrepreneurship\Industrial Production. URL http://samarastat.gks.ru/wps/wcm/connect/rosstat_ts/samarastat/ru/statistics/enterprises/production/. Last accessed: 10 Feb 2019
3. Samara Annual Statistical Report 2017: Stat.sb\Samarastat—P 17. Samara (352p) (2017)
4. Golovanov, P., Livshits, M., Tuponosova, E.: Mathematical model of discharge of specialists by universities. Math. Methods Tech. Technol. (MMTT) J. **2**, 114–119 (2018)
5. Vasil'ev, V., Gurtov, V., Sazonov, B., Surovov, M.: System of monitoring, analysis and forecasting of development of education and educational structures in the regions of Russia. Industria obrazovaniya **5**, 52–60 (2002)
6. Granichina, O.A.: Mathematical models of management of quality of educational process in universities with active optimization. Stoch Optim. Inform. J. **2**(1–1), 77–108 (2006)
7. Borgojakova, T., Lozickaja, E.: Mathematical modeling: definition, application in building models of educational processes. NAUKOVEDENIE Internet J. **9**(2) (2017). Available online: http://naukovedenie.ru/PDF/82TVN217.pdf
8. Tuponosova, E.: Modeling the demand in workforce of highest qualification in the Samara Region. Newsl. Samara Scientific Center Russ. Acad. Sci. **13**(4), 1236–1238 (2011)
9. Zaslonko, O.: Production functions of general education. Econ. Sci. **1**(74), 383–387 (2011)
10. Akinfieva, N.: Comparative analysis of criteria of assessment of quality of higher education. In: Golub, Yu.G. (eds.) Obrazovanie v sovremennom mire (pp. 36–43). Collected articles, Saratov (2012)
11. Klopchenko, V.: Elements of theory of prediction in education. Newsl. TRTU **5**(60), 210–214 (2006)
12. Ronald, C., Lawrence, L.: Comparative performance measurement: a primer on data envelopment analysis. Public Product. Manage. Rev. **22**(3), 348–364
13. Moskovkin, V., Suleiman, B., Lesovik, R.: Mathematical model for the formation of university contingents on the basis of population dynamics equations. Int. J. Appl. Eng. Res. **9**(22), 16761–16775 (2014)
14. Glotova, T., Deev, M., Krevskiy, I., Matukin, S., Sheremeteva, E., Shlenov, Y., Shlenova, M.: Models of supporting continuing education of specialists for high-tech sector. In: 11th Joint Conference on Knowledge-Based Software Engineering (JCKBSE 2014), Volgograd, Russia, 17–20 Sept 2014 (pp. 100–113)
15. Karpenko, O., Bershadskaja, M.: Higher Education Around the World: Analysis of Data of Educational Statistics and Global Ratings in the Sphere of Education (244p). SGU Publishing House, Moscow (2009)
16. Berezhnaya, E., Berezhnoy, V.: Mathematical Methods of Modeling Economical Systems (2nd ed., 432p). Finansy i Statistika, Moscow (2006)

17. Klejner, G.: Production Functions: Theory, Methods, Application (239p). Finansy i Statistika, Moscow (1986)
18. Tan, H.B.: Cobb-Douglas production form. jamador (2008)
19. Zamkov, O., Tolstopyatenko, A., Cheremnykh, Yu.: Mathematical Methods in Economy (386p). MGU, DIS, Moscow (1997)
20. Farebrother, W., Bolch, B., Huang, C.: Multivariate statistical methods for business and economics. Economica **42**, 226 (1975)
21. Ajvazjan, S., Mhitarjan V.: Applied Statistics and Basics of Econometrics (p. 803). Unity, Moscow (1998)
22. Samara State Technical University/Normative Documents/Roadmap of the Development Program. [URL] https://su.samgtu.ru/files. Last accessed: 14 Mar 2019

Decision Support System to Prevent Crisis Situations in the Socio-political Sphere

Andrey Proletarsky, Dmitry Berezkin, Alexey Popov, Valery Terekhov and Maria Skvortsova

Abstract A statement and a general structure for solving the problem of assessing possible crisis situations in the Socio-political sphere is proposed. An approach to analyzing and forecasting the development of crisis situations has been implemented on the basis of continuous monitoring of heterogeneous data from various information sources and summarizing the results of assessing threats obtained using various methods. A model for the development of a crisis situation is presented, which considers the situation as the result of the interaction of various agents in a complex network. The method of historical analogy was applied to the situational forecast. The issues of hardware acceleration of analyzing large data streams are considered through the use of a Leonhard processor that processes large amounts of data due to parallelism. When designing the system, an agent-based development methodology is used. The structure of the system and the results of its application for the analysis of the possible development of crisis situations during political rallies are given.

Keywords Threat · Crisis situation · Decision support system · Hierarchy analysis method · Intelligent agent · Forecasting · Monitoring · Cognitive graphics · Game theory

A. Proletarsky (✉) · D. Berezkin · A. Popov · V. Terekhov · M. Skvortsova
Bauman Moscow State Technical University, 2-nd Baumanskaya st., 5/1, Moscow 105005, Russia
e-mail: pav@bmstu.ru

D. Berezkin
e-mail: berezkind@bmstu.ru

A. Popov
e-mail: alexpopov@bmstu.ru

V. Terekhov
e-mail: terekchow@bmstu.ru

M. Skvortsova
e-mail: magavrilova@bmstu.ru

A. G. Kravets et al. (eds.), *Cyber-Physical Systems: Industry 4.0 Challenges*, Studies in Systems, Decision and Control 260,
https://doi.org/10.1007/978-3-030-32648-7_24

1 Introduction

During the fourth industrial revolution, the rapid development of information and communication technologies, interacting in the physical, digital and biological fields, occurs [1]. The human environment is gradually turning into a cyber-physical system, which leads to a significant change in social and economic relations: society becomes "networked", as a result of which certain crisis phenomena emerge in it with new force and new ones appear [2–4]. There is a gradual transition to the digital economy and the digital army, the widespread introduction of hardware and software and technologies of the "Internet of Things" and "Industry 4.0", the replacement of people by computer systems that act as certain "intelligent agents". The creation of such information systems, which represent a cyber-physical environment in which there is an interaction of people, robots and intelligent agents.

Over time, the structure of socio-political relations can be completely transformed into a global network of agents and will gradually become a network. This will require the creation of fundamentally new systems for managing Socio-political processes since former rigid public administration systems will simply not be able to manage them effectively. In addition, it is necessary to develop new models that describe processes in complex socio-technical systems with a network structure, the purpose of which will be to predict possible threats and crisis situations. Therefore, the actual task is to simulate the processes of socio-political phenomena, to collect and process the information on emerging crisis situations, as well as to formulate proposals for their prevention.

In connection with the above, the purpose of this work is to develop methods and models of a decision support system (DSS) to prevent crises in the Socio-political sphere, taking into account the peculiarities of the development of processes and management of network structures.

2 Problem Statement and Theoretical Substantiation of the Research

In the decision-making process for the prevention of crisis situations in the Socio-political sphere, it is required to choose from a set of possible options (alternatives) a subset of solutions (which, in particular, may contain the best one). In addition, each decision must be assessed from different points of view, taking into account political, economic, social and other aspects. The selection of the required options is made by the decision maker (DM). This may be a person who is a competent specialist in his field, who has the necessary authority and is responsible for the decision made. The concept of the quality of options is characterized by the principle of optimality, which requires the construction of models for optimizing solutions simultaneously using several criteria [5–8].

Thus, the decision-making problem is represented by a pair, (Ω, OP), where Ω is a set of alternatives, OP is the optimality principle. The solution of the problem (Ω, OP), is the set $\Omega_0 \subseteq \Omega$ obtained with the help of the principle of optimality. The mathematical expression of the principle of optimality is the function of choice C0. It matches any part C0(X) of it to any subset X $\subseteq$ Ω. The solution Ω0 to the original problem is the set C0(Ω).

The process of solving the problem (Ω, OP) is organized according to the following scheme: form a set Ω, i.e. prepare alternatives, and then solve the problem of choice. In the process of forming a set Ω, the conditions of possibility and admissibility of alternatives are used, which are determined by the specific constraints of the problem. At the same time, the universal set Ωu of all possible alternatives is considered known. The task of formation Ω is the task of choice (Ωu, OP1), where OP1 is the principle of optimality, expressing the conditions for the admissibility of alternatives in a particular environment. The set $\Omega = C_{0_1}(\Omega_u)$ obtained as a result of solving the specified selection problem is called the initial set of alternatives. Thus, the general decision-making task comes down to solving two consecutive selection problems.

At the first stage of selecting a set of permissible alternatives $\Omega = C_{0_1}(\Omega_u)$, methods of monitoring the current situation are applied in order to identify events and situations that took place in the past. At the stage of choosing the optimal decision maker strategy based on the generated set of alternatives, the hierarchy analysis method (HAM) [5] is used. Game theory models are also used to describe the behavior of interacting players. For additional substantiation and clarification of the decision maker's decision to prevent a crisis situation, methods of cognitive computer graphics are used.

The proposed approach to creating a DSS to prevent crisis situations in the Socio-political sphere is based on the principles outlined earlier in [9].

In Fig. 1 gives a schematic representation of the proposed approach to building information systems for monitoring, analyzing and forecasting threats of crisis situations in the public policy sphere, which is based on the automatic extraction

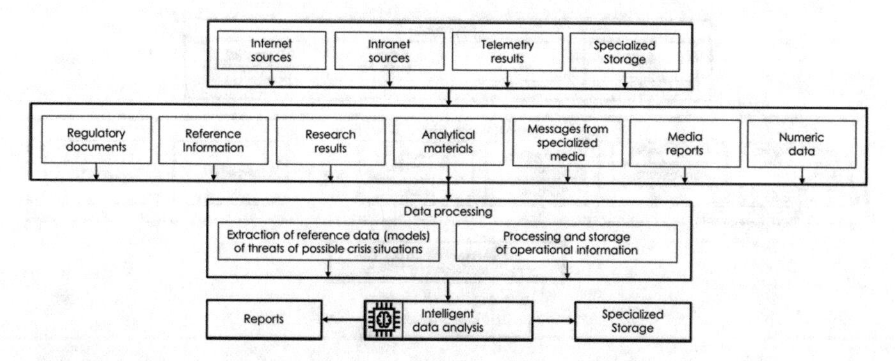

Fig. 1 A schematic representation of the approach in solving the problem of assessing the possible threats of crisis situations in the socio-political sphere

of weakly structured information from various sources, including big data streams, machine text analysis and preparation of proposed solutions based on this analysis.

3 The General Structure of the Solution to the Problem

The general structure of solving the problem of assessing possible threats of crisis situations and developing proposals for their prevention is shown in Fig. 2.

In Fig. 2 schematically shows the means of analyzing the initial information that the user-analyst uses to reasonably form expert assessments in a variety of threat assessment techniques. Techniques can use different methodological tools and input data to develop solutions. Using the HAM to summarize the results allows us to work out a general assessment of the occurrence of possible crisis situations and prepare a draft recommendation that best summarizes the results obtained using particular techniques.

To describe the behavior of interacting participants of Socio-political relations, it is advisable to apply the model of game theory. Recently, models that reflect the network nature of the structure of relations between agents of the analyzed system, in particular, information network models, have become significant. The information network at a qualitative level refers to a structure consisting of a set of agents (subjects—individual or collective, for example, individuals, groups, organizations, states, etc.) and a set of relations defined on it (a set of relations between agents, for example, interactions, cooperation, communication) [9, 10]. Formally, an information network is a graph G(N, E), in which N is a finite set of vertices (agents) and E is a set of edges reflecting the interaction of agents. Players, each of whom seeks to spread some information on the network and influence the agents of other players [9], affect agents.

Let us consider a model of the development of a crisis situation, which represents the situation as the result of the interactions of various agents in a complex network

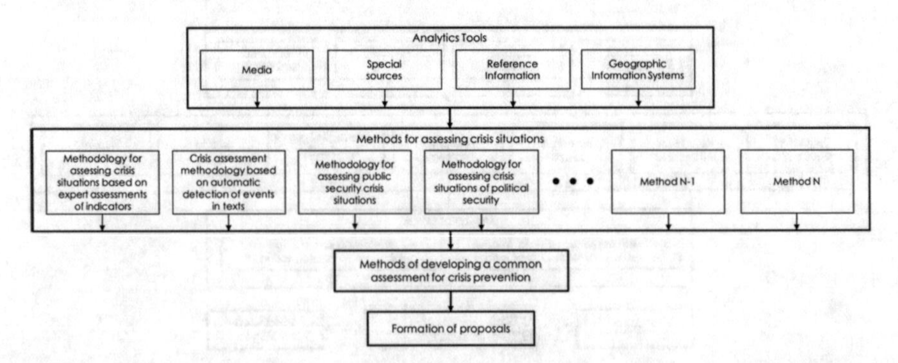

Fig. 2 The general structure of solving the problem of assessing threats of possible crisis situations and developing proposals for their prevention

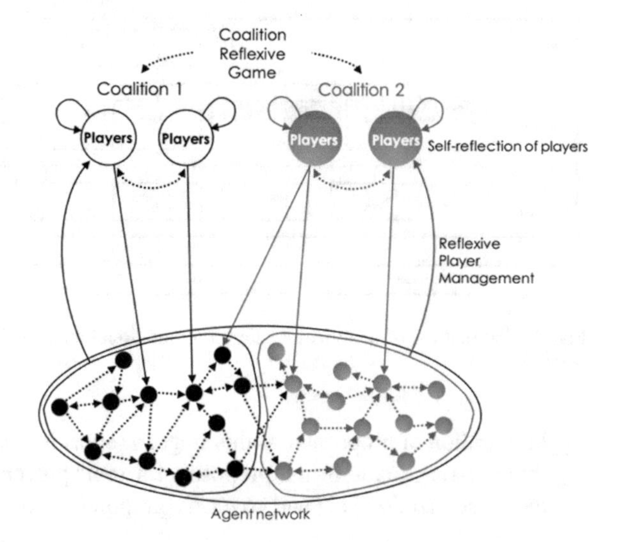

Fig. 3 Model of crisis situations

(Fig. 3). Such agents can be users of social networks that are affected by informational and psychological effects, their virtual counterparts, or some intellectual agents.

Each agent assesses the risks and usefulness of participating or refusing to participate in certain conflicts or coalitions. With the development of crisis situations in such a system, these risks increase and equilibrium states are achieved for all large values of risks. By setting the thresholds, we can evaluate the behavior of the system being modeled as a whole. The choice of criteria for setting the threshold is important; its informed choice requires special study. As the most promising approach for setting the threshold, it is possible to note the use of self-learning methods.

When making decisions, reflection and self-reflection of the parties to the conflict are taken into account, i.e. decisions are formed taking into account how they will be perceived by those agents whose management must be ensured.

4 DSS Methods

During the operation of the proposed DSS for crisis prevention in the Socio-political sphere, four stages of information processing can be distinguished:

1. Collection of information that is performed automatically;
2. Systematization of information on various factors: political, economic, informational and others. The system allows you to configure the factors and signs of systematization of information, as well as perform geo-referencing of the identified crisis situations;
3. Assessment of the likely crisis situations, which are carried out by various methods. The result of the assessment are numerical indicators that characterize the identified crisis situations by factors, countries, and directions;

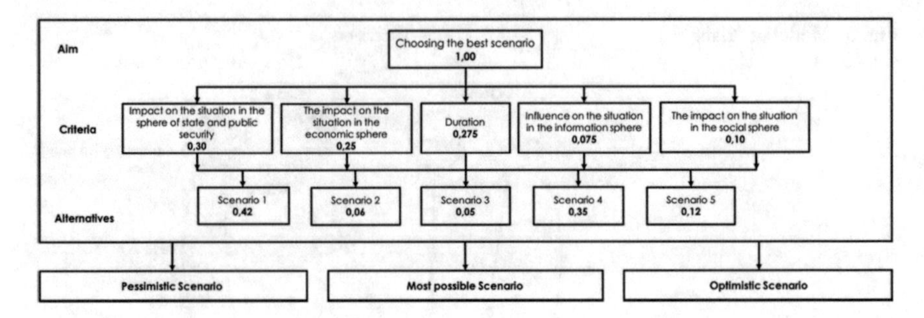

Fig. 4 The method of historical analogy for situational analysis based on the hierarchy analysis method

4. Preparation of proposals. At this stage, a report is generated, including the identified crisis situations and proposals for their prevention [9, 11]. At this stage, data visualization [12] and cognitive graphics are actively used.

For a situational forecast, the system uses an approach based on the automatic search for historical analogies [13, 14] in heterogeneous data streams. It consists of detecting for the current situation similar reference situations that occurred in the past. It is assumed that the detected analogs are possible scenarios for the further development of the current situation. For the purpose of the subsequent formation of proposals, when preparing reference situations, experts should supply them with recommendations, each of which prescribes which person, what actions and in what time period should be performed [14].

The selection of the most likely, optimistic and pessimistic scenarios is assumed. The most likely scenario is based on a reference situation that is closest to the current one. To highlight the most optimistic and pessimistic scenarios, the prioritization of scenarios is performed using the hierarchy analysis method (Fig. 4).

At the last stage of the system's work, the final generation of a report on the results of the analysis of crisis situations and the preparation of proposals for their prevention in the form of a report is performed.

These results can be presented in a form convenient for the decision maker, displayed on a 2-D or 3-D map, and also printed as a document, which can include textual and graphical information.

For a more in-depth analysis of crisis situations, a situational visualizer is used, which uses the hybrid method of dynamic meta anamorphosis is (MDM) developed by the authors.

This approach is based on the use of cognitive graphics—a set of data visualization methods that activate figurative thinking mechanisms of decision makers and facilitate decision making in a complex environment or finding a solution to a complex problem [15]. Such an approach leads to a drastic simplification and acceleration of the time for analyzing crisis situations and, accordingly, reducing the time taken to make an informed decision.

MDM is based on the concept of anamorphosis, which is defined as the transformation of one visual image built on the basis of the Euclidean metric into another visual image based on the metric of the selected indicator [16]. Dynamic meta anamorphosis is called anamorphosis, which is based on an integral indicator. Moreover, each local indicator constituting an integral indicator takes into account one of the characteristics of a crisis situation and evolves in time and space [17].

The image converted using MDM, as a rule, contains more information than before the conversion. Therefore, the original, purely illustrative image is capable of prompting the decision maker new thoughts and ideas for finding a solution. Thus, the original illustrative function of the image becomes a cognitive function.

One can speak of meta anamorphosis as a method of combining indicators (analytical, statistical, probabilistic, etc.) and forms of visual images so that the areas of the original visual images are transformed in proportion to the integral value of the indicator. They allow you to get an idea of the hidden patterns and trends of the analyzed crisis situations, are one of the ways of cognitive visualization of multidimensional data.

The obvious advantages of MDM in the tasks of analyzing crisis situations in the Socio-political sphere are:

- reduction of the dimension of the problem by the number of local indicators of various physical nature rolled up into an integral indicator of meta anamorphosis;
- the possibility of visual modeling of alternative solutions to prevent crisis situations in the Socio-political sphere, taking into account the local indicators of different physical nature that change in time and space, are collapsed into an integral indicator of metamorphosing;
- identification of hidden patterns of evolution in time of crisis situations in the Socio-political sphere, implicitly dependent on local meta anamorphosis is indicators;
- the possibility of constructing scenarios of decision-makers on the basis of visual spatial-temporal analysis of dynamic meta anamorphosis, taking into account the processes associated with a targeted impact, both on individual local indicators and on the integral indicator as a whole;
- forecasting the time series of local indicators of the crisis situation in the Socio-political sphere using a genetic algorithm based on a set of values, an available historical time series of local indicators and a set of basic analytical functions. The resulting syntax tree as a result of the algorithm is a description of the forecast function of a time series of local indicators of a crisis situation in an analytical form.

5 Interpretation and Discussion of Research Results

Creation of information systems, representing a cyber-physical environment in which people, robots and intelligent agents interact, requires the use of special design approaches [18, 19]. It uses the methodology of agent-based development, created in Bauman Moscow State Technical University, based on the best approaches and practices of multi-agent systems design. The methodology includes only what is necessary for the development of heterogeneous software. It provides a set of iterative steps, which include not only general recommendations and basic patterns but also an indication of which models should be developed, contains a description of the overall development process and detailed recommendations for each step of this process.

The structure of the DSS, developed using the methodology of agent-based development, is shown in Fig. 5.

A number of methodologies for assessing the situation and developing proposals to prevent the development of crisis situations in the Socio-political sphere have been implemented. This allowed for an assessment of the situation and the preparation of proposals aimed at countering identified threats, both automatically and in automated modes. All estimates and proposals generated by the system in this mode are given to the operator, who can either accept them or make changes, after which a report is generated and the results are displayed by means of a geographic information system.

The solution to the problem in question may be carried out in multi-user mode. In this case, such work of expert groups is ensured, in which each of them works with its original information using certain techniques in a single information environment.

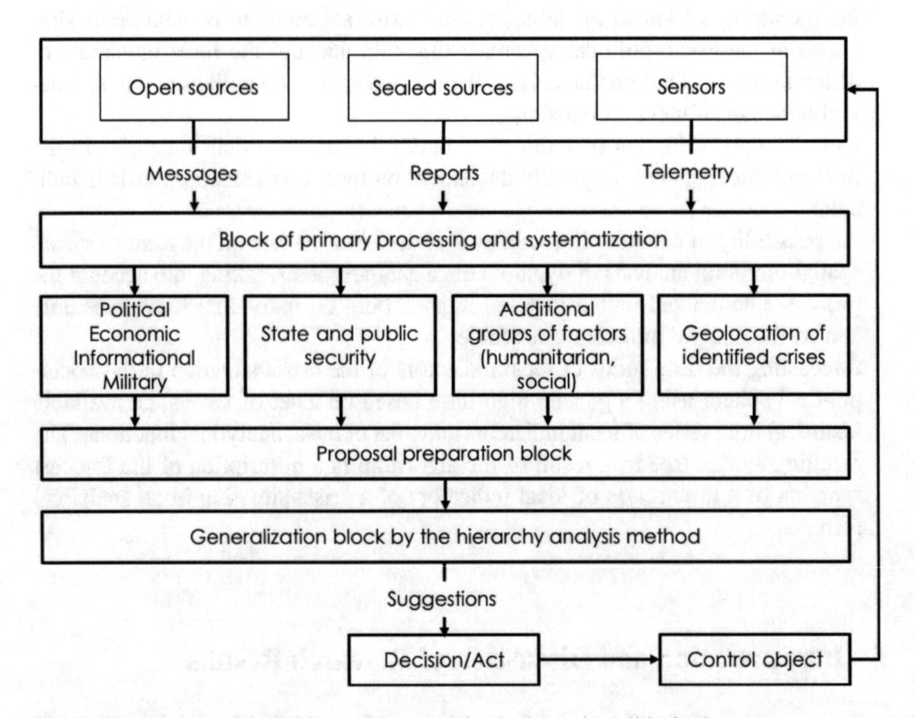

Fig. 5 The structure of the DSS on crisis prevention in the socio-political sphere

The manager monitors the progress of the work from his workplace, can correct the actions of the working groups and, based on the results, makes the final decision.

The proposed model for the development of a crisis situation, methods for predicting the situation and analyzing the flows of big data require significant computational costs. Studies show that a significant acceleration of the work of analytical information systems can be achieved thanks to more advanced methods and algorithms for processing graph data models, as well as by improving the efficiency of the hardware. However, traditional approaches to the design of parallel programs and systems based on the use of a large number of homogeneous universal processors no longer provide the required performance, leading to a significant increase in the complexity and development time and debugging of the program code.

The analysis of the implementation of discrete optimization algorithms in universal microprocessors like Intel's x86, modern GPU (Graphics Processing Unit) accelerators (for example, NVidia Tesla), as well as in special hardware systems (Leonhard microprocessor developed at Bauman Moscow State Technical University) [19–22], which showed a violation of the rhythmic operation of the pipeline, a decrease in the efficiency of hardware prefetch, an increase in the number of pipeline failures due to incorrectly predicted transitions in processing data with strong dependencies and fragmented placing graphs in memory was held. The batch mode adopted for modern types of RAM also does not contribute to the acceleration of calculations, since it contributes to incomplete use of the bus and processor resources when loading list items and trees. Thus, it can be stated that the hardware solutions on which the work of modern computers is based, are aimed at accelerating the processing of vector structures and, on the contrary, slow down the processing of reference data structures and there are no tools that accelerate their processing.

Performance problems in processing such data also arise in processing them on modern GPUs created as part of the NVidia Tesla project. For the GPU, matrices and vectors are the preferred data representation, which can be easily divided into parts corresponding to the cache memory of the streaming processor (SM). With a significant increase in the number of computational cores, the computation speed increases only 1.1–5 times [21].

Therefore, to create a DSS, it is proposed to implement a computing system with many command streams and a single data stream developed by Bauman Moscow State Technical University. This system uses a specialized Leonhard microprocessor with the Discrete Mathematics Instructions Set Computer, DISC, developed specifically for processing dependent data in graph processing tasks. The Leonhard processor allows you to store and independently process large sets and graphs, has a specialized short conveyor. The processor of the computing system DISC showed high efficiency in solving discrete optimization problems on networks and graphs. Acceleration achieved up to 160 times in comparison with universal computing systems.

Machine instructions of the processor for processing structures are complex high-level actions based on operations of discrete mathematics [22].

Comparison of the hardware efficiency of the above types of systems was calculated from the results of tests on Dijkstra and Bellman-Ford algorithms for finding

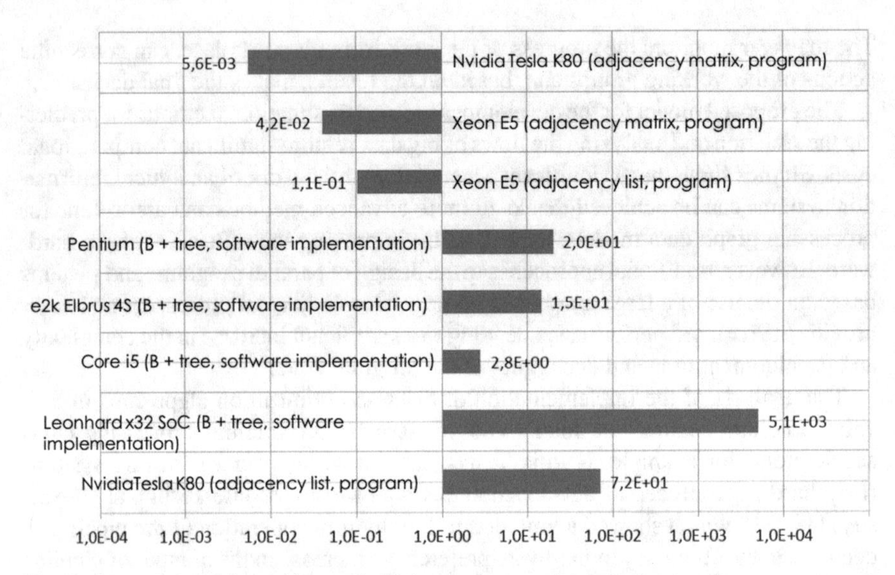

Fig. 6 Comparison of the architectural efficiency of computing systems per valve

the shortest path on graphs (Single Shortest Path Problem, SSSP). The comparison results are shown in Fig. 6, show that the hardware efficiency of graphics accelerators when solving problems on graphs is 102 times lower than the specialized microprocessor Leonhard x32. It is also shown that simpler microprocessors (domestic Elbrus e2k microprocessor, Intel Pentium IV), have better specific architectural efficiency in comparison with more modern multi-core Intel Xeon microprocessors.

The obtained results prove the need for the development and implementation of special tools for accelerating algorithms on graphs, despite the significant achievements of graphics accelerator developers.

From October 2015, the collection of information from various sites of federal and regional authorities, major news agencies, federal and regional mass media, as well as some English-speaking sources of information has begun [9]. It is technically possible to collect and analyze information in other foreign languages.

In Fig. 7 shows the application of the DSS methodology for analyzing the crisis situation of social and political processes related to the activities of non-systemic opposition during the beginning of preparations for the presidential campaign in 2018. The figure on the left shows the results of the analysis of reports on rallies and protests that took place during the previous elections. The right figure shows the results of the analysis and forecast of the development of protest activity in 2017. Comparison of the current situation with existing analogs in the past allows the method of historical analogy to be used to determine possible scenarios for the development of the situation and develop proposals for neutralizing crisis situations and identified threats.

In Fig. 8 shows the results of the automatic selection of events and analysis of the situation based on the processing of the text message flow about the opposition rally

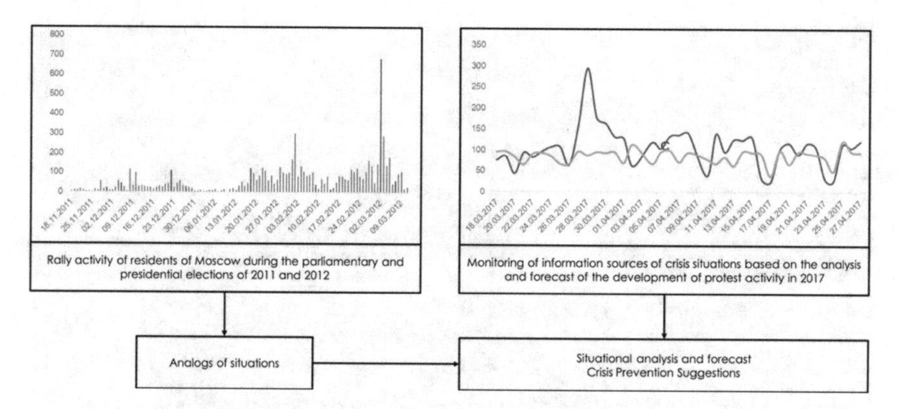

Fig. 7 Use of DSS to analyze social and political processes

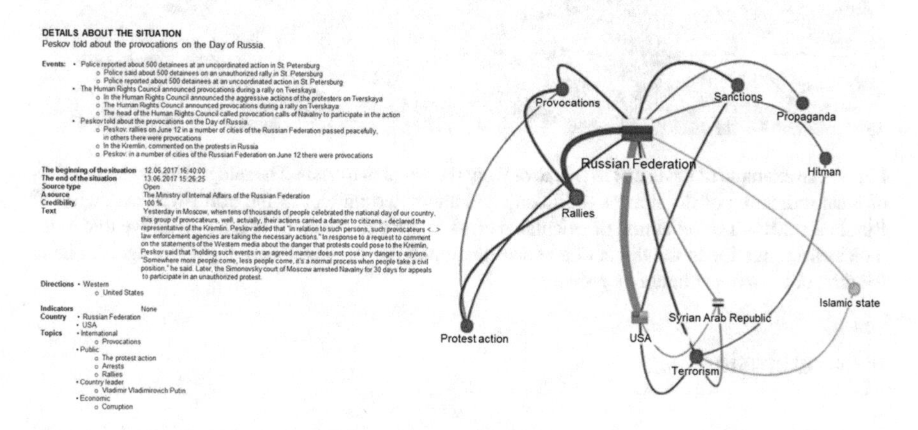

Fig. 8 An example of analyzing the situation in the socio-political sphere

in Moscow on 12.06.2017. Automatically generated graph, which is shown in Fig. 8 on the right, which reflects the main threats identified and related states [23, 24, 25].

At the last stage of the system's work, the final generation of a report on the results of the analysis of crisis situations and the preparation of proposals for their prevention in the form of a report is carried out. Results can be displayed on a 2-D or 3-D map [25], as well as printed as a document, which may include textual and graphical information. An example of the use of MDM to prevent the threat of a crisis is shown in Fig. 9.

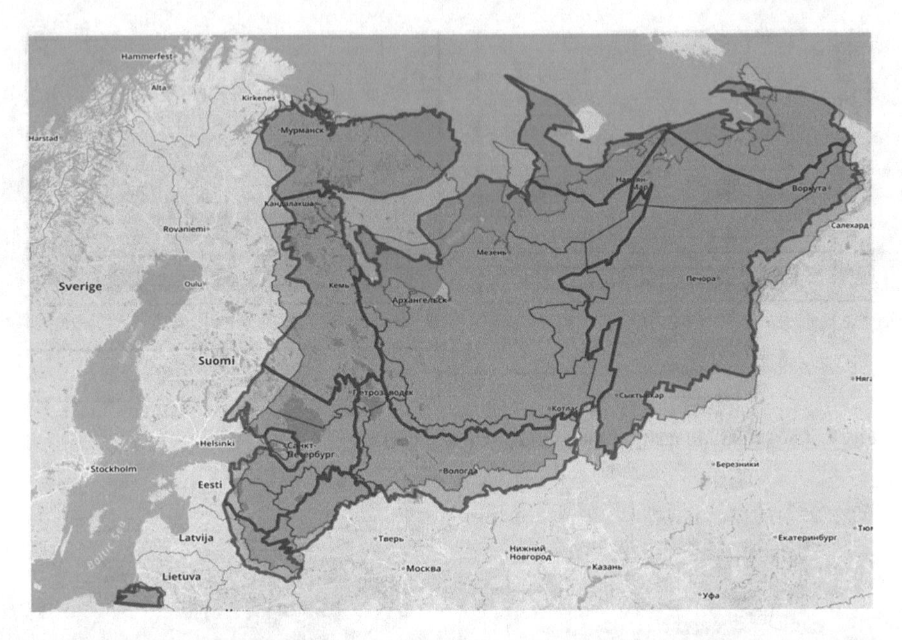

Fig. 9 An example of the use of MDM to prevent the threat of a crisis. The integral indicator consists of local indicators of illicit arms trafficking and attempted murder in the north-western region of Russia for 2014. The reduction of calculated areas relative to the areas of administrative territories indicates a reduction in the threat of a crisis situation. It can be seen that in the Arkhangelsk region the threat of a crisis situation is growing

6 Conclusion

The authors of the chapter have developed methods, models, and algorithms for DSS to prevent crisis situations in the social and political sphere, taking into account the peculiarities of the development of processes and management of network structures. In the process of creating software and system interfaces, web development tools were used that allowed the developed system to be used on various software and hardware platforms (stationary devices, tablets, smartphones, etc.).

The authors continue to work on the creation of a software-technical complex of decision-making support for the prevention of crisis situations in various fields of activity. The proposed method of forecasting the development of situations provides the user with scenarios for further development of the situation and recommendations for actions necessary for their implementation but does not allow managing the development of the situation according to the optimal scenario. The user is required to determine whether the development of the situation corresponds to the previously formed scenario, and receive recommendations if it is necessary to adjust the outlined plan of measures, taking into account the choice of strategies by other agents.

In this regard, the further development of the method is the development of more complex network and game-theoretic models of reference situations that can reflect various options for the possible development of the current situation depending on the actions of DM at each stage of situation management, as well as reflexive control and group interests of interacting agents [23].

References

1. Schwab, K.: The Fourth Industrial Revolution, p. 198. Crown Publishing Group, New York (2016)
2. Luhmann, N.: Soziologie des Risikos. Berlin, New York: Walter de Gruyter, pp. 9–40 (1991)
3. Beck, U.: From industrial society to the risk society. Theor. Cult. Soc. **9**(1), 97–123 (1992)
4. Castells, M.: The information age: economy, society and culture. In: The Rise of the Network Society, vol. 1, 2nd edn, p. 597. Blackwell Publishers, UK (2010)
5. Saaty, T.L.: The Analytic Hierarchy Process, p. 287. McGraw-Hill, New York (1980)
6. Yeung, D.W.K., Petrosyan, L.A.: Subgame Consistent Economic Optimization, p. 396. Springer Science, NY (2012)
7. Strekalovsky, A.S. (2000).: One way to construct a global search algorithm for dc minimization problems. Nonlinear optimization and related topics. In: Di Pillo, G., Giannessi, F. (eds.) Applied Optimization Series, vol. 36, pp. 429–443. Kluwer Academic Publishers, Dordrecht (2000)
8. Saaty, T.L.: Mathematical Models of Arms Control and Disarmament, p. 190. John Wiley & Sons, Inc. (1968)
9. Proletarsky, A.V., Berezkin, D.V., Terekhov, V.I.: Identifying information threats to the security of the Russian Federation, predicting their consequences and developing proposals for their prevention. Dyn. Complex Syst. XXI Century **11**(4), 22–31 (2017)
10. Chkhartishvili, A.G., Gubanov, D.A., Novikov, D.A.: Social Networks: Models of Information Influence, Control and Confrontation, vol. 189. Springer
11. Skvortsova, M., Terekhov V., Grout, V.: A hybrid intelligent system for risk assessment based on unstructured data. In: 2017 IEEE Conference of Russian Young Researchers in Electrical and Electronic Engineering (EIConRus). IEEE, pp. 560–564. https://doi.org/10.1109/eiconrus.2017.7910616 (2017)
12. Lidong, W., Guanghui, W., Cheryl, A.A.: Big data and visualization: methods, challenges and technology progress. Digit. Technol. **1**(1), 33–38 (2015)
13. Aggarwal, C.C., Subbian, K.: Event detection in social streams. In: Proceedings of the 2012 SIAM International Conference on Data Mining. Society for Industrial and Applied Mathematics, pp. 624–635 (2012)
14. Andreev, A., Berezkin, D., Kozlov, I.: Approach to forecasting the development of situations based on event detection in heterogeneous data streams. In: Kalinichenko, L., Manolopoulos, Y., Malkov, O., Skvortsov, N., Stupnikov, S., Sukhomlin, V. (eds.) Data Analytics and Management in Data Intensive DoHAMns. DAMDID/RCDL 2017. Communications in Computer and Information Science, vol. 822, pp. 213–229. Springer, Cham (2017)
15. Keim, D. et al.: Visual Analytics: Definition, Process, and Challenges. Information Visualization, pp. 154–175. Springer, Berlin (2008)
16. Gusein-Zade, S.M., Tikunov, V.S.: A new technique for constructing continuous cartograms. Cartography Geogr. Inform. Syst. **20**(3), 167–173 (1993)
17. Berezkin, D.V., Terekhov, V.I.: Application of the anamorphizing method for modeling and evaluating changes in geopolitical borders. Artif. Intell. Decis. Making **3**, 3–9 (2017)
18. Sakulin, S., Alfimtsev, A., Solovyev, D., Sokolov, D.: Web page interface optimization based on nature-inspired algorithms. Int. J. Swarm Intell. Res. (IJSIR) **9**(2), 28–46 (2018)
19. Nguyen, D.N. et al.: A methodology for developing an agent systems reference architecture. In: Weyns D., Gleizes, M.P. (eds.) Agent-Oriented Software Engineering XI. AOSE 2010. Lecture Notes in Computer Science, vol. 6788, pp. 177–188. Springer, Berlin (2011)

20. Patel, R., Kumar, S.: Visualizing effect of dependency in superscalar pipelining. In: 2018 4th International Conference on Recent Advances in Information Technology (RAIT), pp. 1–5, Dhanbad (2018)
21. Patel, R., Kumar, S.: The effect of dependency on scalar pipeline architecture. IUP J. Comput. Sci. **11**(1), 38–50. Available at SSRN: https://ssrn.com/abstract=3103485 (2017)
22. Popov, A.: An introduction to the MISD technology. In: Proceedings of 50th Hawaii International Conference on System Sciences (HICSS50), pp. 1003–1012 (2017)
23. Chkhartishvili, A.G., Novikov, D.A.: Reflexion and Control: Mathematical Models, p. 373. CRC Press. https://doi.org/10.1201/b16625 (2014)
24. Chernenkiy, V.M., Gapanyuk, Y.E., Kaganov, Y.T., Dunin, I.V., Lyaskovsky, M.A., Larionov, V.S.: Storing metagraph model in relational, document-oriented, and graph databases. In: Proceedings of the XX International Conference "Data Analytics and Management in Data Intensive Domains" (DAMDID/RCDL'2018), pp. 82–89. http://ceur-ws.org/Vol-2277/paper17.pdf (2018)
25. Chernenkiy, V., Gapanyuk, Y., Revunkov, G., Kaganov, Y., Fedorenko, Y.: Metagraph approach as a data model for cognitive architecture. In: Biologically Inspired Cognitive Architectures Meeting, pp. 50–55. Springer, Cham. https://doi.org/10.1007/978-3-319-99316-4_7 (2018)

Development of a Cyber-Physical System for the Specialized On-Track Machine Operators Training

N. A. Staroverova, M. L. Shustrova and M. R. Satdarov

Abstract The cyber-physical systems are a set of mathematical modeling, dynamic, physical, electric, pneumatic characteristics and the systems of simulation now. Its realization allows providing training of the person for an important problem of the modern century, namely competent, fast and high-quality decision-making and complex measures for the safety of the environment and technogenic factors. In simulators, the principles of practical skills development with the integral theoretical preparation are applied. The model of modern networks allows improving constantly systems remotely, to collect information on the quality and extent of threats and to carry out constant completion of the model. These opportunities appeared thanks to the development of such information technologies like virtual reality technologies, machine sight and also the systems of artificial intelligence. One of the most successful branches of the world information industry can note the sphere of simulation technologies. Each profession has some of the most important processes defining the quality and safety of the work. For the train driver, it is the perception of a railway situation in a variety of its manifestations (structures, people on the ways, railway signs, traffic lights, etc.), the analysis and processing of the arriving information and performance of action for control of special rolling stock depending on a surrounding situation. In the article, the development stages of the virtual simulator cyber-physical system intended for drivers-operators training in control of the rectifying and lining and leveling Duomatic 09-32 machine are considered.

Keywords Virtual simulator · Rolling stock · Hardware-Software complex · Cyber-Physical system

N. A. Staroverova (✉) · M. L. Shustrova
Kazan National Research Technological University, 68 Karl Marx Street, Kazan, Tatarstan 420015, Russia
e-mail: nata-staroverova@yandex.ru

M. R. Satdarov
Zarnitza, Kazan, Tatarstan 420006, Russia

A. G. Kravets et al. (eds.), *Cyber-Physical Systems: Industry 4.0 Challenges*, Studies in Systems, Decision and Control 260,
https://doi.org/10.1007/978-3-030-32648-7_25

1 Introduction

Development and use of the virtual simulators modeling the technical systems functioning is the relevant and intensively developing the field of science now. Simulating technologies arose and gained the greatest development in fields where training at real objects errors can lead to extraordinary effects, and their elimination—to big financial expenses.

Training technologies now are the end-to-end systems, the cyber-physical systems of modeling and simulation, the system of visualization, the computer programs, and physical models, special techniques created to prepare the person for acceptance of qualitative and fast solutions that became very serious scientific task and even a problem in the 21st century [1–3].

The use of simulators in the transport industry has rather a wide area now. There is a set of various complexes for training in hardware control skills, beginning from the "simplest" auto simulators to the most difficult hardware-software training complexes of the space industry [4].

Depending on the training purposes, in training of railway transport specialists, apply different types of simulators [5–7]. There are simulators for driving skills training: driver, assistant driver. Also, there is the instructor's workplace who can enter emergency situations, modeling a passing route and also change the movement coefficients, weather characteristics, time of the virtual world [8, 9].

There are specialized simulators of the train dispatcher workplace for training the safe skills of work and to fixing of action skills. This simulator models work of railway stations and depot with various grounds, the network version with the connection in parallel of several participants and ensuring collective work is possible. The virtual stands are used to working off of skills of signs and movements of trains and locomotives.

Training of drivers-operators of special self-moving rolling (SSMR) stock Duomatic 09-32 also includes using of virtual simulators. Such rolling stock participates under repair railroad tracks, rails, cross ties and the tamping of soil and crushed stone. Simulators allow to make complex training of personnel and railway productions employees and to ensure safety and comfortable movement of people, freights and also strategically important objects [9–11].

The purpose of the work is studying of the modeling subject and creation of the virtual simulator imitating functioning of the special self-moved Duomatic 09-32 rolling stock.

In the course of achievement of this purpose the following tasks were formulated and solved:

- the modeling subject studying and analysis,
- development of imitation modules of hydraulic, electric and pneumatic schemes and their components conditions, research and development of a control system and process of negotiations,
- simulation of processes of the world around,

- training of drivers and assistants to drivers in driving, the correct actions in emergency and unusual situations, to the rational expenditure of fuel and energy resources,
- training in the ability to work with safety devices,
- training of operator teams in work and the correct actions during the full cycle of works at construction, diagnostics and repair of railroad tracks and the track equipment, on detection and elimination of all possible malfunctions of the equipment, training in the SSPS device, working off of skills of operation, to service and control of knowledge of service staff [12, 13].

The hardware-software part of simulators uses is architecture, mechanical or virtual simulation of a modeled object on the basis of the client-server touch technologies. It allows providing realistic feedback coupling is applied: the illusion of management, coordination of actions with virtual space by means of acceleration and vibration, audio and visual effects. The space of a server part provides division on managing part and scene.

The emergence of similar simulators was promoted by definition of an algorithm conveyor-based, in particular—the emergence of the same tasks and decisions, the identical sequence of the performed operations.

Developing the simulator equipment has to be corresponding realistic, by addition to the necessary components of office equipment. Besides, the process has to be documented, the program has to carry out the preservation of all performed operations in memory for archiving and access during certain time Also simulators have to be capable to unite on the network as follows [14, 15]:

- by local connection, at restriction in space and finding of clients be near for working off of collective interaction of work on the simulator
- by means of the global network when several people on different continents and corners of the planet are capable to perform collaboration. At the same time the instructor's computer "Server" should be integrated in this network. It carries out information output from monitors and control panels, processes surveillance cameras which are implemented by means of the general interface of the separate modeling systems with the matching device.

For the full development of virtual simulators, it is required to select a convenient environment for an applied part of functionality [14–16].

2 Features of Duomatic 09-32

The use of rolling stocks on railway transport is caused by the need of repair and maintenance of a rail state, soil, safety of a human and technical component of the international community. Duomatic 09-32 carries out bearing and consolidation of the ballast of a cross and longitudinal way profile, dressing of soil, crushed stone and also the analysis entrance and output characteristics. It is applied at the maintenance of railway tracks in working order, to construction and repair (Fig. 1).

Fig. 1 Transport for the performance of railway repair

Operating parts have arrangement under each rail thread: cross ties end faces ballast sealants, the lifting and leveling device (LLD), and lining blocks. Lining blocks are 16 vertical probes which lower ends make cutting in crushed stone and soil by means of the engine and also have vibration effect of the soil strengthening by means of an eccentric shaft. It crashes into ballast in depth of 40–60 cm by means of hydraulic cylinders and carries out rapprochement, compression of cross ties and consolidation of ballast. Consolidation of ballast under two cross ties is at the same time carried out, after the termination of a cycle passes to the following. The block has a possibility of movement in cross-section, carrying out lining operations on curvilinear sites of a way. Represents the vibrating plate with an orientation hydraulic actuator which on both sides carry out the complex improvement.

Lifting and leveling devices have the roller captures providing capture and lifting of a railway line by means of rollers and also the shift of a way for ensuring parallelism and straightening with a hydraulic actuator across. These devices work independently of each other. 5 measuring carts through which there passes the chord which is carrying out measurements of the curvature of a way and being measuring base also are a part.

Measurement is performed on all points. Input and output parameters are compared, and the controller receives a signal of a curvature value and force of the put pressure according to variable potentiometer transducers and degree of an arrow concerning a rope deflection. According to these data, the computer makes a decision on ratios of values and the parties, a ratio of lengths of shoulders of measuring chords.

The program models used in the imitating computer have to display the realistic interaction of components and systems of the modeled process. The completeness of the mathematical model of the developed exercise machine includes models of electric, hydraulic, pneumatic and mechanical systems and also modules of imitation of weather and external conditions.

2.1 Implementation of the Duomatic 09-32 Project

The development of the presented virtual simulator was realized with the application of software: MicrosoftVisio, CorelDraw, C#, Unity 3D, AxureRP, 3DMax, XMind.

Axure RP Pro is the popular software solution for the creation of prototypes and specifications of web projects. The models or prototypes received by means of Axure RP are as close as possible to a final output thanks to a possibility of the interface elements programming. Is helps the program work area prototype with correct transitions, a user interface and necessary functionality was designed.

The MictosoftVisio application allows to describe flowcharts of the program functionality and to give an opportunity for the creation of the correct software product. This application was used, in particular, at the implementation of the flowchart of negotiations regulations for the creation of specialized software by developers and also the organization of a prototype of the interface, type of saving, logging.

The world around and 3D—animation of the project was carried out in Unity 3D. This application allowed to describe 3D—animation in the required sequence and under certain conditions. The project in Unity is divided into stages (levels)—files that contain certain virtual worlds and scenes with a set of functions and objects, various algorithms and scenarios and also settings. Scenes contain various models.

One more advantage of this editor—its support of the C# language which was used for writing of internal functionality, so-called "BackEnd" of the developed client-server application.

In a hardware-software complex 15 touch monitors for full simulation of a road situation in a cabin of the rectifying and lining and leveling machine Duomatic 09-32 are used:

- 9 monitors 40″
- 4 monitors 43″
- 2 monitors 55″

2.2 Implementation of the Hydraulic, Pneumatic and Electric Schemes Duomatic 09-32

In the developed simulator calculation of the electric, pneumatic and also hydraulic scheme, air distributors, cranes, relay, switches, power units is made. The electric current in the electrical circuitry is modelled. Calculation of pressure in the pneumatic system imitates real change of air, animation of this process, the system of search of leaks, their creation tracking of behavior and reaction of the trainee.

The hydraulic drive is a set of devices which are carrying out braking of structures, set cars and hydraulic power of mechanisms in motion. The main elements are the pump and the hydraulic engine. The hydraulic actuator represents a connecting link between mechanisms of the machine and the engine of loading and also performs functions of conrod mechanisms, the belt drive, a reducer. Purpose of a hydraulic actuator, as well as transfer—the transformation of the engine of power shafts according to mechanical loading and to requirements of the converter of an engine output link type, protection against overloads, parameters regulation and control.

The engine gives the rotation moment on a shaft of the pump which working liquid, reports on hydrolines energy via the regulation equipment. Energy comes to the hydraulic engine where it will be transformed to mechanical. Upon termination of liquid on hydrolines comes back to the pump or directly in a tank.

Calculation of behavior model of air in a pneumatic system allows to imitate real passing of air on the system of all structure, operation of distributors, cranes (including animation of this work in what channels will be, to pass air and with what pressure), creation of leaks in the certain places affecting all structure and also to keep track of behavior of a pneumatic system on the long trains.

On the connection diagram of elements show pipelines and elements of the brake highway connections, brake cylinders, the air distributor. At the same time, connections of pipelines show in the form of the simplified external outlines, and pipelines—continuous main lines.

For each car of the rolling stock detailed calculation of a condition of electric parameters is made. When modeling behavior of electric current in electrical circuitry elements (the relay, starting resistors, automatic machines, safety locks, etc.), all elements, work similarly real, that is the dependence of behavior of structure (collecting/analysis of draft, a rise/omitting of pantographs, rheostatic braking, etc.) on any electric element is implemented.

The electric circuit includes the following elements:

- electric devices;
- switches;
- lamps;
- resistors;
- transistors;
- power management.

These systems are realized on the basis of client-server architecture, by means of WPF on the basis of imposition of geometrical objects. Each object characterizes the work of the separate governing body of the modeled elements. When pressing the instructor an element, it is activated and allocated in color. At the same time, this element of the scheme ceases to work according to physical parameters. For maintenance, the driver needs to define the reason of breakage of this element and to carry out a number of necessary actions for its restoration [15, 17–19].

3 The Description of the Functional Characteristics of the Simulator

For the purpose of safety, in order to avoid cases of the undesirable start of the simulator, entering changes into results of a trip, etc., the system of authorization of the instructor is provided. For application launch, the instructor is offered to enter the LOGIN and the PASSWORD. For the management of instructors databases the role of the administrator and user is provided. The password and the login from the administrator account are known to system builders, but not its users. The database is created in PostgreSQL.

Before the start of occupation, the instructor chooses one of the scenarios of a trip created earlier. The scenario defines the complexity of the occupation and its duration, it does not depend on the choice of SSMR and is suitable for any of them. For the scenario creation the following tools are used: choice of the map of the world around, weather conditions control (time of day, rainfall, fog etc.), choice of the beginning point of a trip, choice of the end point of a trip, change of the indication of traffic lights, change of the arrows indication, addition of emergency situations, addition of malfunctions, control of physical quantities coefficients. Some setting up the scenario can be changed during occupation, for example, weather conditions, indications of arrows and traffic lights.

At the beginning of the education the instructor chooses the SSPS type, chooses pupils and appoints roles [the driver (Fig. 2), the assistant, the operator (Fig. 3)], if necessary, prints out necessary documents. During the lesson, it is possible to choose the ready scenario from the list, the SSMR type, its structure, to define or appoint

Fig. 2 Driver view

Fig. 3 Operator view

roles available to pupils, etc. Besides in the course of holding the occupation it is possible to change weather conditions both before the occupation and in process. Such settings as, global time (0–24 h), season, rainfall (rain or snow, seasonally), fog are available (with a possibility of change of density/visibility). The menu of the instructor allows switching values of arrows and traffic lights in 2 ways: by means of the minicard and about the help of the dispatcher of the next logical objects.

The possibility of a task is provided malfunctions. Malfunctions are meant as violations of work in the pneumatic, hydraulic, electric circuit of SSMR They directly affect the operability of separate knots of the train. For example, the safety lock which is responsible for a concrete element will put out of action only it.

Also, the simulator has the possibility of change of dynamics of the train in general (for example, malfunction of the engine will lead to a full stop).

The input of malfunctions does not depend on the position of the locomotive. Malfunction can be set prior to the passing of the scenario, for check, for example, the correctness of the performance of the test of brakes. The input of malfunctions by the instructor is carried out through the special program from the workplace of the instructor. Malfunction elimination by the pupil happens via the Module of malfunctions elimination.

Emergency situations are meant as the situations connected with objects of the world and which are not relating to malfunctions of the pneumatic, hydraulic, electric circuit. For example, a break a rail through 1500 m. Treat emergency situations: obstacles, destructions, fires, signalmen, signs, etc. The emergency situation can be attached to a concrete object on the card or is attached to a concrete kilometer of a way. Coefficients of physical quantities influence on the SSMR loudspeaker. Treat them: coefficient of coupling of wheels (By default 100%), coefficient of brake forces (By default 100%), speed of a side wind (By default 0 m/s). The instructor has an opportunity to control the position of the locomotive by means of the card, the regime card and a band of the traffic light indication.

Also, there is an opportunity to control actions of the pupil with the help: from the surveillance camera and by means of simulation of the panel of the driver. Windows of control over the locomotive and behind actions of the pupil according to the specification are on different screens. The card is the 3D world on which the locomotive moves.

On the card the following actions are available: scaling, switching ways of display (from above, 3D, or from the driver's cabin), switching of browse mode (to tie to the locomotive or free flight). Besides, it is possible to choose subjects of the map and to change their state: to switch shooters/traffic lights, to enter emergency situations, etc. The regime card contains the card of heights, the maximum and current speed of the locomotive, traffic lights, signs. On this card, such actions as scaling, the switching of the traffic lights indication, additions of emergency situations that are not tied to objects are available [17, 20].

After the end of the training, results remain in the database, for each pupil separately. The instructor has an opportunity to browse all results, to look for/sort results, to supplement result with the comment, to save result after adding the comment, to print result. Results represent the list of all violations of the driver, percent of negotiations regulations execution, instructor's personal comments.

4 Conclusion

The developed simulator considers aspects of the railway transport control process (structures, people on the ways, railway signs, traffic lights, etc.) that allows the operator to gain qualitatively skills of the analysis and processing of the arriving information and execution of actions for train control depending on a situation.

Deep simulation of safety devices operation logic, electric and pneumatic circuits, physical processes of train maintaining, physical processes of the world around is implemented.

The training activity is followed with the registration of all of the rolling stock parameters and driver's actions in the special trips analysis program. A similar program of the stationary decoding device is included in a package of the complex locomotive safety control. The program allows wiretapping conversations of the driver with the manager, the driver with the assistant, to control their correctness, also makes the semi-automatic analysis of a trip.

Simulation of the processes which are taking place in the environment allows reproducing most precisely all conditions corresponding to a real trip. For example, there is a possibility of the choice not only time of day, season, the intensity of rainfall, but also the possibility of the choice of ambient air temperature is provided. The program of the exercise machine in this case itself changes the wheel coupling coefficient with a rail, visibility conditions of signals, the train brakes functioning, especially in the conditions of low temperatures, etc.

Training activity includes registration of all driver's and operator's actions and also parameters for special self-propelled train structure and action of the trainee within the special trips analysis program.

This system is put into operation in the course of which proved to be the modern training complex implementing high-quality emulation of processes and allowing to gain correct and relevant knowledge and abilities of train control.

Acknowledgments Authors express gratitude to software design team the Zarnitza (Kazan) for implementation of the project and the provided materials.

References

1. Derler, P., Lee, E.A., Vincentelli, A.S.: Modeling cyber–physical systems. Proc. IEEE **1**(100), 13–28 (2011)
2. Lee, E.A.: Cyber physical systems: Design challenges in 11th IEEE International Symposium on Object and Component-Oriented Real-Time Distributed Computing (ISORC). IEEE, pp. 363–369 (2008)
3. Sridhar, S., Hahn, A., Govindarasu, M.: Cyber–physical system security for the electric power grid. Proc. IEEE **1**(100), 210–224 (2011)
4. Sha, L. et al.: Cyber-physical systems. In: A new frontier in IEEE International Conference on Sensor Networks, Ubiquitous, and Trustworthy Computing, pp. 1–9 (2008)
5. KöglB. Fahrsimulatorenfür die Ausbildung von Triebfahrzeugführern: Elek. Bahnen. **8**, **9** (89), 261–266 (1996)
6. Railway simulators become more diversified. Raihway J. Int. **4**(41), 29–31 (2001)
7. Kim, J., Kim, J.-H.: Development of integrated simulator for AC traction power supply system. Trans. Korean 2010 Inst. Electr. Eng. **59**(1), 75–81 (2010)
8. Takeuchi, Y., Ogawa, T., Morimoto, H.: Development of a train operation power simulator. Japan. Railway Eng. **197**, 13–15 (2017)
9. Miyauchi, T., Imamoto, K., Teramura, K., Takahashi, H.: Evaluating the accuracy of railway total simulator compared with actual, measurement data. IEEE J. Ind. **7**(5), 416–424 (2018)
10. Takeuchi, Y., Ogawa, T., Morimoto, H., Imamura, Y., Minobe, S., Sugimoto, S.: Development of a train operation power simulator using the interaction between the power supply network, rolling stock characteristics & driving patterns, as conditions. Q. Rep. RTRI (Railway Tech. Res. Inst.) **58**(2), 98–104 (2018)
11. Rajkumar, R. et al.: Cyber-physical systems: the next computing revolution. In: Design Automation Conference, IEEE, pp. 731–736 (2010)
12. Madsen, E.S., Bilberg, A., Hansen, D.G.: Industry 4.0 and digitalization call for vocational skills, applied industrial engineering, and less for pure academics. In: Proceedings of the 5th P&OM World Conference, Production and Operations Management, P&OM (2016)

13. Zeyda, F., Ouy, J., Foster, S., Cavalcanti, A.: Formalising cosimulation models. In: Cerone, A., Roveri, M. (eds.) Software Engineering and Formal Methods. SEFM 2017. Lecture Notes in Computer Science, vol. 10729, pp. 453–468. Springer, Cham (2018)
14. Hackenberg, Georg: Test-driven conceptual design of cyber-physical manufacturing systems. Technische Universität München, Diss (2018)
15. Cai, P. et al.: Simulation-enabled vocational training for heavy crane operations. In: Simulation and Serious Games for Education, pp. 47–59. Springer, Singapore (2017)
16. Piccininni, A., Guglielmi, P., Lo Franco, A., Palumbo, G.: Stamping an AA5754 train window panel with high dent resistance using locally annealed blanks. J. Phys. Conf. Ser. **896**(1) (2017)
17. Sutherland, J., Sutherland, J.J.: Scrum: El revolucionario método para trabajar el doble en la mitad de tiempo. Grupo Planeta Spain (2015)
18. Kreg, L.: Application of UML of 2.0 Templates Practical Guidance (Electronic Materials), pp. 736. Williams (2016)
19. Tamburri, D.A., Van den Heuvel, W.J., Lauwers, C. et al.: SICS Softw.-Inensiv. Cyber-Phys. Syst. **34**, 163 (2019)
20. Mueller, W. et al.: Virtual prototyping of cyber-physical systems. In: 17th Asia and South Pacific Design Automation Conference. IEEE (2012)

Building an Knowledge Base of a Company Based on the Analysis of Employee's Behavior

M. V. Vinogradova and A. S. Larionov

The approach is proposed for building the enterprise knowledge base utilizing technologies of cyber-physical systems operating in virtual space. Integration of the enterprise information system (IS) with the knowledge base is considered as the cyber-physical system; employees act as physical objects. The knowledge base model has been developed, whose articles are matched with the objects and business processes of the main information system, based on a three-level semantic network. The algorithm for determining priority materials for writing into the knowledge base driven by the analysis of the user actions track was built. The user actions track is formed according to statistical data obtained from the event logs of the knowledge base and the IS using the sequential analysis to identify patterns of typical operations. The formal model of the interaction process of employees with the knowledge base was built, taking into account the activities of its formation and use. Simulation of the interaction process with the knowledge base was carried out, the results of which confirmed the effectiveness of the proposed approach. With an increase in the number of users, the cost of seeding the knowledge base, taking into account the relevance of its materials, pays its way due to reducing the employees idle time. The proposed models and algorithms contribute to reducing the cost of developing the enterprise knowledge base and increasing its utilization efficiency.

Knowledge management · Enterprise knowledge base · Semantic networks · Analysis of user behavior · Labor cost estimation · Interaction simulation · Sequential pattern

---

M. V. Vinogradova (✉) · A. S. Larionov
Bauman Moscow State Technical University, 5, 2-nd Baumanskaya st., Moscow 105005, Russian Federation
e-mail: vinogradova.m@bmstu.ru

e-mail: andreylar@mail.ru

A. G. Kravets et al. (eds.), *Cyber-Physical Systems: Industry 4.0 Challenges*, Studies in Systems, Decision and Control 260,
https://doi.org/10.1007/978-3-030-32648-7_26

1 Background

Creating one's own knowledge bases containing information on accumulated experience and professional skills for a certain subject area reflects the modern approach to the accumulation of knowledge and competencies within a company or enterprise [1]. The knowledge base turns out to be a useful addition to the main information system and contributes to reducing its operation cost, as well as improving the operational efficiency of the entire enterprise [2, 3]. However, for large and rapidly developing information systems, labor costs for maintaining the knowledge base itself are found to be significant, since it is not known which elements of the system should be described first and at what level of details. Methods and tools are required to determine the most relevant information materials, as well as to provide additional information to the author who seeds the knowledge base. In [4] it was shown that the creation of a useful knowledge base is impossible without an effective strategy for its seeding.

Until the present time, many studies have been conducted on the creating and seeding of knowledge bases, the development of their models and architecture. A critical review of the literature on knowledge management in small and medium-sized enterprises was conducted in [5], which revealed the studies' fragmentariness and the lack of unified approach both to the creating and seeding of knowledge bases and to the knowledge representation and interpretation models used. The absence of a unified and generally accepted approach indicates a niche nature of the results obtained and their focus on the specifics of particular enterprises.

To form a universal approach to the seeding process of an enterprise knowledge base, it is proposed to use technologies from the field of cyber-physical systems.

Cyber-physical systems (CPS), known since 2006, combine physical elements and the information environment into a unified distributed computing system [6]. CPS are used in industry and healthcare, to create smart homes and smart cities, in the Internet of Things and Industry 4.0.

CPS can be divided into two classes: operating in a physical environment and used in a virtual space. The systems of the first class contain sensors and actuators and are aimed at environmental management, the representatives of the second class are used to collect data on user actions in a virtual environment, for example, in social networks or on online store websites in order to predict their future behavior or needs [7].

Some CPS combine the features of both classes. For example, developers of intelligent cities place a person within the system's bounds, pointing out the importance of analyzing his actions and motivation [8]: "Understanding and respecting human behavior is a key component of a CPS; in fact, a CPS should be more accurately referred to as a Cyber-Physical Social System".

To analyze the user actions within the CPS framework, several approaches are proposed. First of all, these are statistical methods. Thus, the authors of [9] use a three-step statistical advisory approach for finding friends in the real world based on information from a social network that supports geospatial navigation. The second, more common approach is based on the sequential analysis methods and on the identification of sequential patterns in user actions. This approach is used in security systems to detect attacks; in e-commerce applications to determine buying preferences [10]; in speech recognition systems [11] and even in tourism to identify the travelers' interests based on their photographs [12].

In this chapter, the integration of the main enterprise information system providing its operation with the knowledge base of this enterprise is considered as a CPS. Physical objects are employees of the enterprise, and information about their actions are automatically collected, analyzed and used to make a decision about the necessity to seed the knowledge base.

The object of the work is to develop a knowledge base model taking into account its interaction with the main enterprise information system, as well as the development of methods for analyzing user actions to identify objects and processes that should be entered into the knowledge base.

To achieve this goal, the following tasks were set:

- to determine the data sources on user actions and the data presentation model for further processing;
- to develop the knowledge base model, allowing to correlate its materials with objects and processes in the main information system;
- to develop the algorithm for estimation of the relevance of writing new materials for the knowledge base in specific areas of the information system;
- to propose the method of obtaining information about the links of objects with other elements of the information system to assist the expert seeding the knowledge base.

The proposed models and methods will improve the efficiency of the enterprise employees, reduce labor costs for seeding the knowledge base and simplify the work of expert authors in the drafting of materials.

2 The Concept of Building a Knowledge Base

In this chapter, the formation of the knowledge base model and methods for analyzing user actions are considered on the example of the computer-aided automotive business management system 1C:"Alfa-Auto", which is included in the product line of "1C: Enterprise" [13] extendable software platform, used to automate various areas of enterprise activity [14]. The 1C:"Alfa-Auto" system is intended for accounting car repair and maintenance services, automating the spare part sales and maintaining warehouse inventory and business accounting. The knowledge base in this example is implemented using the Semantic MediaWiki tools [15].

Three sources can be used to support the analysis of user actions:

1. The log [16] provides information about events in the main information system (for example, recording and posting documents, error messages).

2. The time measuring register [16] records changeovers between screen forms and time spent on this.
3. Google Analytics tools [17] and Semantic MediaWiki's own log files [15] record user-visited articles and search queries.

Each of the three sources provides information on the time of actions and the user who carried them out. Thus, heterogeneous elementary events can be combined into an event sequence [18] suitable for subsequent analysis (action track of the ith user):

$$F^i = \{\ldots, f_p^i, \ldots, f_q^i\}, \text{where}\, p < q. \quad (1)$$

Elementary event sequences, marked on user action tracks, need to be matched with larger meaningful events. To highlight major events, one can rely on user queries they request the knowledge base with. Queries usually relate to the order of business process execution, a detailed description of the system object functionality (for example, "Analyst's automated workstation for ordering goods to the supplier") or a sequence of actions to resolve a specific problem situation that occurs when there is nonstandard interaction of several system objects.

2.1 Knowledge Base Model

In accordance with the most frequently used types of queries, the knowledge base unites three sets of articles. The first of them group information on the system business processes, describing the sequence of actions for performing typical work tasks. The second describe the functionality and rules for working with system objects. At that, within each object, the groups of interlinked elements are separated out that are shared to perform a single operation. The third set of articles describes the work responsibilities of various positions and sets of tasks that their executives regularly face.

The enterprise knowledge base model is defined using ontologies [19], which allow combining information about specific objects, abstract concepts, and links between both of them. In our case, in building the ontological model of the knowledge base, the peculiarities of the structures and links of the elements, the information about which is contained in the pages of the knowledge base, were taken into account. The knowledge base model, which articles are matched with the objects and business processes of the main information system, is described by a three-level semantic network:

$$Q =< V, P >, \quad (2)$$

where

$= A \cup B \cup C$ the set of articles (materials) of the knowledge base;

$= \{a_i\}$ the set of materials on the program objects of the system (documents, reference books, etc.), each of which consists of a set of information elements: $a_i = \{e^i_j\}$, $e^i_j \in E$;

$= \{b_i\}$ the set of materials on business processes, which are described by a sequence of actions to transform information elements: $b_i = \langle \ldots, m^i_p, \ldots, m^i_q \rangle$, $m^i_j : \{e_k\} \to \{e_l\}$;

$= \{c_i\}$ the set of materials on the work responsibilities of users, consisting of a set of standard $\left(\{t^S_i\} \in T^S\right)$ and non-standard $\left(\{t^{NS}_i\} \in T^{NS}\right)$ work tasks T: $c_i = \{t^S_i\} \cup \{t^{NS}_j\}$.

$\{p_{ij}\}$ the set of links between knowledge base materials of all types: $p_{ij} = \langle v_i, \mathrm{v_j} \rangle$, $v_i, \mathrm{v_j} \in V$

Standard work tasks consist of multiple business processes $t^S_i = \langle \ldots, m^i_p, \ldots, m^i_q \rangle$. The nonstandard task is described by the triple $t^{NS}_i = \langle b_i, G_i(\{e_k\}), \{\phi_i\} \rangle$ and defines the situation when the system is in the nonstandard state during the execution of the work process b_i (as specified by the predicate $G_i(\{e_k\}), e^i_j \in E$) and therefore additional actions $\phi_j : \{e_k\} \to \{e_l\}$ need to be taken.

If there are complex semantic dependencies between the elements of the same level, then it is advisable to go over from a graph model to a metagraph model [20]. Using the metagraph will reduce the number of objects considered, which will simplify the model processing. To go over to the metagraph model, the formal approach considered in [21] is used.

Taking into account the proposed knowledge base model (2), the element f^i_p of the user's action track (1) can be represented as follows:

1. for the action $d_k \in D$ on seeding or reading a knowledge base article $v_k \in V$ at the instant of time t_k:

$$f^i_p = \langle t_k, d_k, v_k \rangle; \tag{3}$$

2. when working with the main information system for the action $d'_l \in D'$, that uses the information item $e_l \in E$ and obtained the result $r_l \in R$ at the instant of time t_l:

$$f^i_q = \langle t_l, d'_l, e_l, r_l \rangle. \tag{4}$$

Schematic representation of the knowledge base model and its links with user action tracks are presented in Fig. 1. The cross-reference model leads to the fact that for events in the system the description can be constructed containing indicators to linked elements from all three sets. The description can be constructed for elementary events, such as pushing a specific button, or to enlarge the event up to entire blocks of business processes. To change between the detailed and enlarged representations of the event, the rules for the transformation of graph and metagraph models considered in [22] are used.

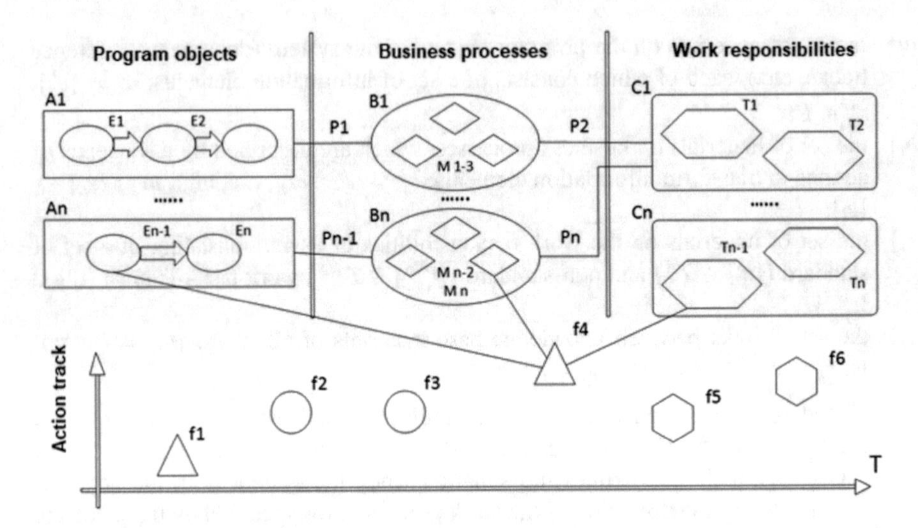

Fig. 1 Illustration of a knowledge base model and its links to user action tracks

2.2 *Knowledge Base Seeding Algorithm*

The built knowledge base (2) and user actions track (3, 4) models allow to automate the process of determining the priorities of writing articles and to obtain additional information about the linked objects. Let's consider the appropriate algorithm. But beforehand, experts need to manually mark the article section headings and match them with the objects and business processes of the main information system.

- Step 1. Identify failed operations (events that resulted in an error) based on user tracks (4).
- Step 2. For each type y of failed operations, estimate the error probability P_0^y for the period T by the tracks of all users: $P_0^y = N_{\mathrm{N}}^y / N_A^y$, where N_{N}^y—the number of events of type y that failed; N_A^y—total number of events of this type.
- Step 3. For each type of operation y, identify the associated information item $e_y \in E$.
- Step 4. For the element e_y calculate the time between the failed operation and the similar successful one that solved the problem. If both operations are performed by the same user, then the time between them we call the unit idle period $T_i^{\mathrm{IDL}(y)}$ of the ith user.
- Step 5. For the ith user analyze his track (3, 4) in the idle period. Assign weights k_i^y depending on whether the user completely stopped working or switched to other tasks.
- Step 6. Calculate the weighted average idle time of all n users over period T:

$$T_{av}^{\mathrm{IDL}(y)} = \left(\sum_{i=1}^{n} \left(k_i^y * T_i^{\mathrm{IDL}(y)} \right) \right) / n.$$

- Step 7. Make a forecast how much users will spend cumulatively in idle mode for the next T′ period, taking into account the number of events and the error probability. For short periods of time (up to a week), it makes no sense to take into account seasonal fluctuations in the forecast; it is enough to linearly transfer the data of the previous period:

$$T_{\text{fore}}^{\text{IDL}(y)} = P_0^y * N_A^y * T_{\text{av}}^{\text{IDL}(y)} * T'/T.$$

- Step 8. Based on the analysis of user tracks (3), calculate how much time users spent on average creating articles in the knowledge base for m logically related items: $T_{av}^{\text{SP}} = \left(\sum_{i=1}^{m}(T_i^{\text{SP}})\right)/m$, where T_i^{SP}—the time spent on writing the ith article.
- Step 9. Calculate for each type y of failed events the relevance coefficient as the ratio of the forecast of the total user idle time to the average time of writing the article: $K_{\text{REL}}^y = T_{\text{fore}}^{\text{IDL}(y)}/T_{av}^{\text{SP}}$.
- Step 10. Sort the resulting relevance coefficients for all types of failed operations in descending order. The resulting list shows the priority of writing articles for the knowledge base.

Writing a new article on a problem in the knowledge base is considered relevant if the relevance coefficient K_{REL}^y exceeds the value K_{REL}: $K_{\text{REL}}^y > K_{\text{REL}}$ determined by the expert.

Having determined the priorities for the creating of articles for the knowledge base, it is necessary to define the set of the system objects $v_i \in V$ (2), which should be mentioned in the article. For this, an analysis of the user action track (4) is performed and its description is made as a Petri net [23]:

$$N = \langle \text{E}, D', L \rangle, \tag{5}$$

where E—finite set of states (corresponds to a set of information items); D′—finite set of transitions (corresponds to elementary actions); $L \subseteq (ExD') \cup (D'xE)$—set of directed arcs.

Petri nets are built for each user's track. On the basis of the built Petri nets, a search of their fragments, which can be associated with business processes, is carried out using the CloSpan algorithm for mining closed sequential patterns [24, 25]. The found fragments (patterns) corresponding to the failed operation and associated with the business process are subject to further analysis.

If a pattern associated with the failed operation is found, one should review the sequence of user actions in both directions along the time axis. Since elementary events are linked with the knowledge base materials, their belonging to objects and business processes of the system can be used to filter out noise in the user actions. As a result, for each type of failed operation, the following patterns are identified: S_1—the initially correct execution of the operation; S_2—performing the operation that ended with failure; S_3—performing the operation, that has corrected a previously unsuccessful attempt.

These work scenarios are used to construct a finite automaton using the Markov algorithm [18]. The probabilities of state transitions will vary for each S_i. Comparing the sequence of events in different scenarios can provide an expert with important information about the trends that need to be reflected in the article. For example, the difference between the patterns S_1 and S_3 will allow to identify information items and additional actions for non-standard situations, and the difference between S_1 and S_2—objects and business process, which are insufficiently described in the knowledge base.

3 Estimation of the Proposed Approach Efficiency

In order to estimate the effectiveness of the proposed algorithm for seeding the knowledge base, we will simulate the employees work with and without our approach. Let's check whether the relevance estimate of the preparation of materials for the knowledge base can reduce idle time in the work of employees.

On the basis of statistical data obtained from the knowledge base event log and the main information system, we build a model of interaction between employees and the knowledge base. Using the model, we estimate the time spent by users and technical support staff when errors occur.

The model assumes that an employee of the enterprise performs work operations in the main information system, and these operations can be completed either successfully or with an error. If an error occurs, the employee contacts the knowledge base and tries to find a hint there. At that, he spends time searching the knowledge base and reading its articles. The model introduced the assumption that reading the article unambiguously allows to solve the user's problem and eliminate the error that occurred.

If the required article is absent in the knowledge base, the user contacts the support service, whose employee helps to solve the problem. In this case, the support service may initiate the writing of the missing article. The time spent by the user for searching the knowledge base and waiting, and technical support for troubleshooting is idle time.

The model of a queue system was compiled, which demonstrates how users refer to the knowledge base materials and how the technical support staff works when errors occur. The model diagram is shown in Fig. 2. For the simulation, a program was written in the GPSS simulation language.

The movement of the transaction in the built model simulates the processing of a single event by users and support staff. Consider the life cycle of a transaction:

- Step 1. A transaction emerges in the Str block, where according to a given probability P_i from the set of probabilities $P = \{P_i\}$, where $\sum_{1=1}^{M} P_i = 1$, the identifier of the ith event type is assigned to it. The event type corresponds to a typical user action (employee work task). Counter N_i to account for the number of all events of this type is increased by one.

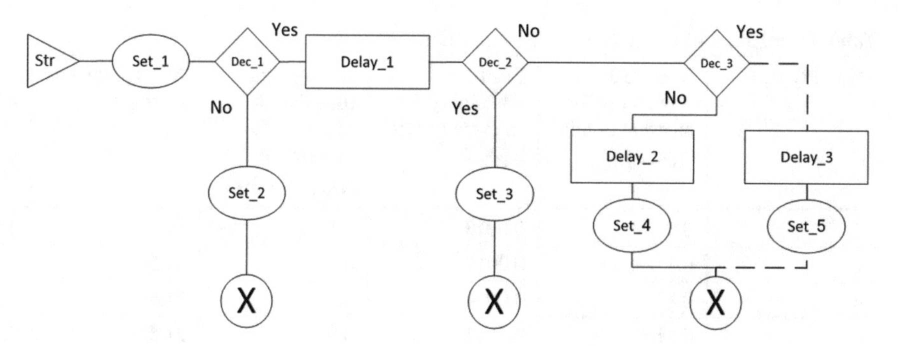

Fig. 2 Model of the interaction process of enterprise employees with the database

- Step 2. The transaction goes into the Set_1 block, where a flag is being assigned to it with a probability P_i^{ER}, that an error has occurred.
- Step 3. In the Dec_1 block, the presence of an error flag is checked. If there was no error, then the transaction goes into the Set_2 block (step 4), otherwise to the Delay_1 block (step 5).
- Step 4. In the Set_2 block, the counter of successful events of this type N_i^{OK} is incremented. The transaction without errors leaves the model without increasing the total idle time.
- Step 5. The transaction with an error goes into the block Delay_1. This block corresponds to the time T_i^{RD}, which the user spends on trying to understand the error and preliminary search for solutions in the knowledge base.
- Step 6. In the Dec_2 block, the presence of a knowledge base article on the given problem is checked. Knowledge base materials are considered sufficient, i.e. reading the article allows to eliminate the error associated with it. If the article is found, then the transaction goes into the Set_3 block (step 7), otherwise into the Dec_3 block (step 8).
- Step 7. In the Set_3 block, the counter of events with errors N_i^{ER} is incremented. For this transaction, the idle time is determined by the time of searching and reading an article in the knowledge base T_i^{RD}. Total idle time increases by T_i^{RD}. The transaction with an error leaves the model.
- Step 8. In the Dec_3 block, the decision is made about the need to write an article on this topic. If the relevance coefficient K_{REL}^i of this type of events i exceeds the threshold value set by the expert K_{REL}, then the transaction goes into the Delay_3 block (step 9), otherwise into the Delay_4 block (step 11).
- Step 9. In the Delay_3 block, writing a new article to the knowledge base is simulated, which takes time T_i^{WR} both from the user (he is waiting for the article) and from the support staff member. The transaction goes into the Set_4 block.
- Step 10. In the Set_4 block, the error counter with the created articles N_i^{ER} is incremented. Total idle time increases by $(2 * T_i^{WR})$. The transaction leaves the model.

Table 1 Fragment of source data for the simulation

Event type, i	Occurrence probability for an event of this type, P_i	Error probability for an event of this type, P_i^{ER}	Average idle time during search in the knowledge base, T_i^{RD}[h]	Average time of writing the article, T_i^{WR} [h]	Averag manua the pro T_i^{SS} [h
1	0.2	0.0013	0.5	21.5	1.7
2	0.15	0.0017	0.5	21.5	2
3	0.13	0.0021	0.7	21.5	3.2
4	0.11	0.0013	0.9	21.5	2.1
5	0.11	0.0005	0.8	21.5	3.4
6	0.04	0.0003	0.8	21.5	4
7	0.03	0.0001	0.4	21.5	3.5
8	0.03	0.002	2	21.5	7
9	0.02	0.001	1.3	21.5	5
…	…	…	…	…	…
50	0.000018	0.000001	2.1	21.5	2
51	0.000014	0.000001	0.7	21.5	3.1
52	0.00001	0.000002	0.9	21.5	3.8

- Step 11. In the Delay_4 block, the manual solving of the problem by support service is simulated, which takes time T_i^{SS} both from the user and from the support staff member. The transaction goes into block Set_5.
- Step 12. In the Set_5 block, the error counter with the created articles N_i^{ER} is incremented. Total idle time increases by $(2 * T_i^{SS})$. The transaction leaves the model.

The result of the simulation is total idle time. Statistics on the total period of user idle time is accumulated when multiple transactions are passed through the model. The result consists of the time the user searches for information in the knowledge base (T_i^{RD}) and the time spent by the support staff member on manually solving the problem $\left(T_i^{SS}\right)$ or writing articles $\left(T_i^{WR}\right)$.

The simulation model of the system was built using the GPSS World tool. Source data is obtained on the basis of the analysis of the work of a particular enterprise using the "1C: Alfa-Auto" system [13]. The data sources are the log, time measuring register [16] and Google Analytics tools [17]. The initial data for the simulation are: the occurrence probability of an event of the *i*th type P_i; the error probability for an event of the ith type P_i^{ER}; the average idle time when searching in the knowledge base T_i^{RD}; the average time of writing the article T_i^{WR}; the average time of manual solving the problem T_i^{SS}.

Table 1 shows the specific values of the source data obtained from the event logs and used for modeling. Values are obtained from statistics on the module of warehouse inventory accounting and spare part sales. Indicators for 9 types of the

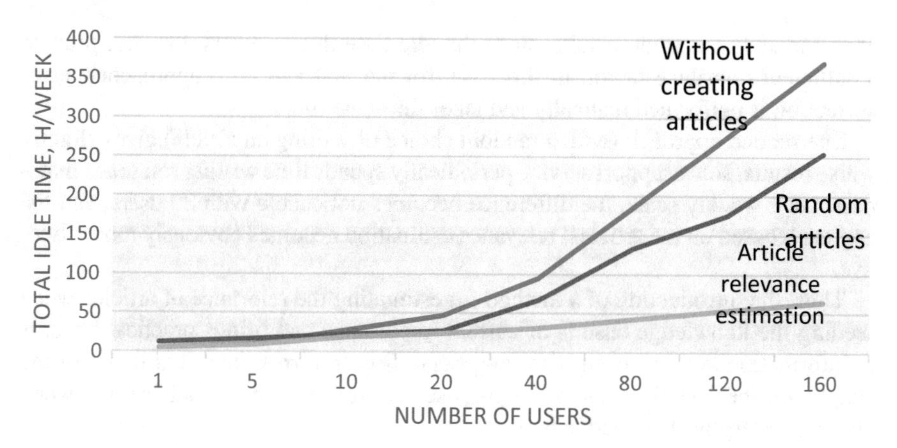

Fig. 3 Dependence of total idle time on the number of users

most frequent events ($P_i > 0.02$) which are being recorded in the Log are given. The probability of all other 43 operations types is cumulatively less than 0.18. Here is a part of the table of probability distribution by types of events below. The average time of writing articles is common to all types (average value), since each event corresponds to only one article.

To prove the effectiveness of the method to estimate the relevance of the articles, three approaches to seeding the knowledge base were considered:

1. The knowledge base is not used; all errors are fixed by the support service manually. When modeling this situation, the transaction from the Dec_3 block always goes into the Delay_2 block.
2. The knowledge base materials are filled in random order. When modeling this situation, the transaction from the Dec_3 block goes into the Delay_2 or Deley_3 blocks with equal probability.
3. The knowledge base is seeded in accordance with the proposed algorithm. For the ith event type, the coefficient of relevance K^i_{REL} is calculated and compared with the threshold value K_{REL}. If $K^{\mathrm{i}}_{\mathrm{REL}} > K_{\mathrm{REL}}$, then one should write an article, otherwise it is more efficient to solve the rare problem manually. When modeling this situation, the transaction from the Dec_3 block goes into the Delay_2 block, if $K^i_{\mathrm{REL}} > K_{\mathrm{REL}}$, and into the Delay_3 block otherwise.

The most effective approach is the one, in which the total user idle time will be the smallest. A simulation of the three mentioned situations was carried out. The simulation results are shown in Fig. 3.

Figure 3 shows the graphs of the total idle time per week as a function of the number of users for each of the mentioned approaches to seeding the knowledge base.

Based on real statistical data, user performs on average 6.5 actions per hour, which are marked in the Log as events. It can be seen from the simulation results that while few users and, correspondingly, errors occur in the system, all approaches show approximately the same results, since the idle time due to errors does not provide a sufficient relevance factor. In this case, for the first and third approaches, error correction is performed manually and takes the same time.

The second approach (with a random choice of writing an article) gives slightly worse results, since support service periodically spends time writing reference materials. On a weekly scale, the difference becomes noticeable with 20 users, and the approach based on the articles' relevance estimation becomes obviously more effective.

Thus, the introduction of a method for estimating the relevance of articles when seeding the knowledge base is of current importance and brings practical benefits to information systems even with few users. This confirms the initial thesis of the chapter on the effectiveness of the proposed approach to reduce labor costs when building enterprise knowledge base.

4 Conclusion

The models and algorithms proposed in the chapter were created on the example of particular, albeit common software, however, they are rather abstract for wide application and can be adapted for other software products. The knowledge base model consisting of three sets of articles with cross-references allows building a projection of its materials on objects and business processes of the main information system and is applicable to most enterprises. The tracking of user elementary actions is performed by any logging tool, both built-in and universal.

The use of CPS approaches to analyze the track of user actions allows to estimate the relevance of writing specific articles. Identifying patterns of events in the action track allows to detect trends in the subject area for inclusion them in the articles. The combination of CPS capabilities with knowledge base models gives an opportunity to create an effective knowledge management system in the enterprise that is closely integrated with the main information system and is capable of responding in real time to the needs and requirements of employees to acquire knowledge.

Using the proposed models and methods saves time for experts seeding the knowledge base and reduces idle time in enterprise employees work.

References

. Alavi, M., Leidner, D.E.: Knowledge management and knowledge management systems: conceptual foundations and research issues. MIS Q. 107–136 (2001)
. Gorlacheva, E.N., Gudkov, A.G., Omelchenko, I.N., Drogovoz, P.A., Koznov, D.V.: Knowledge management capability impact on enterprise performance in russian high-tech sector. In: 2018 IEEE International Conference on Engineering, Technology and Innovation (ICE/ITMC), IEEE, pp. 1–9 (2018)
. Migdadi, M.M., Abu Zaid, M.K.S.: An empirical investigation of knowledge management competence for enterprise resource planning systems success: insights from Jordan. Int. J. Prod. Res. **54**(18), 5480–5498 (2016)
. Mohapatra, S., Agrawal, A., & Satpathy, A.: Designing knowledge management strategy. In Designing Knowledge Management-Enabled Business Strategies, pp. 55–88. Springer, Cham (2016)
. Massaro, M., Handley, K., Bagnoli, C., Dumay, J.: Knowledge management in small and medium enterprises: a structured literature review. J. Knowl. Manag. **20**(2), 258–291 (2016)
. Sadiku, M.N., Wang, Y., Cui, S., Musa, S.M.: Cyber-physical systems: a literature review. Eur. Sci. J. ESJ **13**(36), 52 (2017)
. Zanni, A. (2015). Cyber-physical systems and smart cities. IBM Big data and analytics, 20
. Cassandras, C.G.: Smart cities as cyber-physical social systems. Engineering **2**(2), 156–158 (2016)
. Yu, X., Pan, A., Tang, L.A., Li, Z., Han, J.: Geo-friends recommendation in gps-based cyber-physical social network. In: 2011 International Conference on Advances in Social Networks Analysis and Mining, IEEE, pp. 361–368, July 2011
. Chernova, V.Y., Tretyakova, O.V., Vlasov, A.I.: Brand Marketing Trends in Russian Social Media. https://mediawatchjournal.in/wp-content/uploads/2018/09/Brand-Marketing-Trends-in-Russian-Social-Media.pdf (2018). Accessed 20 Apr 2019
. Popolov, D., Callaghan, M., Luker, P.: Conversation space: visualising multi-threaded conversation. In: Proceedings of the Working Conference on Advanced Visual Interfaces, ACM, pp. 246–249, May 2000
. Vu, H.Q., Li, G., Law, R., Zhang, Y.: Travel diaries analysis by sequential rule mining. J. Travel Res. **57**(3), 399–413 (2018)
. C-Rarus: 1C: Enterprise Platform. Business Automation, Consulting and Support. https://rarus.ru/en. Accessed 20 Apr 2019
. Vazhdaev, A.N., Chernysheva, T.Y., Lisacheva, E.I.: Software selection based on analysis and forecasting methods, practised in 1C. In: IOP Conference Series: Materials Science and Engineering, vol. 91, no. 1, p. 012067. IOP Publishing (2015)
. Semantic MediaWiki (2019) Semantic MediaWiki. http://mediawiki.org/wiki/Semantic_MediaWiki. Accessed 20 Apr 2019
. C-Rarus: Objects of Configuration at 1C: Enterprise 8. https://its.1c.ru/db/metod8dev/content/2579/hdoc. Accessed 20 Apr 2019
. Clifton, B.: Advanced Web Metrics with Google Analytics. John Wiley & Sons (2012)
. Aggarwal, C.C.: Data Mining: The Textbook. Springer (2015)
. Fedotova, A.V., Tabakov, V.V., Ovsyannikov, M.V., Bruening, J.: Ontological modeling for industrial enterprise engineering. In: International Conference on Intelligent Information Technologies for Industry, pp. 182–189. Springer, Cham, September 2018
. Chernenkiy, V., Gapanyuk, Y., Terekhov, V., Revunkov, G., Kaganov, Y.: The hybrid intelligent information system approach as the basis for cognitive architecture. Procedia Comput. Sci. **145**, 143–152 (2018)
. Kanev, A., Cunningham, S., Valery, T.: Application of formal grammar in text mining and construction of an ontology. In: 2017 Internet Technologies and Applications (ITA), IEEE, pp. 53–57, September 2017
. Chernenkiy, V.M., Gapanyuk, Y.E., Kaganov, Y.T., Dunin, I.V., Lyaskovsky, M.A., Larionov, V.S.: Storing Metagraph Model in Relational, Document-Oriented, and Graph Databases (2018)

. Alfimtsev, A.N., Loktev, D.A., Loktev, A.A.: Comparison of development methodologies for systems of intellectual interaction. In: Proceedings of Moscow State University of Civil Engineering, no. 5, pp. 200–208 (2013)
. Barbara, D., Kamath, C. (eds.) Proceedings of the 2003 SIAM International Conference on Data Mining. Society for Industrial and Applied Mathematics (2003)
. Cook, J.E., Du, Z., Liu, C., Wolf, A.L.: Discovering models of behavior for concurrent workflows. Comput. Ind. **53**(3), 297–319 (2004)

Printed in the United States
By Bookmasters